THE CHEMISTRY AND TECHNOLOGY OF PECTIN

FOOD SCIENCE AND TECHNOLOGY
A Series of Monographs

A complete list of the books in this series appears at the end of the volume.

THE CHEMISTRY AND TECHNOLOGY OF PECTIN

EDITED BY

Reginald H. Walter

Department of Food Science
Cornell University
Geneva, New York

ACADEMIC PRESS, INC.

Harcourt Brace Jovanovich, Publishers
San Diego   New York   Boston
London   Sydney   Tokyo   Toronto

Copyright © 1991 by ACADEMIC PRESS, INC.

All Rights Reserved.

No part of this publication may be reproduced or transmitted in any form or by any means, electronic or mechanical, including photocopy, recording, or any information storage and retrieval system, without permission in writing from the publisher.

Academic Press, Inc.
San Diego, California 92101

United Kingdom Edition published by
Academic Press Limited
24–28 Oval Road, London NW1 7DX

Library of Congress Cataloging-in-Publication Data

The chemistry and technology of pectin [edited by] Reginald H. Walter.
    p. cm. -- (Food science and technology series)
    Includes bibliographical references.
    ISBN 0-12-733870-5
    1. Pectin.   I. Walter, Reginald H.   II. Series.
    TP248.P4C49   1991
    664'.25--dc20   91-4287
                                                   CIP

Transferred to digital printing 2005

91 92 93 94   9 8 7 6 5 4 3 2 1

CONTENTS

4. Tropical Fruit Products

A. S. Hodgson and L. H. Kerr

5. The Chemistry of High-Methoxyl Pectins

D. G. Oakenfull

6. The Chemistry of Low-Methoxyl Pectin Gelation

M. A. V. Axelos and J.-F. Thibault

7. Gelation of Sugar Beet Pectin by Oxidative Coupling

J.-F. Thibault, F. Guillon, and F. M. Rombouts

CONTRIBUTORS

Numbers in parentheses indicate the pages on which the authors' contributions begin.

M. A. V. **Axelos** (109,228), Institut National de la Recherche Agronomique, Laboratoire de Physico-Chimie des Macromolécules, 44026 Nantes Cedex 03, France

J. K. **Burns** (165), Citrus Research and Education Center, IFAS, University of Florida, Lake Alfred, Florida 33850

H.-U. **Endress** (251), Herbstreith and Fox KG, Pektin-Fabrik Neuenbürg, DW-7540 Germany

F. **Guillon** (119), Institut National de la Recherche Agronomique, Laboratoire de Technologie Appliquee a la Nutrition, Nantes Cedex 03, France

A. S. **Hodgson** (67), Department of Food Science and Human Nutrition, University of Hawaii at Manoa, Honolulu, Hawaii 96822

A. C. **Hoefler** (51), Hercules Incorporated, Middletown, New York 10940

L. H. **Kerr** (67), Food Technology Institute, Kingston, Jamaica

J. **Lefebvre** (227), Institut National de la Recherche Agronomique, Laboratoire de Physio-Chimie des Macromolécules, Nantes Cedex 03, France

D. G. **Oakenfull** (24, 87), Food Research Laboratory, CSIRO Division of Food Processing, New South Wales 2113, Australia

G. W. **Pilgrim** (24), The Red Wing Company, Fredonia, New York 14063

L. A. **Pitifer** (135), Department of Food Science and Technology, Cornell University, Geneva, New York 14456-0462

C.-G. **Qui** (228), Department of Food Science and Technology, Cornell University, Geneva, New York 14456

M. A. **Rao** (228), Department of Food Science and Technology, Cornell University, Geneva, New York 14456

F. M. **Rombouts** (119), Agricultural University of Wageningen, Department of Food Science, Bomenweg 2, The Netherlands

T. **Sajjaanantakul** (135), Department of Food Science and Technology, Faculty of Agro-Industry, Kasetsart University, Thailand

J.-F. **Thibault** (109, 119), Institut National de la Recherche Agronomique, Laboratoire de Biochimie et Technologie des Glucides, Nantes Cedex 03, France

J. P. Van Buren (1), Department of Food Science and Technology, Cornell University, Geneva, New York 14456

R. H. Walter (24, 189), Department of Food Science and Technology, Cornell University, Geneva, New York 14456

PREFACE

During the approximately four decades that elapsed after publication in 1951 of Dr. Kertesz's now classic book, *The Pectic Substances*, a fundamental understanding of polymers has evolved concurrently with advances in analytical instrumentation. The theories and methodologies developed during this interval are broadly applicable to the galacturonans. These biopolymers, collectively called pectin, have seldom been discoursed comprehensively in the context of the new knowledge; nor has this knowledge hitherto been systematically organized into a sequel to Dr. Kertesz's text. This is the object of *The Chemistry and Technology of Pectin.*

The current text attempts to explain the scientific and technical basis of many of the practices followed in processing and preparing foods fabricated with or containing pectin. The subject matter was engendered of the research programs and practical experiences of the authors. It was compiled with a minimum of abstruse discussion and mathematical treatment in the hope of facilitating its comprehension by the diverse personnel involved in the preparation and processing of fruit products for food. Thus, it provides information useful to the manufacturer and the researcher, the student and the teacher, the scientist and the technologist. Nonfood uses are included to illustrate how pectin, a multifunctional macromolecule, is experiencing wide applicability in a myriad of industries.

Given the limitations on the practical size of the current text, the authors made difficult choices in selecting data. Indeed, a compilation of equal size is possible from the omitted information. Nevertheless, many important aspects of pectin have been elaborated, from its isolation from the cell-wall middle lamella to its ultimate use in modern fabricated foods; the intention, as much as possible, has been to accelerate a redirection of its behavior toward contemporary scientific principles and methodologies.

CHAPTER 1

Function of Pectin in Plant Tissue Structure and Firmness

J. P. Van Buren

Department of Food Science and Technology
Cornell University
Geneva, New York

I. INTRODUCTION

Pectin substances are complex mixtures of polysaccharides that make up about one third of the cell-wall dry substance of dicotyledonous and some monocotyledonous plants (Hoff and Castro, 1969; Jarvis et al., 1988). Much smaller proportions of these substances are found in the cell walls of grasses (Wada and Ray, 1978). The location of pectin in the cell-wall–middle lamella complex has been known since the earliest work on this material (Kertesz, 1951). Highest concentrations are seen in the middle lamella, with a gradual decrease as one passes through the primary wall toward the plasma membrane (Darvil et al., 1980). Digestion of tissues with pectolytic enzymes leads to dissolution of the middle lamella and cell separation (Vennigerholz and Wales, 1987).

The pectic substances contribute both to the adhesion between the cells and to the mechanical strength of the cell wall, behaving in the manner of stabilized gels (Jarvis, 1984). They are brought into solution more easily than other cell-wall polymers, although their extractability varies widely from species to species. They have a higher degree of chemical reactivity than do other polymeric wall components. Physical changes, such as softening, are frequently accompanied by changes in the properties of the pectic substances.

Tissue firmness can be described as the resistance to a deforming force. Resistance to small deforming forces results from turgor and cell-wall rigidity. When forces are great enough, they can lead to irreversible changes in the conformation of a tissue. At the cellular level, several things may take place. There can be a breakage of cell walls accompanied by release of vacuolar contents. When this type of failure takes place with cells that consist of a large proportion of vacuoles, the tissues can often be described as juicy. Another type of conformation change is a separation of one cell from another. There is a failure at the adhesive layer between the cells. When this loss of adhesion is pronounced, there are extensive cell separations resulting from applied forces, and this type of tissue is frequently described as having a mealy or slippery character.

The type of failure that actually takes place is determined by the relative strengths of the cell wall and of the adhesive layers between the cells. Before the formation of the secondary cell wall, the strength of adhesion between cells is often greater than the strength of the primary cell wall; consequently, cell-wall breakage is the usual result of excessive deforming forces. The primary cell wall is composed of interwoven cellulose fibrils embedded in an amorphous polysaccharide matrix. Its strength is related to its thickness.

Force-induced failures of intercellular adhesion are seen in some ripe or overripe fruits, in heated tissues, and in special areas such as abscission zones. In these cases there have usually been changes that resulted in a weakening of the middle lamella and degradation of its pectin component. Once there has been extensive cell separation, it becomes difficult to cause cell-wall-breakage.

II. PECTIN

A. Classes

Pectins have frequently been classified by the procedures used to extract them from cell walls. In general, three types have been distinguished: water-soluble pectins extractable with water or dilute salt solutions; chelator-soluble pectins extractable with solutions of calcium chelating agents such as ethylenediaminetetraacetic acid (EDTA), (cyclohexanediaminotetraacetic acid (CDTA), or hexametaphosphate; and protopectins that are brought into solution with alkali solutions or hot dilute acids. The difficulty in extracting protopectin may be owing to acid-and/or alkali-labile bonds that secure the protopectin in the primary cell wall matrix. A large part of the protopectin can be solubilized by $0.05M$

Na_2CO_3, but a small fraction remains insoluble after the use of extractants as strong as $4M$ KOH (Massiot *et al.*, 1988; Ryden and Selvendran, 1990). Selvendran (1985) has suggested that the water-soluble and chelator-soluble pectins are derived from the middle lamella. It is possible that a part of the protopectin chain is imbedded in the cell wall, with the rest extending into the middle lamella.

The proportions of these pectin types vary considerably between different tissues. In carrots and snap-bean pods (Sajjaanantakul *et al.*, 1989) most of the pectin is of the chelator-soluble type. In ripe and even senescent apples, most is of the protopectin type (Massey *et al.*, 1964; O'Beirne *et al.*, 1982). In some other ripe fruits, such as freestone peaches (Postlmayr *et al.*, 1956), most of the pectin is of the water-soluble type, while in ripe clingstone peaches, approximately equal proportions of all three types were found. In tissues such as carrots, potatoes, and snap-bean pods with high proportions of chelator-soluble pectin, the infusion of chelators into the tissue results in dramatic losses of cohesion (Linehan and Hughes,1969; Van Buren *et al.*, 1988). Tissues such as beet root, with a high proportion of protopectin, show little loss of cohesion when treated with chelating agents (S. Shannon, unpublished data, 1975).

The water-soluble and chelator-soluble pectins are typically composed mainly of galacturonic acid residues with about 2% rhamnose and 10–20% neutral sugar. The distribution as well as the number of free carboxyl groups may be important in affecting whether a pectin is water soluble or chelator soluble. The protopectins, particularly if they are extracted with alkali, have high neutral sugar content (Selvendran, 1985), mainly galactose and arabinose. Commercially prepared pectins often resemble water-soluble and chelator-soluble pectins in their composition, but it is likely that many of their neutral sugars have been removed by hydrolysis during extraction.

It seems that the major contribution to intercellular adhesion comes from the chelator-soluble fraction and the protopectin. In general, softening during ripening (Massey *et al.*, 1964; Gross, 1984) or heating (Van Buren *et al.*, 1960b) is accompanied by a loss of protopectin and an increase in water-soluble pectin.

B. Structure

The principal constituent of the pectin polysaccharides is D-galacturonic acid, joined in chains by means of α-(1→4) glycosidic linkages. Inserted into the main uronide chain are rhamnose units, joined to the reducing end of the uronide by (1→2) linkages and the nonreducing end of the next uronide unit by (1→4) bonds. Rhamnose introduces a kink into the otherwise straight chain. The mole-percent of rhamnose in potato chelator-soluble pectin was much lower than that found in potato protopectin (Ryden and Selvendran, 1990).

Often, arabinan, galactan, or arabinogalactan side-chains are linked (1→4) to the rhamnose. In the side-chains, the arabinose units have (1→5) linkages while galactoses are mutually joined mainly by (1→4) linkages, but (1→3) and

(1→6) linkages also occur. Other sugars, such as D-glucuronic acid, L-fucose, D-glucose, D-mannose, and D-xylose are sometimes found in side-chains.

The pectin material solubilized in water or in solutions of calcium chelators shows variation with regard to the distribution of the rhamnose units along the main chain. For large parts of the chains, the rhamnose may be distributed in a fairly regular fashion, since acid hydrolysis gives segments 25–35 units long (Powell *et al.*, 1982). In other regions, the L-rhamnose units are closer together, and in these regions there are two to three moles of neutral sugar per mole of galacturonic residue (DeVries *et al.*, 1982).

The size of the neutral sugar side-chains appears to differ between the sparsely rhamnosylated regions and the densely rhamnosylated regions. Using the assumptions that all the rhamnose units have neutral sugar chains, and these chains join only to rhamnose, one can conclude that the sparse regions have neutral sugar-chain lengths of 4 to 10 residues, while the dense regions have chain lengths of 8 to 20 residues (Selvendran, 1985; DeVries *et al.*, 1982). DeVries *et al.* (1982) have designated these dense regions as *hairy* regions, and the sparse regions as *smooth*.

Pectins from some species, such as beet, have significant amounts of acylation on the uronide residues (Kertesz, 1951). The most common substituent is acetate. The pectic fractions of carrot were acetylated to a degree similar to that of apricot pectins, 7–13%, but less than that for sugar beet pectins (Massiot *et al.*, 1988). Komalavilas and Mort (1989) have shown that acylation occurs at the O-3 position of the uronide residues in rhamnose-rich portions of pectin polymers. Ferulate and coumarate are found attached to neutral sugars (Fry, 1986).

C. Degree of Esterification

An important factor characterizing pectin chains is the degree of esterification (DE) of the uronide carboxyl groups with methyl alcohol. Pectins might be formed initially in a highly esterified form, undergoing some deesterification after they have been inserted into the cell wall or middle lamella. There can be a wide range of DEs dependent on species, tissue, and maturity. In general, tissue pectins range from 60 to 90% DE. Water-soluble pectins and protopectins have slightly higher DEs than do chelator-soluble pectins. It seems that the distribution of free carboxyl groups along the pectin chains is somewhat regular, and the free carboxyl groups are largely isolated from one another (DeVries *et al.*, 1986).

The DE has a bearing on the firmness and cohesion of plant tissues. Reductions in DE result in greater cohesion, which is particularly apparent in heated tissues. The pectin methylesterase enzyme (see Chapter 8), present in most tissues, can slowly bring about demethoxylation. This enzyme has a rather low activity in normal tissue, but it becomes much more active when tissue is damaged (Robinson *et al.*, 1949) by procedures such as heating to 50 to 80°C, bruising, chilling, or freezing. These conditions are often experienced during processing.

Table I Effect of preheating[a] on the methoxylation of potato tissue pectin

Preheat °C	Pectin DE	Ca in cell wall mM/g dry wt	Relative softness[b]
27	67	43	7.5
50	56	28	5.3
60	39	46	2.2
70	33	88	2.0
80	60	45	5.4

[a]60-min preheating of 1.2 cm diced potato.
[b]After boiling for 30 min (measured as penetrometer values). From Bartolome and Hoff (1972).

The effect of preheating diced potatoes on the pectin DE and softness and after cooking is shown in Table I. Low-temperature alkaline demethoxylation results in firmer heated tissue (Van Buren, unpublished data, 1990). Treatment of apple tissues to convert carboxyl groups to methylesters causes loss of intercellular cohesion (Knee, 1978).

The firming effect involves two separate phenomena. In fresh tissue, the formation of free carboxyl groups increases the possibilities and the strength of calcium binding between pectin polymers. In heated tissue, there is a combination of increased calcium binding and a decrease in the susceptibility of the pectin to depolymerization by β-elimination (Sajjaanantakul et al., 1989).

Some commodities that have been found to show firmer texture after activation of pectin methylesterase and consequent decrease in pectin DE are snap beans (Sistrunk and Cain, 1960; Van Buren et al., 1960a), cauliflower (Hoogzand and Doesburg, 1961), tomatoes (Hsu et al., 1965), cherries (Buch et al., 1961; LaBelle, 1971; Van Buren, 1974), potato (Bartolome and Hoff, 1972), apple (Wiley and Lee, 1970), cucumber (Sistrunk and Kozup, 1982), sweet potato (Buescher and Balmoori, 1982) and carrot (Lee et al., 1979).

In many tissues such as apples (O'Beirne et al., 1982) and tomatoes (Burns and Pressey, 1987), there are normal decreases in DE that are not accompanied by firming during ripening; this is because simultaneous pectin solubilizing and degradation reactions are occurring as part of the ripening and senescence processes.

III. INTERACTIONS

A. Entanglement and Cohesion

Pectin is present in the cell wall in a highly concentrated condition. Cell walls contain approximately 60% water and 40% polymers (Jarvis, 1982). Pectins make up 20–35% of the polymers; therefore, pectins are present at 8 to 14% (w/w) overall concentrations. Since pectin predominates in the middle lamella,

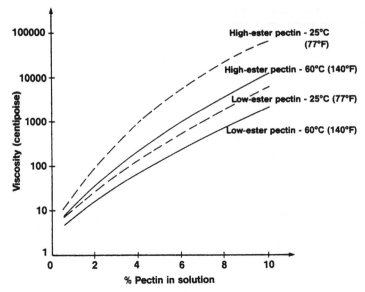

Figure 1 Typical viscosities of commercial high-ester and low-ester pectins at 25 and 60°C. From Copenhagen Pectin Factory (1985).

pectin in this structure may have a concentration in the order of 10 to 30%. Under these conditions, the pectin can behave as a highly coherent and strongly adhesive layer.

Properties of this pectin layer can be considered a combination of a viscous liquid and a stable cross-linked network (Fery, 1980). The viscous behavior will be strongly affected by the entanglements inherent in concentrated polymer systems (Kaelble, 1971), showing a steep rise in viscosity with increasing concentrations (Fig. 1).

The effect of entanglement on viscosity and effective cross-linking has been given a mathematical treatment by Kaelble (1971). If M is the molecular weight of the polymer and M_e is the average molecular weight between entanglements, then x_e is the mole fraction of the polymer enclosed within an entanglement network, and is given as

$$x_e = 1 - (2M_e/M).$$

As polymer concentration increases, so does x_e, up to a limit of 1.

A fundamental character of entanglement networks is a slippage factor S, which has a value from 1 to 0, with $S = 0$ when no slippage takes place, and $S = 1$ where entanglement does not affect chain movement through the network. Movement of the chains through the network is always constrained to some degree by the entanglement. A chain entangled with other chains can be regarded as being confined within a tunnel defined by the loci of its intersections with neighboring molecules (DeGennes, 1976). Transverse motions are prevented by neighbors, so the chain can move only by small displacements along the tunnel. This is called reptation. As the chain moves away from the tunnel formed by its contour conformation at some initial time, it continually finds itself in a new tunnel, thereby acquiring a new conformation. In the absence of cross-linking, the movement along the tunnel will be retarded by friction at the points of entanglement. This retardation and friction will be increased by side-chains on the main chain and results in decreased values for S.

Kaelble's analysis indicates that the effect of entanglement on the polymer viscosity is to increase viscosity by a factor equal to

$$(M/2M_e)(x_e^{(1-S)/S}),$$

thus the viscosity increases as x_e increases, as S decreases, and as M_e decreases with increasing concentration. Entanglement also decreases the rate of polymer diffusion, slowing the movement of chains from regions of high chain concentrations to regions of low chain concentrations.

Chemical cross-linkage creates additional constraints on movement. These cross-links may be easily reversible, such as hydrogen bonds, or nearly irreversible, such as diferulate linkages. As cross-linkages increase, the polymer molecules are brought more and more into a single effective molecule or branched structure in which the effective molecular weight (MW) is limited only by the summed weights of all the polymer chains.

The establishment of stable cross-linkages greatly retards reptation. The entanglements then behave as though they are also cross-links and contribute to a value describable as effective cross-links, which is the sum of chemical cross-linkages and entanglements. Therefore the entanglements contribute to the cohesion and dimensional stability of the branched structure.

The stress failure value for a cohesive network is a function of the density of the polymer, P, divided by the average molecular weight, M_x, between the effective cross-links; (P/M_x). Thus the more effective cross-links per network, the higher the stress failure value. At high polymer concentrations and extensive entanglement, near-maximal stability is achieved when only a few stable chemical cross-links per polymer chain have been established.

In the middle lamella, with pectin concentrations in the order of 10 to 30%, the large numbers of entanglements together with enough chemical cross-linking to prevent reptation will result in strong cohesiveness in the structure. In many species, this chemical cross-linking is carried out through Ca^{2+} (Van Buren, 1968; Demarty et al. 1984), while in others, such as beets, diferulate ester linkages may be important (Fry, 1983); Rombouts and Thibault, 1986).

Cell-to-cell adhesion requires components embedded in the primary walls that can be effectively entangled and cross-linked with the middle lamella pectin. These components may be pectins or hemicelluloses. How firmly these components are attached in the primary wall may also be determined by both entanglement and chemical cross-links. In the case of pectins, extremely low slippage factors are likely, because pectins associated with the primary wall have high proportions of neutral sugars (Selvendran, 1985) and consequently, extensive side-chains, some of which have their own sub-side-chains.

B. Calcium Cross-Links

The ability of calcium to form insoluble complexes with pectins is associated with the free carboxyl groups on the pectin chains. There is an increased tendency for gel formation as the DE of the pectins decreases (Anyas-Weisz and Deuel, 1950). Calcium linkages involve other functional groups in addition to the carboxyl groups (Deuel et al., 1950). The strong interaction between calcium and other oxygen atoms on the pectin has been described by Rees et al. (1982). Calcium complexes with neutral as well as acidic carbohydrates (Angyal, 1989). These complexes involve coordination bonds utilizing the unfilled orbitals of the calcium ion. The calcium ion is particularly effective in complexing with carbohydrates (Angyal, 1989), in large part because its ionic radius, 0.1 nm, is large enough that it can coordinate with oxygen atoms spaced as they are in many sugars, and because of a flexibility with regard to the directions of its coordinate bonds.

For the calcium-induced coagulation and gelation of pectin, a so-called egg box structure has been proposed (Rees et al., 1982), in which calcium ions ionically interact and coordinate with the oxygen functions of two adjacent chains, giving rise to a cross-linking of the chains. The calcium cross-linkages becomes more stable by the presence of cooperative neighboring cross-linkages, with maximal cross-link stability being reached when 7–14 consecutive links are present (Kohn and Luknar, 1977). Consecutive calcium links in plant cell walls are indicated by the electron spin resonance studies carried out by Irvin et al. (1984). Calcium chloride did not coagulate pectins with DEs over 60%, and the concentrations needed to coagulate low DE pectins increased as the viscosity MW decreased (Anyas-Weisz and Deuel, 1950).

It has long been known that hard water induces firmness in tissues (Bigelow and Stevenson, 1923). The effect is due to the calcium in the water. Interactions between calcium ions and cell-wall pectin play key roles in stabilizing wall structure (Demarty et al., 1984). Calcium has been used to maintain firmness in canned tomatoes (Loconti and Kertesz, 1941), apples (Wiley and Lee, 1970), carrots (Sterling, 1968), snap beans (Van Buren, 1968), cauliflower (Hoogzand and Doesburg, 1961), and brined cherries (Van Buren et al., 1967). Calcium pectate has been identified by Loconti and Kertesz (1941) as the wall component responsible for the firming effect. Most tissue calcium is associated with the cell wall–middle lamella (Demarty et al., 1984). Molloy and Richards (1971) have shown that pectin is the major calcium-binding constituent of the cell walls. Acylation of the uronide residues decreases the binding of calcium (Kohn and Furda, 1968).

In edible tissues the concentration of calcium is low; for example, potatoes have 0.07–0.13% calcium on a dry-weight basis, while apples are reported to have an average of 0.005% on a fresh-weight basis (Perring, 1974). There is considerable variation in the concentration of in different parts of a plant, e.g., the potato (Addiscott, 1974). A comprehensive coverage of calcium contents in edible plant parts has been compiled by Watt and Merrill, (1963).

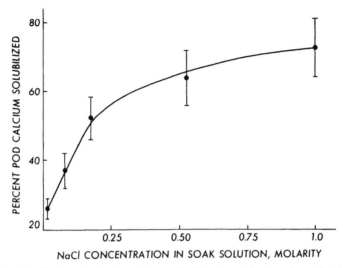

Figure 2   Solubilization of cooked snap bean pod calcium by soaking in sodium chloride solution. From Van Buren (1984).

In plant tissue, about 90% of the calcium is present in a bound or insoluble condition. Fifty to seventy percent is bound in a form easily displaced by molar NaCl concentrations (Fig. 2) (Jarvis, 1982, Van Buren, 1984). Such displacements result in a moderate loss in firmness (Fig. 3), but the most dramatic decreases in cohesion take place when tightly bound calcium is removed by chelating agents (Fig. 4) (Linehan and Hughes, 1969).

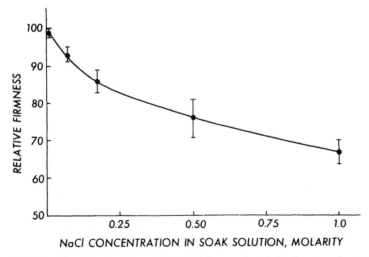

Figure 3   Effect of soaking in sodium chloride solutions on the relative firmness of canned snap beans. From Van Buren (1984).

J. P. Van Buren

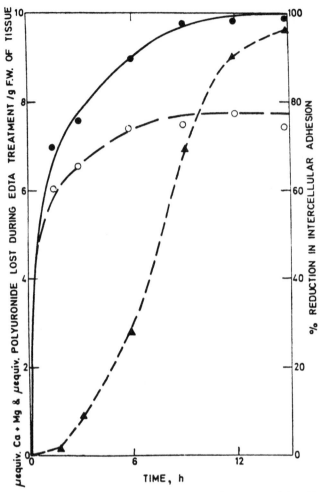

Figure 4   Effect of soaking in 0.05 M EDTA (pH 6.5) at 35°C on intercellular adhesion ▲---▲; calcium plus magnesium ○---○; and polyuronide ●---● lost from potato tuber sections. From Linehan and Hughes (1969).

The firmness of a heated tissue will be influenced by the relative concentrations of calcium and monovalent cations. Contour lines of constant relative firmness with canned snap beans for a range of calcium chloride and sodium chloride concentrations are shown in Fig. 5.

The use of chelating agents to remove tissue calcium and thereby increase the solubility of pectic materials has long been practiced; ammonium oxalate was one of the first used (Sucharipa, 1925). In later years, EDTA, CDTA, and sodium hexametaphosphate have been more widely employed (Selvendran, 1985). When these chelating agents are applied to plant tissues, cohesiveness is markedly decreased. This provides strong evidence that calcium–pectins are the principal

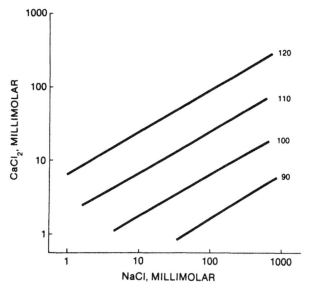

Figure 5   Contour lines of equal firmness for cooked snap bean pods held in solutions containing different concentrations of CaCl₂ and NaCl. Numbers next to lines give relative firmness. From Van Buren *et al.* (1988).

materials contributing to intercellular adhesion. It does not tell us the relative importance of chelator-soluble pectin and protopectin, since the use of chelating agents removes calcium from both these classes of pectins.

C. Ferulate Cross-Links

Possibilities for ester cross-linking of pectin chains have been raised by the discovery of the participation of ferulate and coumarate groups in promoting polysaccharide gel formation (Geissmann and Neukom, 1971; Rombouts and Thibault, 1986b). Ferulic acids form esters with neutral sugar hydroxyl groups. Cross-linking takes place, with the aid of peroxidase and H_2O_2, by the formation of a covalent bond between two ferulate phenyl rings (Fry, 1986). Ferulate residues associated with polysaccharides are more prominent in some species, such as beets and spinach, but they have not been found in potatoes and cherries (Rombouts and Thibault, 1986a). In the beet, little of the pectin is extractable with chelating agents, with most being extracted by alkali and hot acid. These acid and alkali extracts are relatively rich in ferulate residues, containing an estimated one diferulate ester group per pectin molecule (Rombouts and Thibault, 1986a).

In spinach, Fry (1983) has shown that feruloylation occurs on the nonreducing termini of the arabinose and/or galactose residues of the pectin fraction. Thus the diferulate linkages are between neutral sugar side-chains of the rhamnoga-lacturonans. There was about 1 feruloyl group per 100 uronide residues. En-zymatic digestion yielded arabinose or galactose feruloyl disaccharides, which accounted for more than 60% of the original ferulate. In the sugar beet, 20–30% of the pectic feruloyl groups are attached to terminal nonreducing arabinose residues, and the rest are attached to terminal nonreducing galactose residues (Guillon and Thibault, 1989). $NaClO_2$, which cleaves cinnamate derivatives, increases the amount of pectin solubilized from sycamore and kale (Fry, 1986).

D. Hydrogen Bonding and Hydrophobic Interaction

Two important types of noncovalent interchain associations are those mediated by hydrogen bonding and those enhanced by hydrophobic interaction (Tanford, 1961). They play an important part in pectin aggregate formation in the absence of cross-linking agents (Jordan and Brant, 1978; Davis *et al.*, 1980; Walkinshaw and Arnott, 1981).

Multiple hydrogen-bonding possibilities are present along the pectin chain owing to their oxygen functions and hydroxyl hydrogens. Walkinshaw and Arnott (1981), using models based on x-ray difraction studies, have shown that hydrogen bonds are favored by the conformations of adjacent uronide residues in the pectin chains. While hydrogen bonds are weak and easily broken, the large number of such bonds tends to stabilize polymer aggregates.

The methoxyl groups abundantly present in normal tissue pectins can partic-ipate in hydrophobic interactions. These interactions stabilize aggregates by reducing surface area at the interface between polar and nonpolar regions, thereby reducing the energy of the system. Conformational analysis shows that such regions can be formed by the methyl groups of adjacent pectin residues. Plash-china *et al.* (1985) showed that attractive forces exist between pectin molecules, owing to the methoxyl groups.

In isolated pectin systems, such as in pectin–acid–sugar gels, these weak bondings and interactions are believed to account for the formation of the junction zones that stabilize the gels (Oakenfull and Scott, 1984). Their participation in

Table II   The effect of urea on snap bean firmness[a]

First-soak solution	Second-soak solution	Relative firmness[b]
0.1 *M* NaCl	None	100 a
0.2% NaHMP[c] and 0.1 *M* NaCl	None	53 b
0.1 *M* NaCl	6 *M* urea and 0.1 *M* NaCl	98 a
0.2% NaHMP and 0.1 *M* NaCl	6 *M* urea and 0.1 *M* NaCl	36 c

[a]Measured after the last soak as resistance to compression.
[b]Means not followed by the same letter were significantly different at $P < 0.05$.
[c]Sodium hexametaphosphate (polyphosphate). From Van Buren *et al.* (1988).

cell adhesion is strongly suggested by the ability of urea to decrease firmness of some plant tissues (Linehan and Hughes, 1969; Van Buren *et al.*, 1988). High concentrations of urea have the ability to interface with noncovalent interactions between polymer chains (Von Hippel and Wong, 1965). These bonds may play only a minor role in normal tissues, since the urea effect is seen only when the tissues have been previously weakened by the removal of calcium (Table II).

E. Electrostatic Effects

Electrostatic repulsion plays a role in reducing chain aggregation. This is most readily seen in the effects of pH and electrolyte concentrations on the viscosity of solutions in which pectin behaves as a typical polyelectrolyte (Fuoss and Strauss, 1949). At pH higher than 7, there are maximal negative charges on the pectin chain owing to carboxyl group dissociation. At low electrolyte concentrations, there is little or no shielding of these charges, leading to a maximum of chain repulsion (Delahay, 1965), a minimum of aggregation, and a maximum elongation of the chains (Cesaro *et al.*, 1982). The result is a higher viscosity for the pectin solution compared to that found at moderate salt concentrations, such as 0.1 *M* NaCl (Michel *et al.*, 1982). In the same way, salt concentrations affect firmness in bean tissue (Van Buren, 1984) (Fig. 6). In this comparison,

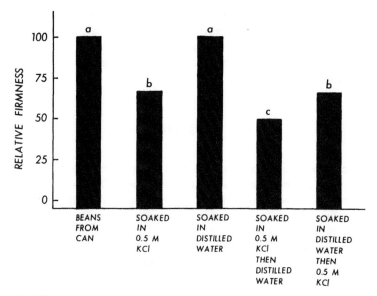

Figure 6. Effect on canned-bean firmness of different sequences of 0.5*M* KCl and distilled water soaks. Distilled water soaks were for 3 days with daily changes of 10 volumes of water. KCl soaks were for 3 days at 1°C. From Van Buren (1984).

Table III   The recementation of EDTA-treated potato tuber section by various cations

Cation (2.0 Eq/liter)	Percentage increase in intercellular adhesion compared with samples held in water[a]
Sodium	150
Potassium	190
Magnesium	280
Calcium	520

[a]Compressive strength of samples held in water was 0.7 kg/cm^2. From Linehan and Hughes (1969).

with beans that had been soaked in both distilled water and 0.5 M KCl, but in different sequences, less firmness was seen when the liquid present at the time of measurement was distilled water instead of 0.5 M KCl. Since both sets of treatments lead to similar calcium displacement, the increased firmness in the presence of KCl can be ascribed to decreased electrostatic repulsion. After calcium has been almost completely displaced by EDTA, the firming effect of monovalent ions can also been seen (Table III) in the case of potato sections. The high increase with calcium is probably attributable largely to the reformation of calcium cross-links in the protopectin network.

Increasing the pH of canned bean pods in the range of 4 to 7 caused a decrease in firmness; the magnitude of the decrease was dependent on the prior treatment of the pods or the presence of salts in the pH adjustment soak solutions (Van Buren et al., 1988). Loss of firmness as a result of higher pH was most pronounced when the pods had been previously depleted of calcium by the use of chelating agents, and the electrolyte concentration was at very low levels. The least change in firmness was seen when calcium was included in the pH adjustment solutions.

IV. RIPENING

Softening and loss of tissue cohesion has been studied more extensively in fruits than in any other higher plant organ. Extensive reviews (Poovaiah et al., 1988; Dey and Brinson, 1984; John and Dey, 1986; Knee and Bartley, 1981) have appeared dealing with the changes that accompany fruit softening. Changes in pectins have been a focus of interest, as well as polysaccharide-degrading enzymes that are present or develop at maturity or during senescence. Histological studies confirm that ripening-related fruit softening involves extensive middle-lamella disruption (Ben-Arie et al., 1979).

In general, and there are exceptions (Postlmayr et al., 1956), as fruits soften and lose cohesion, there is an increase in water-soluble pectins accompanied by a loss of protopectin (Bartley and Knee, 1982; Massey *et al.*, 1964; Knee *et al.*, 1977). This increase in water-soluble pectin is usually ascribed to the action of polygalacturonases, (Pressey *et al.*, 1971) acting in concert with other enzymes, such as pectin methylesterases and various glycosidases. In climacteric fruits, the glycosidases normally increase in amount or activity at the time of increased ethylene production.

There are two general types of polygalacturonases. Exo-polygalacturonases remove uronic acid residues from the nonreducing end of the pectin chains (Pressey and Avants, 1976). Endo-polygalacturonases cleave the α-1,4 glycosidic linkages randomly along the uronide chain. Both types of enzymes require deesterified residues. The action of the exo-polygalacturonase may be limited, in that its action is terminated by other residues, such as rhamnose units, along the chain. Both endo and exo enzymes have been found in many fruits such as pear, peach, cucumber, and tomato (Dey and Brinson, 1984). Nonspecified polygalacturonases are reported in avocado, pineapple, cranberry, grape, and dates.

Polygalacturonase enzymes increase in all the above fruits during ripening. Pectin methylesterase (see Chapter 8) is usually present in abundance in fruit tissue well before softening takes place. It is rather inactive *in situ,* unless the tissue is traumatized by such processes as heat shock, freezing, or maceration. Possibly there is normally a separation of the esterase from the methoxylated pectin, preventing access of the enzyme to the substrate. Some action of this enzyme appears necessary in order to have sufficient amounts of demethoxylated uronide residues to render pectin chains susceptible to attack by polygalacturonases (John and Dey, 1986).

The loss of galactose residues from cell-wall polysaccharides during ripening was first described by Tavakoli and Wiley (1968). More recently a survey of 15 species (Gross and Sams, 1984) showed 7 having a preferential loss of galactose and 7 having a preferential loss of arabinose. A good part of these neutral sugar residues can come from pectin side-chains. Implicated in the removal of these neutral sugars are a variety of fruit glycosidases (Wallner and Walker, 1975) detected using nitrophenyl glycoside substrates. To the extent that the removal of neutral sugars is from pectic material, there can be an influence of the neutral sugar glycosidases on pectin breakdown and solubilization. Removal of neutral sugar side-chains could increase the susceptibility of the pectin to attack by polygalacturonases and pectin methylesterases. Loss of side-chains would decrease the extent of entanglement and increase the slippage factor for the pectin chains.

A great deal of research has been done on the softening of tomato fruit. These fruits have unusually high levels of enzymes that attack pectins, both endo- and exopolygalacturonases and pectin methylesterase (Buescher and Tigchelaan, 1975; Tucker *et al.*, 1980). These enzymes are bound to the cell walls and can carry out autolysis of cell-wall pectins (Wallner and Bloom, 1977; Rushing and Huber, 1990). As ripening progresses, there are dramatic increases in the proportion of water-soluble pectin with MW in the range of 20,000 (Gross and Wallner, 1979).

Their low neutral sugar content is related to the loss of galactose residues from the cell-wall polysaccharides.

In contrast to the tomato, apples contain very low levels of polygalacturonase, and thus far only the exo-polygalacturonase has been identified (Bartley and Knee, 1982). While many varieties of apples become more juicy as they ripen, they never reach the degree of juiciness found in tomatoes and freestone peaches, and, at the overripe stage, most cultivars have a mealy character.

Simultaneously during softening, there is a decrease in the DE of the middle-lamella pectins, due to the action of pectin methylesterase (Hobson, 1963; Burns and Pressey, 1987). This renders the pectin more susceptible to endo-polyga-lacturonase. Several forms of pectin methylesterase are present (John and Dey, 1986).

The increase in water-soluble pectin is balanced, in part, by a decrease in protopectin. In the *nor* (nonripening) and *rin* (ripening inhibited) genotypes (Gross, 1984), there is little increase of water-soluble pectin, and polygalactu-ronase is largely absent.

These ripening-inhibited tomato genotypes have made possible extensive studies on the genetics of ripening and the relation of DNA to the ripening phenomena (Schuch *et al.*, 1989). The incorporation of polygalacturonase antisense genes leads to decreased levels of polygalacturonase and less decline in the average MW of the fruit pectin. Incorporation of the polygalacturonase gene into ripening-inhibited lines resulted in the formation of polygalacturonase in the transformed lines (Giovannoni *et al.*, 1989).

The presence of a variety of glycosidases in tomato has been shown by Wallner and Walker (1975). The galactosidases are present in several isozyme forms (Pressey, 1983), one of which hydrolyzed galactose from a tomato cell-wall polysaccharide fraction. In the *rin* mutant, galactose is lost during maturation (Gross and Wallner, 1979; Gross, 1984), while softening does not take place.

In the ripening tomato, two types of softening take place. Cell walls are weakened, and the placental gel is liquified, leading to a juicy character. At the same time, intercellular adhesion is decreased in the pericarp, leading to a tender mealiness in this region.

During ripening, there is a loss of cell-wall galactose. With the Cox's Orange Pippin apple cultivar, the loss of galactose takes place before significant pectin solubilization; it has been suggested that the removal of galactose side-chains is a precondition for the degradation of the main uronide chain. Apples contain an active galactosidase; however, it has not yet been demonstrated that this enzyme acts on apple cell-wall components (Bartley, 1977).

There is a great deal of variation in the behavior of different apple cultivars with regard to softening. Some, such as Yellow Transparent, soften very quickly, faster than tomatoes. Others, such as McIntosh and Gravenstein, soften at a moderate rate and can be stored for several months at low temperatures. Still others, such as York and Red Delicious, can be stored for 6 months or more and retain a crisp texture. No consistent relationships have been found between softening and measured amounts of water-soluble pectin, polygalacturonases, and galactosidases (DeVries *et al.* 1984; Poovaiah *et al.*, 1988).

Despite the difficulty in relating cell-wall compositional changes with differing rates of softening, histological evidence points to the involvement of pectin breakdown. Ben-Aries *et al.* (1979) showed a progressive splitting and dissolution of the pectin-rich apple middle lamella during softening conditions. Poovaiah *et al.* (1988) showed fracture at the middle-lamella regions when tissues are subjected to tensile stress. Apple softening may be greatly affected by changes in very restricted regions or planes of the middle lamella, so that detection of chemical changes in these critical regions is obscured by the main components of the cell wall–middle lamella complex.

A few general statements can be made. During softening there is a weakening of the cell walls and a decreased adhesion between the cells. Which process dominates varies between species of fruit and cultivars, but the two modes of softening appear to involve degradation and solubilization of pectin. The most likely mechanism is through the action of polygalacturonase on pectic substrates already suitably deesterified by pectin methylesterase. The loss of cell wall neutral sugars such as galactose and arabinose is consistently seen before and during softening, but there is as yet no established causal relationship to softening.

V. ABSCISSION

Abscission takes place in specialized abscission layers, where, at the onset of abscission, the cells are observed to become rounder, and the walls separate from each other (Sexton and Roberts, 1982). The result is a loss of adhesion. Plant hormones such as ethylene, auxin, and abscisic acid influence the course of the process. Pectic changes have been found to be involved, particularly the solubilization and loss of both methoxylated and acidic pectins (Berger and Reid, 1979). Accompanying these losses, the middle lamella swells, becomes porated, and then disappears.

Polygalacturonase is active in the abscission zone (Morre, 1969; Riov, 1975). Some reports suggest that a latent polygalacturonase is activated by ethylene, while others have evidence that ethylene produces its rapid effect by promoting enzyme formation (Abeles *et al.*, 1971).

Tissue fracture at abscission usually occurs long the planes of the middle lamella, leaving a high proportion of the cells on the fracture face still intact (Davenport and Marinos, 1971). At the time of separation, calcium, as detected by electron-probe X-ray analysis, is lost from the wall (Poovaiah and Rasmussen, 1973).

VI. SUMMARY

Pectin contributes to the firmness and structure of plant tissue as a part of the primary cell wall and as the main middle-lamella component involved in intercellular adhesion. It forms highly entangled networks further stabilized by calcium and alkali-labile cross-links. The strength of adhesion is usually greater than that of the primary cell walls; therefore, the force needed to cause irreversible structural changes is often determined by the strength of the wall. In those cases in which adhesion is weak, there normally has been a degradation of the pectin by enzymes or heat.

References

Abeles, F. B., Leather, G. R., Forrence, L. E., and Craker, L. E. (1971). Abscission: Regulation of senescence, protein synthesis, and enzyme secretion by ethylene. *Hort. Sci.* **6**, 371–376.

Addiscott, T. M. (1974). Potassium and the distribution of calcium and magnesium in potato plants. *J. Sci. Food Agric.* **25**, 1173–1183.

Angyal, S. J. (1989). Complexes of metal cations with carbohydrates in solution. *Adv. Carbohydr. Chem. Biochem.* **47**, 1–43.

Anyas-Weisz, L., and Deuel, H. (1950). Uber de Koagulation von Natriumpektinaten. *Helv. Chim. Acta* **33**, 559–562.

Bartley, I. M. (1982). A further study of β − galactosidase activity in apples ripening in store. *J. Exptl. Bot.* **28**, 943–948.

Bartley, I. M., and Knee, M. (1982). The chemistry of textural changes in fruit during storage. *Food Chem.* **9**, 47–58.

Bartolome, L. G., and Hoff, J. E. (1972). Firming of potatoes: Biochemical effects of preheating. *Agric. Food Chem.* **20**, 266–270.

Ben-Arie, R., Kislev, N., and Frenkel, C. (1979). Ultrastructural changes in the cell walls of ripening apple and pear fruit. *Plant Physiol.* **64**, 197–202.

Berger, R. K., and Reid, P. D. (1979). Role of polygalacturonase in bean-leaf abscission. *Plant Physiol.* **63**, 1133–1137.

Bigelow, W. D., and Stevenson, A. E. (1923). "The Effect of Hard Water in Canning Vegetables." National Canners Association Research Laboratories Bull, No. 20-L, Washington, D.C.

Buch, M. L., Satori, K. G., and Hills, C. H. (1961). The effect of bruising and aging on the texture and pectic constituents of canned red tart cherries. *Food Technol.* **15**, 526–531.

Buescher, R. W., and Balmoori, M. R. (1982). Mechanism of hardcore formation in chill-injured sweet potato (*Ipomea batatas*) roots. *J. Food Biochem.,* **6**, 1–11.

Buescher, R. W., and Tigchelaan, E. C. (1975). Pectinesterase, polygalacturonase, and Cx-cellulase activity in the *rin* tomato mutant. *Hort. Sci.,* **10**, 624–625.

Burns, J. K., and Pressey, R. (1987). Ca^{2+} in cell walls of ripening tomato and peach. *J. Amer. Soc. Hort. Sci.,* **112**, 783–787.

Cesaro, A., Ciana, A., Delben, F., Manzini, G., and Paoletti, S. (1982). Thermodynamic evidence of a pH-induced conformational transition in aqueous solution. *Biopolymers* **21**, 431–449.

Copenhagen Pectin Factory Ltd. (1985). "Handbook for the Fruit Processing Industry." Lille Skensved, Denmark.

Darvill, A., McNeil, M., Albersheim, P., and Delmer, D. P. (1980). The primary cell walls of flowering plants. *In* "The Biochemistry of Plants" (P. K. Stumpf and E. E. Conn, eds), Vol. 1, pp. 91–161. Academic Press, New York.

Davenport, T. I., and Marinos, N. G. (1971). Cell separation in isolated abscission zones. *Aust. J. Biol. Sci.* **24**, 709–714.

Davis. M. A. F., Gidley, M. J., Morris, E. R., Powell, D. A., and Rees, D. A. (1980). Intermolecular association in pectin solutions. *Int. J. Biol. Macromol.* **2**, 330–332.

DeGennes, P. G. (1976). Dynamics of entangled polymer solutions. The Rouse model and hydrodynamic interactions. *Macromolecules* **9**, 587–598.

Delahay, P. (1965). "Double Layer and Electrode Kinetics." Wiley, New York.

Demarty, M., Morvan, C., and Thellier, M. (1984). Calcium and the cell wall. *Plant, Cell Environ.* **7,** 441–448.

Deuel, H., Huber, G., and Anyas-Weisz, L. (1950). Uber 'Salzbrucken' zwischen Makromolekulen von Polyelektrolyten, besonders bei Calciumpektinaten. *Helv. Chim. Acta* **33,** 563–567.

DeVries, J. A., Hansen, M., Soderberg, J., Glahn, P.-E., and Pedersen, J. K. (1986). Distribution of methoxyl groups in pectins. *Carb. Polymers* **6,** 165–176.

DeVries, J. A., Rombouts, F. M., Vorgen, A. G. J., and Pilnik, W. (1982). Enzymic degradation of apple pectins. *Carbohydr. Polymers* **2,** 25–33.

DeVries, J. A., Vorgen, A. G. J., Rombouts, F. M., and Pilnik, W. (1984). Changes in the structure of apple pectic substances during ripening and storage. *Carbohydr. Polymers* **4,** 3–13.

Dey, P. M., and Brinson, K. (1984). Cell wall and fruit ripening, *Adv. Carbohydr. Chem. Biochem.* **42,** 339–382.

Fery, J. D. (1980). "Viscoelastic Properties of Polymers." Wiley, New York.

Fry, S. C. (1983). Feruloylated pectins from the primary cell wall: Their structure and possible functions. *Planta* **157,** 111–123.

Fry, S. C. (1986). Cross-linking of matrix polymers in the growing cell walls of angiosperms. *Annu. Rev. Plant Physiol.* **37,** 165–186.

Fuoss, R. M., and Strauss, V. P. (1949). The viscosity of mixtures of polyelectrolytes and simple electrolytes. *Ann. N.Y. Acad. Sci.* **51,** 836–851.

Geissmann, T., and Neukom, H. (1971). Vernetzung von Polysacchariden durch oxydative phen-olisch Kupplung. *Helv. Chim. Acta* **54,** 1108–1112.

Giovannoni, J. J., DellaPenna, D., Bennett, A. B., and Fisher, R. L. (1989). Expression of chimeric polygalacturonase gene in transgenic *rin* (ripening inhibitor) tomato fruit results in polyuronide degradation but not fruit softening. *Plant Cell* **1,** 53–59.

Gross, K. C. (1984). Fractionation and partial characterization of cell walls from normal and nonripening mutant tomato fruit. *Physiol. Plant* **62,** 25–32.

Gross, K. C., and Sams, C. E. (1984). Changes in neutral sugar composition during fruit ripening: A species survey. *Phytochem.* **23,** 2457–2461.

Gross, K. C., and Wallner, S. J. (1979). Degradation of cell-wall polysaccharides during tomato ripening. *Plant Physiol.* **63,** 117–120.

Guillon, F., and Thibault, J.-F. (1989). Enzymic hydrolysis of the "hairy" fragments of sugar beet pectins. *Carbohydr. Res.* **190,** 97–108.

Hobson, G.E. (1963). Pectinesterase in normal and abnormal tomato fruit. *Biochem. J.* **86,** 358–365.

Hoff, J. E., and Castro, M. D. (1969). Chemical composition of potato cell wall. *Agric. Food Chem.* **17,** 1328–1331.

Hoogzand, C., and Doesburg, J. J. (1961). Effect of blanching on texture and pectin of canned cauliflower. *Food Technol.* **15,** 160–163.

Hsu, C. P., Deshpande, S. N., and Desrosier, N. W. (1965). Role of pectin methylesterase in firmness of canned tomatoes. *J. Food Sci.* **30,** 583–588.

Irwin, P. L., Sevilla, M. D., and Shieh, J. J. (1984). ESR evidence for sequential divalent cation binding in higher plant cell walls. *Biochim. Biophys. Acta* **805,** 186–190.

Jarvis, M. C. (1982). The proportion of calcium-bound pectin in plant cell walls. *Planta* **154,** 344–346.

Jarvis, M. C. (1984). Structure and properties of pectin gels in plant cell walls. *Plant Cell Environ.* **7,** 153–164.

Jarvis, M. C., Forsyth, W., and Duncan, H. J. (1988). A survey of the pectic content of nonlignified monocot cell walls. *Plant Physiol.* **88,** 309–314.

John, M. A., and Dey, P. M. (1986). Postharvest changes in fruit cell wall. *In* "Advances in Food Research" (C. O. Chichester, ed.), Vol. 30, pp. 139–193. Academic Press, Orlando, Florida.

Jordan, R. C., and Brant, D. A. (1978). An investigation of pectin and pectic acid in dilute aqueous solution. *Biopolymers* **17,** 2885–2895.

Kaelble, D. H. (1971). "Physical Chemistry of Adhesion." Interscience, New York.

Kertesz, Z. I. (1951). "The Pectic Substances." Interscience, New York.

Knee, M. (1978). Properties of polygalacturonate and cell cohesion in apple fruit cortical tissue. *Phytochem.* **17,** 1257–1260.

Knee, M., and Bartley, I. M. (1981). Composition and metabolism of cell wall polysaccharides in ripening fruit. *In* "Recent Advances in the Biochemistry of Fruits and Vegetables" (J. Friend and M. J. C. Rhodes, eds.), pp. 133–148. Academic Press, New York.

Knee, M., Sargent, A. J., and Osborne, D. J. (1977). Cell-wall metabolism in developing strawberry fruits. *J. Exp. Bot.* **28,** 377–396.

Kohn, R., and Furda, I. (1968). Binding of calcium ions to acetyl derivatives of pectin. *Coll. Czech. Chem. Commun.* **33,** 2217–2225.

Kohn, R., and Luknar, O. (1977). Intermolecular calcium ion binding on polyuronates. *Coll. Czech. Chem. Commun.* **42,** 731–744.

Komalavilas, P., and Mort, A. J. (1989). The acetylation at O-3 of galacturonic acid in the rhamnose-rich portion of pectins. *Carbohydr. Res.* **189,** 261–272.

LaBelle, R. L. (1971). Heat and calcium treatments for firming red tart cherries in a hot-fill process. *J. Food Sci.* **36,** 323–326.

Lee, C. Y., Bourne, M. C., and Van Buren, J. P. (1979). Effect of blanching treatments on the firmness of carrots. *J. Food Sci.* **44,** 615–616.

Linehan, D. J., and Hughes, J. C. (1969). Texture of cooked potatoes. III. Intercellular adhesion of chemically treated tuber sections. *J. Sci. Food Agric.* **20,** 119–123.

Loconti, J. D., and Kertesz, Z. I. (1941). Identification of calcium pectate as the tissue-firming compound formed by treatment of tomatoes with calcium chloride. *Food Res.* **6,** 499–508.

Massey, L. M., Parsons, G. F., and Smock, R. M. (1964). Some effects of gamma radiation on the keeping quality of apples. *Agric. Food Chem.* **12,** 268–274.

Massiot, P., Rouau, X., and Thibault, J.-F. (1988). Characterization of the extractable pectins and hemicelluloses of the cell wall of carrot. *Carbohydr. Res.* **172,** 229–242.

Michel, F., Doublier, J. L., and Thibault, J.-F. (1982). Investigations on high-methoxyl pectins by potentiometry and viscometry. *Prog. Food Nutr. Sci.* **6,** 367–372.

Molloy, L. F., and Richards, E. L. (1971). Complexing of calcium and magnesium by cell-wall fractions and organic acids. *J. Sci. Food Agric.* **22,** 397–402.

Morre, D. J. (1969). Cell-wall dissolution and enzyme secretion during leaf abscission. *Plant Physiol.* **43,** 1545–1559.

Oakenfull, D., and Scott, A. (1984). Hydrophobic interactions in gelation of high-methoxyl pectins. *J. Food Sci.* **49,** 1093–1098.

O'Beirne, D., Van Buren, J. P., and Mattick, L. R. (1982). Two distinct pectin fractions from senescent Idared apples extracted using nondegradative methods. *J. Food Sci.* **47,** 173–176.

Perring, M. A. (1974). The mineral composition of apples. Method for calcium. *J. Sci. Food Agric.* **25,** 237–245.

Plashchina, I. G., Semenova, M. G., Braudo, E. E., and Tolstoguzov, V. B. (1985). Structural studies of the solutions of anionic polysaccharides. IV. Study of pectin solutions by light scattering. *Carbohydr. Polymers* **5,** 159–179.

Poovaiah, B. W., Glenn, G. M., and Reddy, A. S. N. (1988). Calcium and fruit softening: Physiology and biochemistry. *Hort. Rev.* **10,** 107–152.

Poovaiah, B. W., and Rasmussen, H. P. (1973). Calcium distribution in the abscission zone of bean leaves. *Plant Physiol.* **52,** 683–684.

Postlmayr, H. L., Luh, B. S., and Leonard, S. J. (1956). Characterization of pectin changes in freestone and clingstone peaches during ripening and processing. *Food Technol.* **10,** 618–625.

Powell, D. A., Morris, E. R., Gidley, M. J., and Rees, D. A. (1982). Conformation and interaction of pectins. II. Influences of residue sequence on chain association in calcium pectate gels. *J. Mol. Biol.* **155,** 517–531.

Pressey, R. (1983). β-galactosidases in ripening tomatoes. *Plant Physiol.* **71,** 132–135.

Pressey, R., and Avants, J. K. (1976). Pear polygalacturonases. *Phytochem.* **15,** 1349–1351.

Pressey, R., Hinton, D. M., and Avants, J. K. (1971). Development of polygalacturonase activity and solubilization of pectin in peaches during ripening. *J. Food Sci.* **36,** 1070–1073.

Rees, D. A., Morris, E. R., Thom, D., and Madden, J. K. (1982). Shapes and interactions of carbohydrate chains. *In* "The Polysaccharides" (G. O. Aspinall, ed.), pp. 195–290. Academic Press, New York.

Riov, J. (1975). Polygalacturonase activity in citrus fruit. *J. Food Sci.* **40**, 201–202.

Robinson, W. B., Moyer, J. C., and Kertesz, Z. I. (1949). Thermal maceration of plant tissue. *Plant Physiol.* **24**, 317–319.

Rombouts, F. M., and Thibault, J.-F. (1986a). Feruloylated pectic substances from sugar-beet pulp. *Carbohydr. Res.* **154**, 177–187.

Rombouts, F. M., and Thibault, J.-F. (1986b). Sugar beet pectins: Chemical structure and gelation through oxidative coupling. *In* "Chemistry and Function of Pectins" (M. L. Fishman and J. J. Jen, eds.), pp. 49–60. American Chemical Society, Washington, D.C.

Rushing, J. W., and Huber, D. J. (1990). Mobility limitations of bound polygalacturonase in isolated cell wall from tomato pericarp tissue. *J. Am. Soc. Hort. Sci.* **115**, 97–101.

Ryden, P., and Selvendran, R. R. (1990). Structural features of cell-wall polysaccharides of potato (*Solanum tuberosum*). *Carbohydr. Res.* **195**, 257–272.

Sajjaanantakul, T., Van Buren, J. P., and Downing, D. L. (1989). Effect of methyl ester content on heat degradation of chelator-soluble carrot pectin. *J. Food Sci.* **54**, 1272–1277.

Schuch, W., Bird, C. R., Ray, J., Smith, C. J. S., Watson, C. F., Morris, P. C., Gray, J. E., Arnold, C., Seymour, G. B., Tucker, G. A., and Grierson, D. (1989). Control and manipulation of gene expression during tomato fruit ripening. *Plant Mol. Biol.* **13**, 303–311.

Selvendran, R. R. (1985). Developments in the chemistry and biochemistry of pectic and hemicellulosic polymers. *In* "The Cell Surface in Plant Growth and Development" (K. Roberts, A. W. B. Johnston, C. W. Lloyd, P. Shaw, and H. W. Woolhouse, eds), pp. 51–88. Company of Biologists, Cambridge, England.

Sexton, R., and Roberts, J. A. (1982). Cell biology of abscission. *Annu. Rev. Plant Physiol.* **33**, 133–162.

Sistrunk, W. A., and Cain, R. F. (1960). Chemical and physical changes in green beans during preparation and processing. *Food Technol.* **14**, 357–362.

Sistrunk, W. A., and Kozup, J. (1982). Influence of processing methodology on quality of cucumber pickles. *J. Food Sci.* **47**, 949–953.

Sterling, C. (1968). Effect of solutes and pH on the structure and firmness of cooked carrot. *J. Food Technol.* **3**, 367–371.

Sucharipa, R. (1925). "Die Pektinstoffe." Serger and Hempel, Braunschweig, Germany.

Tanford, C. (1961). "Physical Chemistry of Macromolecules." Wiley, New York.

Tavakoli, M., and Wiley, R. C. (1968). Relation of trimethylsilyl derivatives of fruit-tissue polysaccharides to apple texture. *Proc. Am. Soc. Hort. Sci.* **92**, 780–787.

Tucker, G. A., Robertson, N. G., and Grierson, D. (1980). Changes in polygalacturonase isoenzymes during ripening of normal and mutant tomato fruit. *Eur. J. Biochem.* **112**, 119–124.

Van Buren, J. P. (1968). Adding calcium to snap beans at different stages in processing. Calcium uptake and texture of the canned product. *Food Technol.* **22**, 790–793.

Van Buren, J. P. (1974). Heat treatments and the texture and pectins of red tart cherries. *J. Food Sci.* **39**, 1203–1205.

Van Buren, J. P. (1984). Effects of salts added after cooking on the texture of canned snap beans. *J. Food Sci.* **49**, 910–912.

Van Buren, J. P., Kean, W. P., and Wilkison, M. (1988). Influence of salts and pH on the firmness of cooked snap beans in relation to the properties of pectin. *J. Texture Stud.* **19**, 15–25.

Van Buren, J. P., LaBelle, R. L., and Splittstoesser, D. F. (1967). The influence of SO_2 level, pH, and salts on color, texture, and cracking of brined Windsor cherries. *Food Technol.* **21**, 1028–1030.

Van Buren, J. P., Moyer, J. C., Wilson, D. E., Robinson, W. B., and Hand, D. B. (1960a). Influence of blanching conditions on sloughing, splitting, and firmness of canned snap beans. *Food Technol.* **14**, 233–236.

Van Buren, J. P., Moyer, J. C., Robinson, W. B., and Hand, D. B. (1960b). Pectische

Veranderungen bei der Verarbeitung gruner Bohnen. *Hoppe-Seyler's Zeit. physiol. Chemie* **321,** 107–113.

Vennigerholz, F., and Wales, B. (1987). Cytochemical studies of pectin digestion in epidermis with specific cell separation. *Protoplasma* **140,** 110–117.

Von Hippel, P. H., and Wong, K. Y. (1965). On the conformational stability of globular proteins. *J. Biol. Chem.* **240,** 3909–3923.

Wada, S., and Ray, P. M. (1978). Matrix polysaccharides of oat celeoptile cell walls. *Phytochem.* **17,** 923–931.

Walkinshaw, M. D., and Arnott, S. (1981). Models for junction zones in pectinic acid and calcium pectate gels. *J. Mol. Biol.* **153,** 1075–1085.

Wallner, S. J., and Bloom, H. L. (1977). Characteristics of tomato cell wall degradation *in vitro*. Implications for the study of fruit-softening enzymes. *Plant Physiol.* **60,** 207–210.

Wallner, S. J., and Walker, J. E. (1975). Glycosidases in cell-wall degrading extracts of ripening tomato fruits. *Plant Physiol.* **55,** 94–98.

Watt, B. K., and Merrill, A. L. (1963). "Composition of Foods." U.S. Department of Agriculture Handbook 8 (rev.), Washington, D.C.

Wiley, R. E., and Lee, Y. S. (1970). Modifying texture of processed apple slices. *Food Technol.* **24,** 1168–1170.

CHAPTER 2

Jams, Jellies, and Preserves

G. W. Pilgrim
The Red Wing Company, Inc.
196 Newton Street
Fredonia, New York

R. H. Walter
New York State Agricultural Experiment Station
Cornell University
Department of Food Science
Geneva, New York

D. G. Oakenfull
Food Research Laboratory
CSIRO Division of Food Processing
North Ryde
New South Wales, Australia

I. Introduction
II. Commercial Pectin Extraction
III. Form and Function of Ingredients
  A. Pectin Stability
  B. Methylester Content and Setting Rates
  C. Physical State of Pectin
  D. pH Effects
  E. Soluble-Solids Content
  F. Calcium
IV. Processing Variables
  A. Fruit Preprocessing
  B. The Cooking Temperature
  C. The Setting Time and Temperature
  D. Size of Containers
  E. Order of Ingredients Addition
V. Texture Measurements
VI. Miscellaneous Pectins and Products

A. Confectionery Pectins
B. Fluid High-Methoxyl Gels
VII. Practical Problems and Solutions
  A. Floating Fruit Pieces
  B. Opacity
  C. Discoloration
  D. Pregelation
  E. Syneresis Control
  F. Miscellaneous Problems
VIII. Quality Assurance and Control
  A. Color of Jelly
  B. Calculation of Unknown Ratio of Fruit to Sugar
  C. Calculation of Jelly Yield
  D. Calculation of Acid
  E. Calculation of Liquid Additions
References

I. INTRODUCTION

Jams, jellies, preserves, and other pectin products are an example of art preceding science. From the time of Goldthwaite (1909–1917) who, it is said (Charley, 1970), first attempted to put jelly making on a scientific basis, manufacturing of pectin food products has progressed to a truly scientific and technological enterprise. The process of pectin gelation begins with extraction of the pectin under controlled conditions to effect certain outcomes, and ends with a visco-elastic or elastic body whose behavior is now interpreted within the framework of contemporary polymer principles. Current knowledge has permitted the use of pectin far beyond its original application in jelly manufacturing.

II. COMMERCIAL PECTIN EXTRACTION

Most of the pectin of commerce is extracted from citrus peels that contain 25% pectin (Keller, 1983), and dried apple pomace, which contains 15–18% pectin (Hang and Walter, 1989). Initially, these sources undergo pretreatment involving cleaning to remove foreign particles, washing to remove sugar and acid, distilling (of lemon oil from citrus), inactivating the demethylating and depolymerizing enzymes, drying, comminuting, and storage. The substrate material is refluxed for several hours, with agitation, with one of various concentrations of sulfurous, sulfuric, nitric, or hydrochloric acid. The acid acts on the insoluble protopectin to give the useful moiety called pectin. Sulfurous acid as SO_2 offers the advantage of being removable by aeration after the treatment. This gas is sparged in the refluxing matrix to bring it to pH 1.8 to 2.7. Sometimes, a residue of sulfurous acid, because of its antimicrobial activity, is beneficial to storage of the crude extract.

   The greater the acidity in the extraction solvent, the higher the yield of pectin. However, high acidities might result in a deterioration of certain desirable physical properties (Joseph and Havighorst, 1952). After the refluxing interval of 2 to 10 hr from 50 to 100°C, the liquid fraction is decanted or filtered and cooled. When the substrate material is apple pomace, it is necessary to treat the extract with amylase for the purpose of hydrolyzing starch to water-soluble maltose (Keller, 1984), since congeneric starch would contaminate pectin in its final form. Afterwards, ammonium chloride is dissolved in the extract, and aluminum pectinate, a colloidal precipitate, is formed. If sulfurous acid was used, the extract is aerated until the pH rises to 3.2 to 4.0. Acid other than sulfurous acid would also cause the refluxing to terminate below pH 3.2 to 4.0, so an appropriate quantity of sodium carbonate would be added until the mixture is neutralized to pH 3.2 to 4.0, where aluminum pectinate becomes hydrolyzed to pectin and aluminum hydroxide. The crude pectin isolate is freed of aluminum hydroxide and aluminum ions by washing with acidified isopropanol (HCl, pH

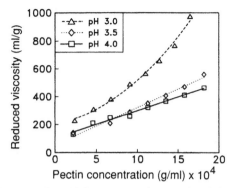

Figure 1   Effect of pectin (DE = 72.9%) concentration on reduced viscosity at different pH at 25°C. Redrawn from Michel *et al.* (1985).

1.0). The immersion must be of short duration, or extensive demethylation might occur. Subsequently, the alcoholic mixture is partially neutralized with ammonia (to pH 3.6–4.2), and the solvent is separated by decantation and filtration or distillation. The reagents used in the extraction have a significant impact on the character of the pectin isolated.

Alternatively, pectin may be isolated and purified after refluxing by direct precipitation with alcohol, followed by washing. The moist pectin isolate, containing a high percentage of methyl esters, is dried *in vacuo* from approximately 60% to 7 to 10% moisture. Isolated pectin exhibits viscous flow that is typical of hydrocolloids (Fig. 1) in the range of pH in which it is often used as a thickener.

Pectins are classified as high-methoxyl (HM) and low-methoxyl (LM) pectins, depending on the degree of esterification (DE). The cut-off DE between HM and LM pectins is arbitrarily 40 (Fig. 2) to 50% (Ahmed, 1981; Hercules, Inc., 1985). Pectins isolated by the methods outlined above are HM types. LM pectins are the product of further regulated acid, alkaline, or enzyme treatment of HM pectins. Most of the pectins of commerce are ground and screened versions of these two kinds of pectin, having a bulk density of approximately 0.7 g/cm^3.

III. FORM AND FUNCTION OF INGREDIENTS

Pectin was defined by Kertesz (1951) as those water-soluble pectinic acids of varying methylester content and degree of neutralization that are capable of forming gels with sugar and acid under suitable conditions. This definition is herein extended to include gelation with calcium of pectinic acids, defined by Kertesz (1951) as the colloidal polygalacturonic acids isolated from plants, containing more than a negligible proportion of methyl ester groups. Herein also,

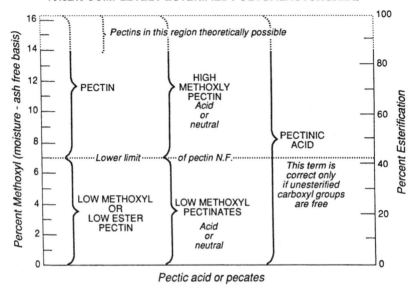

Figure 2 Classification of pectins on the basis of methylester content. Reproduced with permission from Joseph (1953).

pectin is used collectively to include pectic acid, the completely deesterified form of pectin. The pectin forms are generally recognized as safe (GRAS) by the U.S. Food and Drug Administration (Food and Nutrition Encyclopedia, 1983). Legal specifications for them are enunciated internationally (Hercules, Inc., 1985).

Pectin gelation depends on the chemical nature of the pectin, the soluble-solids content, and pH or calcium. The pectin may be in either a liquid dispersion or in powder form. Tressler and Woodroof (1976) outlined a recipe for liquid pectin for household use, consisting of 3 lb of 150-grade, slow-set pectin, 3 lb of citric acid, 10 lb 9 oz of granulated sugar and 7 gal 104 fl oz of water. The combination makes about 9 gal of stock dispersion. Recipes for its use are given. Such aqueous pectin dispersions age rapidly in storage, and it is best not to prepare more than what will be consumed in 24 hr.

From a scientific standpoint, pectin gelation is a continuous series of physical–chemical events whereby a fluid dispersion is converted at a temperature, or over a short temperature range, to a definite, three-dimensional structure at the end of some finite interval. The gel reaches its maximum strength (setting) during the following 24 hr. In this chapter, gelation and setting, differing only in a time context, are used interchangeably.

It has been suggested that deesterification of pectic substances, initiated *in situ* by the action of pectin methylesterase on damaged tissue, firms fruits, and vegetables by strengthening the cell walls and enhancing intercellular cohesion *via* a mechanism involving calcium (Van Buren, 1974).

Strawberries were once packed in syrup with LM pectin at 0.10 to 0.15% of the weight of the berries. The principle underlying the custom was that calcium exuding in the juice from broken cells would react with the pectin to form a calcium pectate seal, thus blocking the capillary exit channels of the exudate on the broken surface (Joseph, 1953). There is no documentation to show that this procedure was effective.

A. Pectin Stability

LM pectins are more chemically stable to moisture and heat than are HM pectins, because of the latter's tendency to deesterify in a humid atmosphere. This is especially true at elevated temperatures. HM pectins must consequently be thoroughly dried before sealing in moisture-proof containers. The containers should be stored in a cool, dry place. The two kinds of pectin are relatively stable at the low pH levels existing in jams and jellies, but prolonged heating in a strongly acidic or alkaline medium might lead to lower-molecular-weight pectins and should therefore be avoided. The higher the DE, the greater the probability of demethylation. HM and LM pectins are equally susceptible to acidic depolymerization under severely high temperature conditions. A combination of high pH, elevated temperature and long holding time increases the probability that liquid pectin dispersions will lose viscosity through β-elimination of water. This occurrence is evidenced by splitting of the pectin molecule into two moieties, one of which develops monounsaturation.

B. Methylester Content and Setting Rates

Rapid-set and slow-set designations of pectin refer to the rate at which a jelly's incipient structure develops at or near the gelation temperature. Their rate of gelation influences the product's texture. HM pectins are either rapid- or slow-setting. The rate of setting declines as many as threefold with declining DE. Intermediate rates lead to designations such as medium rapid-set, medium slow-set, etc. HM pectins gel at a faster rate than do LM pectins. HM pectins with higher DE gel at a higher temperature than do HM pectins with lower DE, under the same cooling gradient.

Standard jellies are normally made with HM slow-set pectin. The slow rate of gelation allows enough time (25–30 min) for air bubbles trapped in the cooling sol during pouring to escape from the container, and for other operations, e.g., filling, capping, and labeling, to be completed before setting, so that the process is not disturbed. In preserves, rapid-set or medium rapid-set pectins ensure uniform distribution of fruit pieces throughout the gel by holding the pieces in place before they have time to settle or float. Rapid-set pectin will allow jelly products to be made in the pH range 3.30 to 3.50. Slow-set pectins perform best at approximately pH 2.8–3.2. A mixture of HM and LM pectins will sometimes impart a degree of thixotropy to a jelly sol.

Some natural pectins, e.g., beet pectins, are acetylated. The exact acetyl locations are not always known with certainty, but the C-2 and C-3 hydroxyl groups, as well as the C-6 carboxyl group, have been implicated. The C-2 and C-2 hydroxyl groups form reverse esters, in that the larger pectin molecule is the alkyl group. As a consequence of the deesterification reaction with ammonia, the C-6 carboxyl group of some LM pectins may be amidated.

A jelly is considered to be normal by U.S. standards, if it has a ridgelimeter sag of 23.5%. A ridgelimeter is the device with which all high-ester pectins are standardized in the United States (IFT, 1959). British manufacturers have reconciled their standards, established under their conditions, with U.S. standards, by increasing the amount of pectin by 1.09 times the U.S. equivalent (B.F.M.I.R.A., 1951). The amount of pectin in a jelly that is to be set to the standard firmness depends primarily on the pectin grade. Soluble-solids (ss) content and pH may, however, be adjusted to meet textural specifications other than that of standard jelly. The pectin requirement is fairly consistent when depectinized juice is used but is subject to fluctuations for preserves, because the pectin present in fruit varies with climate and season.

The amount of LM pectin to be used in LM pectin jellies is determined by the quality and properties of the expected end-product. These jellies are relatively independent of pH, and are mostly dependent on calcium. Amidated and conventional LM pectins differ in their physical characteristics. At a constant pH and sugar concentration, amidated pectins tend to give a rigid gel similar to that of HM pectins, while the conventional LM pectins tend to give a softer, spreadable gel. At a constant ss content, the amidated pectins require a slightly lower use-level than the conventional LM pectins, to reach a given set condition. The amidated pectins are far less shear reversible at pH 3.5 and below than are the conventional LM pectins. The use-level of conventional LM pectins is 10–20% higher than that of amidated pectins, to achieve equal firmness (Hercules, Inc., 1985).

Pectin jellies are sometimes an ingredient of baked goods. The structure of the jelly incorporated in pastries should not disintegrate when exposed to heat. There are two ways to circumvent this potential problem; first, since the flow of HM pectin gels is not normally thermally reversible at pastry internal baking temperatures, and since the rapid-set types have a higher gelation temperature than do the slow-set types, HM rapid-set jelly is preferable to HM slow-set jelly. Second, the flow of LM pectin jelly is normally thermally reversible at baking temperatures, and as a result, the gel structure is not likely to be damaged by heat. There is, however, the problem of containment of LM jellies within the sharp boundaries delineated by the dough or batter. In this instance, a high calcium concentration will make a cohesive jelly with a high sol–gel conversion temperature. When containment is not essential, either HM or LM pectin may be used satisfactorily. The choice is immaterial for prebaked items like jelly donuts, in which heating precedes the injection of jelly.

C. Physical State of Pectin

Pectin for the manufacture of jams and jellies may be added in the solid (powdery) state or dispersed in water. Either form should be completely dispersed by the time the boiling temperature is reached, and certainly before the acid or calcium is added. In the powdery state, dispersion is facilitated by preblending it with no more than 20–25% its weight in sugar. Dispersion is retarded above this limit. Syrup and honey, exercising the dual function of particle separation and water dispersion, can also perform as diluents of the powdered pectin. The dry blend, with or without sugar, has the advantage of remaining functional in storage for an extended duration. The aqueous form can be added before or after the boiling temperature has been reached.

D. pH Effects

HM pectins are not particularly responsive to pH, because of their decreased charge density as a result of the methyl-ester groups. With declining DE, carboxyl ionization increases to the level in LM pectins at which pH control is crucial to their nongelling, dispersion behavior. The sensitivity of LM pectins to pH is minimized or eliminated by hydrogen counter-ions (low pH), inasmuch as the ionization is totally depressed in a strongly acidic medium (Table I).

Table I pH dependence of the charge density of pectin[a]

pH		Ionized residues (% total uronate)
	38% DE	72% DE
5.0	60	27
4.5	55	25
4.0	44	20
3.5	27	12
3.0	12	5.6
2.5	4.6	2.1
2.0	1.5	0.7

[a]From Morris et al. (1980).

In the manufacture of jams and jellies, the acid is customarily added just before filling the containers, so that the heating interval is as short as possible, thereby minimizing the chance of pectin degradation in the acidic medium.

Tartness is an acid-related attribute. The common food acidulants each have a characteristic tartness. Tartness intensity may be lowered by small quantities of fumaric, tartaric, or phosphoric acid. These acids possess a higher dissociation constant than do malic and citric acid, and should therefore provide equal tartness at lower concentrations. Citric acid is preferred, because it is very soluble and it is ubiquitous. In the United States, the use of phosphoric acid is restricted to imitation jams and jellies.

Flavor intensity in pectin jelly products is modified by pH, and many manufacturers include buffer solutions in the product formula, when there is need to maintain flavor through a constant level of titratable acidity in products with fruit pieces. Without buffering, some variation in flavor and flavor intensity might be expected, since acid will migrate from the fruit pieces to the medium. This problem is addressed by selecting the appropriate buffer system or increasing the quantity of fruit in the formula.

Excess acid in an HM jam or jelly can lead to an undesirably firm texture. In contrast, a higher-than-optimum pH might cause the gel to develop more slowly, and a longer-than-normal interval would be necessary for a proper gel to form. If the product is to be restricted to a moderately high pH, it will be necessary to increase the amount of pectin in the formulation. This will impart the same jelly texture as that obtained at pH 2.8–3.5. Too high a pH may completely prohibit gelation. Figure 3 shows the effect of pH on HM pectin gels made with different sugars.

The LM pectin systems respond differently to a higher pH (less acid) by forming a less rigid and a more thixotropic structure.

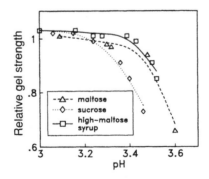

Figure 3   Effect of pH on the relative strength of pectin gels (0.5%) formed in the presence of different sugars (68% w/w total solids). Redrawn from May and Stainsby (1986).

E. Soluble-Solids Content

The overwhelming percentage of ss in a jelly comes from the added sugar ingredient, usually sucrose. Solution properties of this sugar are listed in Table II. A regular HM pectin jelly contains approximately 65% sucrose. Occasionally, a rapid-set pectin with 55% ss, or a slow-set pectin with 68% ss is substituted. The useful range of ss content of the various products is 50–80% with pectin DE 58–60. At a constant pH (3.1) and temperature (85°C), more sugar is required (62% ss) for slow-set pectin to set to the same gel strength (22% sag) as a rapid-set pectin jelly with 57% ss (Fig. 8 of Keller, 1983). At varying pH, the sugar determines the setting temperature (Fig. 4). The ss range for LM pectin (DE 30–34) jellies is 0–45%.

Alternative sweeteners are permitted in standardized pectin products. There are advantages and disadvantages to their use. For example, honey might

Table II Equivalency of sucrose concentration[a]

°Brix	Lb/gal	Sp. Gr.	Density
45	4.51	1.205	10.03
46	4.63	1.210	10.07
47	4.76	1.215	10.12
48	4.88	1.221	10.16
49	5.00	1.226	10.21
50	5.12	1.226	10.25
51	5.25	1.237	10.30
52	5.38	1.243	10.34
53	5.51	1.248	10.39
54	5.64	1.254	10.44
55	5.77	1.260	10.49
56	5.90	1.265	10.53
57	6.03	1.271	10.58
58	6.16	1.277	10.63
59	6.30	1.283	10.68
60	6.44	1.289	10.73
61	6.58	1.295	10.78
62	6.72	1.301	10.83
63	6.85	1.307	10.88
64	7.00	1.313	10.93
65	7.14	1.319	10.98
66	7.28	1.325	11.03
67	7.42	1.331	11.08
68	7.57	1.337	11.13
69	7.71	1.343	11.18
70	7.86	1.350	11.23

[a]In water at 20°C (68°F). Condensed from The National Bureau of Standards, No. 457 (1946).

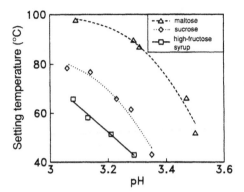

Figure 4 Effect of pH on the setting temperature of pectin gels (0.5%, 68% w/w total solids). Redrawn from May and Stainsby (1986).

introduce an extraneous flavor that might be intensified by heating. Dextrose is less soluble and less sweet than sucrose, and its crystallization is therefore highly probable at high ss content. On the sweetness scale, dextrose is 74% the sweetness of sucrose. Crystallization is not a problem with sucrose, because it is converted by acid to invert sugar (an equimolar mixture of glucose and fructose) that does not crystallize so readily. Common sweeteners of pectin gel products are listed in Table III, along with observations on them at the Red Wing Company. The corn syrups possess unique flavors that tend to mask natural fruit flavor. Some require heating before bulk storage for long periods, because their low ss content is not conducive to biological stability. The apparent discrepancy between percentage of dextrose and dextrose equivalent (d.e.) in Table III results from the fact that reducing starch derivatives are included in the latter value. The liquid sweeteners, to varying degrees, are prone to develop a yellow discoloration on

Table III Common sweetening agents for jams, jellies, and preserves

Sweetening agent	Total solids (%)	Dextrose (%)	Relative sweetness
Sucrose	100	0	100
Fructose	100	0	120
42 d.e.[a] corn syrup	80	19–20	30
62 d.e.[a] corn syrup	80	37	45
High-maltose corn syrup	80	7–10	35
High-fructose corn syrup 42% fructose	71, 80	50	92
High-fructose corn syrup 55% fructose	76–77	41	99

[a]Dextrose equivalent.

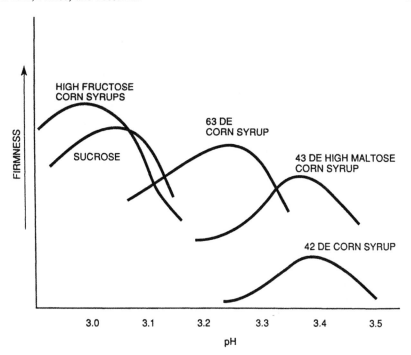

Figure 5   Relative effect of sweeteners in different pH ranges on the firmness of a pectin gel. From Ehrlich (1978).

extended storage. Yellowing is more prominent in the syrups containing higher total solids.

The firmness of a pectin gel is fundamentally a property of the ss content. The sugar used has a secondary effect on texture (Kawabata *et al.*, 1976; May and Stainsby, 1986). The effect of the various sweeteners on pectin gel firmness at different pH is illustrated in Fig. 5. It is seen that only high-fructose corn syrups, sucrose, and 63 d.e. corn syrup are interchangeable, within narrow limits, in the vicinity of pH 3.1. By itself, 42 d.e. corn syrup is incapable of reaching the texture of the others.

The IFT-prescribed jelly formula calls for a finished net weight of 1015 g (IFT, 1959). The jelly boiling point of the pectin : sugar : acid ratios given is 103–105°C (217–221°F). Arrival of the heated sol at this temperature is an indication that the appropriate ss content after cooling has been reached. The boiling point decreases by approximately 1°F for every 550 ft of elevation. The concentration of sweetener in the finished product may be ascertained by evaporating water to a predetermined weight, refractometer reading, or boiling point. Only a drop of the cooked jelly is needed for the refractometer measurement. The drop should be cooled to room temperature before taking the reading. Modern refractometers for hand-held factory use, standardized at 25°C but with temperature correction, are available with direct calibration in % sugar ± 0.3% (as sucrose).

F. Calcium

Ionic calcium is an essential element in the gelation mechanism of LM pectin sols. The lower the pectin DE, the lower the calcium demand to reach a given texture (of acid-precipitated pectinates) (Owens *et al.*, 1949). According to Joseph *et al.* (1949), sensitivity to calcium is also a property of the method of obtaining LM pectin. The latter authors stated that 4–10 mg and 15–30 mg per gram is required of LM pectin obtained by enzyme- and ammonia-deesterification, respectively, and that the demand rises twofold (15–30 mg/g) for acid-hydrolyzed LM pectin.

Calcium ions are prone to complexation with polyacids as insoluble salts. Not surprisingly, they have this effect on LM pectins, increasing their dispersion; hence, the order of addition should be calcium after the pectin has been dispersed. Calcium may be added simultaneously with fruit.

Naturally occurring calcium in fruits for jams, preserves, marmalades, etc., ranging between 90 ppm (apple) and 700 ppm (lemon peel) (Hercules, Inc., 1985), enables a reduction in the amount of calcium added for LM pectin products. The addition must be slowly and carefully executed with stirring, inasmuch as a localized concentration of calcium will cause pregelation (Cole *et al.*, 1930). The gelation temperature of LM pectin jelly sols is directly proportional to calcium as well as to ss content.

1. Source of Calcium

It has been found that gelled products of different fruits have different calcium demands. For example, grape jelly (32% ss) has a higher demand than strawberry and raspberry jellies similarly constituted. Before adding calcium salts to a formula, it is therefore necessary to determine the natural calcium content of the fruit in question and the calcium requirement of the selected pectin. This information is usually obtainable from fruit processors and pectin manufacturers, respectively.

2. The Role of Calcium

LM pectins will gel in the presence of an adequate concentration of calcium in a wide range of ss content. The calcium requirement averages 20 mg/g of LM pectin. An insufficiency will accelerate a separation of liquid from the gel (syneresis) (*vide infra*). The more calcium-reactive pectins will gel at a higher temperature than the less calcium-reactive pectins. Highly calcium-reactive pectins are intended for use in formulae containing 20–40% ss. LM pectins with low calcium reactivity are intended for use in formulae containing 60–80% ss. These LM pectins approach HM pectin in behavior.

One area of research in which more information is required is in the effects of electrolytes on the gelation of HM pectins. Calcium ions are required for gelation of LM pectins (Chapter 6), and there are indications in the early literature that calcium (and other metal ions) also influence the gelation of HM pectins. There appear to have been no recent systematic studies on such effects. Different anions may be of significance, since electrostatic factors are important in limiting junction-zone formation.

IV. PROCESSING VARIABLES

Pectin DE and grade, jelly pH, and to some degree, calcium content, are intrinsic variables of pectin products manufacturing that are usually fixed by the formulation. Their variability is therefore only marginal in the production scheme. On the other hand, the manufacturing process is influenced by the extent to which such extrinsic variables as fruit preprocessing, transportation, cooking temperature, container size, order of ingredients addition, etc., affect the final product.

A. Fruit Preprocessing

The best fruits make the best products, because color and flavor cannot be improved by processing. For this reason, it is not possible to manufacture a superior preserve, for example, from culls. Sorting, washing, and removing calyx tissue are examples of unit operations that translate into premium-quality fruit stock. The same criterion applies to juice extracts, whether single-strength or concentrate.

In whole-fruit preserves, some rupturing of tissue will occur, but the overall structural integrity will prevent the release of much natural pectin into the surrounding sugar solution. Pectin addition will therefore be necessary. In acquiring pulp for pectin products, the screens through which the material passes will shear the tissues, and hence will affect the flow of pectin into the sugar solution. The finer the screens, the greater the shear, and the higher will be the quantity of released pectin. Whole fruits with an inherently tough or leathery pericarp should be precooked.

Fruits contain active enzymes. Mindful that clarified juice for making clear jellies might have been enzyme treated, the manufacturer should guard against active residues that might survive low-heat processes, with the consequence that added pectin might be degraded on storage. If the history of the fruit pieces or the juice is not known, the extra precaution of heating them to 190°F is advised, before incorporating them in a vacuum cook.

B. The Cooking Temperature

The difference between a pectin product cooked at atmospheric pressure and under vacuum is one of quality retention and appearance. A standard formula of jelly ingredients boils at approximately 105°C (221°F) in an atmospheric cook. Vacuum pans operate at 38 to 60°C (100–140°F). Sugar solutions boil at a higher temperature at atmospheric pressure than under vacuum. The milder heat treatment *in vacuo* conserves more of the original texture, color, and flavor of the fruit pieces and of the color and flavor of the juice. Atmospheric cooking is performed in open systems (Fig. 6); vacuum cooking is performed in essentially closed systems (Fig. 7) from which the air is exhausted. Thus, atmospheric cooking offers an opportunity for oxidative deterioration that is absent in vacuum cooking. The reduced air pressure in vacuum cooking has the ancillary advantage of exhausting air from the interior of the fruit pieces, enabling sugar and calcium to diffuse easily into the tissue. This diffusion, with time, equalizes with the external solution the density of fruit pieces that are predisposed to float.

Sugar caramelization is a virtual certainty in atmospheric cooking, if the engineering and manufacturing setup is not specifically designed to prevent it. This sugar reaction to heat is nonexistent with vacuum cooking.

Figure 6   Kettles for atmospheric cooking of pectin products in a modern processing plant.

Figure 7   Pans for vacuum cooking of pectin products in a modern processing plant.

A 65% sugar solution is much more viscous at 38 to 60°C than at 105°C; so the engineering prerequisites for pumping and transporting cooked pectin fluids must take into consideration post-cooking temperatures. More-viscous products are accommodated by more-powerful pumps and larger-diameter pipes and spigots of the delivery systems. The increased viscosity of low-temperature fluids has ramifications for accurate measuring systems.

Slow-set pectins are preferable for vacuum cooking, because they lessen the chance of pregelation at vacuum-cooking temperatures. Preheating the ingredients shortens the vacuum cooking time. In all circumstances the heat accumulated in the cooked pectin products must be removed as rapidly as possible.

Jellies rely on their high sugar concentration (water activity approximating 0.8) for microbiological stability. However, at a jelly–air interface, particularly in humid climates under home conditions, the sugar concentration at the surface of frequently opened bottles may often be below the osmotic concentration necessary to impart biological stability. This condition invites mold growth on the surface. An atmospheric cook exceeds the pasteurization temperature of the product, and as a result, the post-cook microbial population is small or non-existent. If the filled, capped containers are quickly inverted, the caps will be pasteurized. The maximum temperatures of vacuum cooking do not ensure microbiological stability. These considerations also pertain to reduced-calorie imitation-pectin products in which the ss content is below the critical osmotic concentration.

C. The Setting Time and Temperature

Doesburg and Grevers (1960) defined setting time as the time between the moment that all ingredients necessary for forming the jelly are present in the heated solution in the correct proportion, and the moment that the jelly develops into a coherent mass. Setting time and setting temperature are of practical interest to the manufacturer who must schedule the unit operations (filling, cooling, and packing), to economize on time.

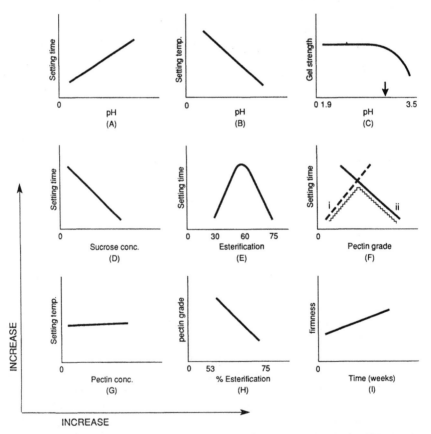

Figure 8 Generalized pectin jelly relationships. From Doesburg and Grevers (1960); Fig. F(i) from Baker and Goodwin (1944).

   It is self-evident that setting time is fixed by the cooling rate, i.e., the faster the cooling, the shorter the setting time. Also, the faster the cooling, the lower the setting temperature (Hinton, 1950). Rapid cooling promotes an unstable gel state called *undercooling* by Hinton (1950). The phenomenon is pH dependent. These time–temperature–pH dependencies make the physical state of a jelly not exclusively a property of the formula but also of manufacturing conditions. The numerous interrelationships between intrinsic and extrinsic factors of jelly quality are graphically summarized in Fig. 8. It is seen that setting time, to a limited degree, is directly related, and setting temperature is inversely related, to pH (Fig. 8A and 8B); below a critical pH interval (2.5–3.1), gel strength is independent of pH, but declines rapidly above it (arrow in Fig. 8C). Setting time is inversely related to sugar concentration (Fig. 8D) and pectin grade (Fig. 8E). Interestingly, setting time has also been reported to be directly related to pectin grade (Fig. 8F, i) (Baker and Goodwin, 1944). Figure 8D and 8F(i) are similar, in that pectin grade is the ability of pectin to gel a quantity of sugar. Figure 8E illustrates that there is a maximum DE (approximately 60%), below which setting time is directly proportional to DE, and above which it is inversely proportional to DE. Pectin grade is inversely related to DE (Fig. 8H). Setting temperature is raised by increasing concentrations of sugar, and only slightly by pectin concentration (Fig. 8G). This last relationship is a reasonable expectation, in view of the constant molecular characteristics and critical minimum requirement of pectin to effect gelation in a normal process. The firmness of any jelly will increase with age (Fig. 8I) to a definite maximum (setting).

D. Size of Containers

Large containers (above 16-oz jars) necessitate an increased amount of pectin in the formula. Considering the extra support afforded a gel by the walls of a small container, the principle of adding more pectin is to increase firmness to the extent that the elastic limit of the gel in a large container, effectively unsupported away from the container walls, will not be overcome by the forces generated during handling and transport. A ruptured gel will undergo syneresis, and syneresis is a quality defect (*vide infra*). Cooking to a higher temperature or for a longer interval concentrates the sugar solution with the same end result, except that a shorter duration is more conducive to retention of fruit color and flavor. However, the boiling temperature of a standard jelly cannot be increased by more than 1 to 2 degrees at atmospheric pressure. Sunkist Growers (1964) recommends a 10% pectin increase when the product is to be shipped in 5-lb containers, and a 30% increase for 30-lb containers.

   Small containers, allowing for rapid heat transfer through a high temperature gradient with the surroundings, may be air- or water-cooled. Large containers lose heat from the interior more slowly by conduction, and water-cooling might not be suitable for them, for fear that a product having a gel with an outer layer at the container surface, cooled at a faster rate than the interior volume, might have a nonuniform texture.

E. Order of Ingredients Addition

This section refers to the addition of pectin and acid. Pectin has a low hydrophilicity, and consequently, it does not easily disperse in media in which conditions are favorable to gelation, e.g., high ss concentration, high ionic strength, low temperatures. Pure water is the best medium in which to disperse dry pectin.

As previously mentioned, sustained heat at or near the boiling temperature in an acidic medium of jelly can induce depolymerization and demethylation of pectin with resulting impairment of jelly texture. Heat transfer in a 65% sugar solution is slow, and the time the mixture takes to reach the boiling temperature, or to cool spontaneously to room temperature, can be enough to cause pectin decomposition in the presence of an adequate concentration of acid. This event may be minimized by adding acid at the last possible moment, or by elevating the sugar–acid solution temperature to near boiling before adding liquid pectin. The dry form is not amenable to the second option, because of the time it would take for wetting and dispersion of the granules. The dry pectin should be dispersed very early in the sugar–water solution, mindful of the sugar limitation previously stated. Recall that liquid pectin may also be added after boiling, which would lessen the chance of pectin decomposition. Some manufacturers claim that a water–sugar–pectin mixture without pectin heats faster than one with pectin.

In a normal process the acid, dissolved in water, should be added to the sugar–pectin mixture after the boiling temperature has been reached. Home-canners who pack small volumes sometimes add the acid solution to individual containers. This alternative enhances the prospect of a nonuniform distribution of the acid throughout the jelly matrix in the container, and localized pregelation may ensue as a consequence, where the acid concentration is highest. Where the concentration is lowest, gelation and setting may fail to occur. Slow-set pectin allows time between cooking and filling the containers to mix the acid solution more thoroughly than does rapid-set pectin.

V. TEXTURE MEASUREMENTS

Texture, as it relates to pectin gels, has been repeatedly mentioned without so far ascribing meaning to it. Consistency is a related term embodying the total physical characteristics that consumers respond sensorially to, *viz.*, cohesion, adhesion, hardness, flow, elasticity, rubberiness, etc.

In the trade, texture is used synonymously with gel strength. For low-solids gels, the moduli of elasticity appear to be related to the calcium:pectin ratio (Owens *et al.*, 1947).

Many attempts have been made to harmonize consumer sensorial responses to pectin products with gel strength measured on a variety of empirical instruments (Angalet, 1986; Beach, *et al.*, 1986; Johnson and Breene, 1988). The basic instrumentation is of two kinds—destructive and nondestructive. The former ruptures the test jelly's bonding; the latter does not. Destructive tests are claimed to simulate human sensory perception of jelly texture more so than do nondestructive tests (Crandall and Wicker, 1986).

Measurements of commercial pectin jellies have shown a correlation between breaking strength and viscosity (Beach *et al.*, 1986). The index does not correlate with pectin grade. The factors that affect breaking strength are pH, type of pectin, pectin molecular weight (MW) and gel-testing conditions (Crandall and Wicker, 1986). The yield stress depends on DE also, passing through a maximum in the range of 70 to 80% (Fig. 9). The higher the MW, the stronger the gel (Morris *et al.*, 1980). However, since pectins come to the manufacturer in fixed chemical states, MW is a constant not relevant to the jelly-manufacturing process.

There has been no unanimity on the acceptance of a single, official instrument for measuring jelly texture. The one most closely associated with an official status is the ridgelimeter (Cox and Higby, 1944; Joseph and Baier, 1949; IFT, 1959). It is nondestructive and simple to operate, but it is incapable of a comprehensive evaluation of gel structure, because it cannot evaluate spreadability (the fluid state of the jelly that allows it to spread two-dimensionally, as does butter), which is the most important criterion from a consumer's standpoint (Beach *et al.*, 1986); neither is it suitable for already packaged products or products with large pieces of fruit.

The Voland–Stevens Texture Analyzer (Voland Corporation, Hawthorne, N.Y. 10532) is a recent device for measuring the gel strength of an assortment of foods, including pectin products. The device has facilities for recording and

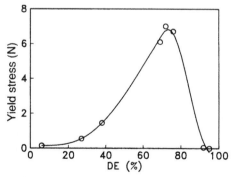

Figure 9   Yield stress of pectin gels (1%, w/v) in 60% (v/v) ethylene glycol formed at 25°C, pH 1.9, as a function of degree of esterification. From Morris *et al.* (1980).

displaying the penetrating and withdrawal forces acting on probes of different geometries and sizes. It draws a texture profile from which information on hardness, adhesion, brittleness, etc., other than gel strength, may be gleaned.

Qui *et al.* (1990) measured fundamental parameters (intrinsic viscosity, yield stress, shear, and instantaneous moduli) on pectin gels with a controlled-stress rheometer (Deer Rheometer III, Niewleusen, Holland). They found that the jelly made with a slow-set apple pectin was the most rigid, while that made with a rapid-set pectin was the least rigid. They quantified the plasticizing effect of fructose substituted for sucrose.

VI. MISCELLANEOUS PECTINS AND PRODUCTS

Jams, jellies, preserves, and marmalades are the traditional items for which a wide variety of recipes have been documented (Sunkist Growers, 1964; Hercules, Inc., 1985). Other items are less known but equally delightful to consume. These processed fruit products intentionally or unintentionally exploit the universal delectability of fruit and the physical–chemical response of added water–sugar–acid–pectin systems to heat.

A. Confectionery Pectins

Confectionery pectins are slow-setting pectins approximating DE 60, capable of setting on the addition of acid. Candied pectin gels rely on an optimization of the pectin, the cooking temperature, the invert sugar and glucose syrup concentrations, and the pH to create a durable, viscoelastic body. Evaporation during cooking raises the ss content to 80% or more. The critical pH 3.5–3.8 is buffered with phosphate. The hot, acidified sol is cast in various molds and cooled, before coating with chocolate or other flavors. The gel candy is finally wrapped.

B. Fluid High-Methoxyl Gels

HM pectin gels containing the IFT-prescribed quantities of ingredients will gel irreversibly to a solid mass after heating to the boiling temperature (105°C) and cooling. They will soften with heat, but they will not ordinarily revert to a fluid condition. The same formula exposed to minimum heat at no higher than 40–50°C will yield a soft, reversible gel, more in the nature of a paste (Deuel and Eggenberger, 1956). This gel is metastable, not particularly reproducible with the same viscosity characteristics from batch to batch (Walter and Sherman,

1986). It is mostly of theoretical interest, since it will revert to a sol in such a short time that it would be commercially infeasible to use it in retail packs.

VII. PRACTICAL PROBLEMS AND SOLUTIONS

The manufacture and storage of pectin gel products are occasionally attended with problems whose solutions become obvious, once they are investigated, as in the example of a too-soft jelly (add sugar, pectin and/or calcium; lower pH). Many solutions are implicit in the origin of the problem, as in discoloration (avoid caramelization, enzyme action, metal ions). There are other potential quality defects (Table IV) that are worthy of separate discussion, because corrective measures are not always apparent.

Table IV   Defects in pectin products and their causes

Defects	Causes
Too-rigid gel	(i) Too-high ss content
	(ii) Too-high pectin content
	(iii) Too-high calcium content (LM pectin)
Too-soft gel	(i) Too-high pH
	(ii) Too-low ss content
	(iii) Too-low pectin content
	(iv) Too-low calcium content (LM pectin)
	(v) Aged pectin (rapid-set converted to slow-set; HM converted to LM)
	(vi) Degraded pectin (heat + acid + time)
Excessive foaming	(i) Traces of protein
	(ii) Saturation with air
Haze formation	(i) Impure ingredients (dust, fiber, etc.)
	(ii) Undispersed pectin (too-large mesh size)
	(iii) Unclarified juice
Discoloration	(i) Caramelization (nonuniform heating)
	(ii) Metal ions (Fe, Cu contamination)
Crystallization	(i) Surface evaportation of water
	(ii) Too-high glucose content
	(iii) Too-high ss content
	(iv) High tartrate content (grape juice)
Flaccid fruit tissue	(i) Overripe fruit
	(ii) Excessive precooking
	(iii) Heat-labile fruit
	(iv) Freeze-labile fruit
	(v) Oversized fruit (berries)
Pregelation	(i) Nonuniform distribution of acid and calcium
	(ii) Cold additions of calcium or acid solution
	(iii) Temperature and concentration gradients
Syneresis	(i) Improper pectin choice
	(ii) Too-high ss content
Microbial contamination	Microscopic leaks (negative container pressure)

A. Floating Fruit Pieces

Matter floats when it is lighter than the surrounding medium. Fruit and fruit pieces in preserves float when they are less dense than the surrounding 65% sugar solution. The densities may be equalized by pre-equilibrating the fruit and fruit pieces in a 65% sugar solution, or by preheating or vacuum-cooking to expel air. In these instances, water and sugar displace gas in the vacuolar and intercellular spaces. A simpler alternative is to add enough pectin to increase viscosity, which in turn, slows the rate of ascension of the floating fruit and fruit pieces.

B. Opacity

A cloudy appearance in jelly is an indication of the presence of colloidal-size particles throughout the medium. These particles may have been introduced with unclarified juice. Dispersed air is a less-probable source of the cloudiness, because the small air bubbles either become reabsorbed upon cooling or coalesce to larger bubbles that rise and escape at the surface, before gelation and capping. The solution to opacity might then be using depectinized juice or evacuating the system more thoroughly, before or after bottling. Here, the vacuum should not be greater than the cooking vacuum, or air bubbles might emerge from those fruit pieces that might not have been completely evacuated during cooking. Remedial action like precooking the fruit pieces or prolonging the boiling may be precluded by the quality demands of the product.

C. Discoloration

In a normal cook, the color of the final product is largely determined by the color of the juice or fruit ingredient and its inherent response to heat. As stated previously, prolonged atmospheric cooking is almost always accompanied by incipient caramelization of sugar on the walls of the cooking vessel, unless guarded against in the engineering design of the equipment or in the cooking procedures. Modern kettles are equipped with stirrers, agitators, and scrapers that ensure uniform mass and heat distribution.

Metal ions catalyze the decomposition of heat-labile fruit pigments, generating melanoidin-like or humin-like discolorations. Stainless steel is inert to organic acids, is easy to clean, and is therefore the preferred surface for all paraphernalia expected to come in contact with pectin ingredients and products not bottled in glass.

D. Pregelation

It has already been made clear that pregelation in pectin jellies, the localized, premature solidification of small, discrete volumes of a pectin–sugar mixture in a larger volume of sol, initiated by a nonuniform distribution of the acid, or by temperature or ss concentration gradients, is a quality defect. The condition may also arise from a localized concentration of unusually cold acid, and a too-high ss and/or acid content. With LM pectins, a too-high calcium concentration is also to be considered. Excess calcium in the sol may simply be sequestered with phosphoric or citric acid. Given its flow reversibility, defective LM pectin jelly may occasionally be heated and reincorporated in a new batch. The flow of HM pectin jelly, being thermally irreversible, cannot be similarly treated. However, an HM pregel can be completely redispersed in water or in a heated, dilute jelly sol.

E. Syneresis Control

The exudation of liquid consisting of an equilibrium fraction of soluble components from a ruptured jelly surface or from a jelly mass upon standing is called syneresis. It is a manifestation of a gel's predisposition to reach a state of minimum energy. The phenomenon is somewhat pectin, pH, and ss dependent. Under different circumstances, it occurs with equal frequency with LM and HM pectins. Intact HM gels will undergo syneresis less than will intact LM gels under optimum conditions of production and storage. Gels with a high ss content will undergo syneresis less than will gels with a low ss content. A too-high calcium content of LM jellies will accelerate the problem (Baker and Goodwin, 1944). Whether or not pH has any effect seems to depend on the ss content (of slow-set apple) (Swajkajzer, 1962). If exudation is not severe, addition of a small quantity of a humectant to succeeding batches may mask the problem. Otherwise, a different pectin should be tried.

Syneresis is a problem with LM pectin gels containing less than 25% ss. On the assumption that other factors, e.g., calcium level, ss content, etc., control the choice and amount of pectin, it is possible to mask or completely retard syneresis, by incorporating in the formula a larger percentage of hydrocolloid with the appropriate humectant or viscosity-increasing function. Carboxymethylcellulose, guar, and locust bean gum are common additions to a pectin gel formula for this purpose.

F. Miscellaneous Problems

An otherwise dependable source of processing water, for some indeterminate reason, e.g., municipal breakdown of the water-treatment plant, might suddenly transport an unusual level of mineral elements to the processing plant. Water from deep wells is likely to be harder than water from shallow wells. Fe and Cu ions might be inadvertently introduced into the formulation with a new utensil or a new batch of ingredient. Protein contamination, always a possibility with plant extracts and food materials, might induce abnormal foaming. This last problem is resolved by the use of an antifoam or by sudden interruptions of vacuum, where applicable. It is not uncommon for grape juice to have a high concentration of tartaric acid. Juice pumped from the lower strata of a storage tank may contain close to saturation levels of potassium acid tartrate, causing crystals to form in low water activity gels. Micropores occurring around the lid of containers are a conduit of air and cooling water, as the temperature (and pressure) of the packaged product falls to that of its surroundings. Such leaks are a source of post-process contamination, identifiable by a loss of vacuum in the container. All these defects and problems are correctible by proper troubleshooting.

VIII. QUALITY ASSURANCE AND CONTROL

The quality control laboratory has the technical responsibility of monitoring conformance with management and production standards. The analytical methodologies involve mostly elementary chemistry, such as measurements of acidity and ss content. These procedures are discussed elsewhere. Some calculations are usually necessary, and examples are included in this section.

A. Color of Jelly

The color of apple jelly may be determined with Virtis color standards (The Virtis Co., Gardiner, N.Y. 12525). Four rectangular bottles provided with the device are filled with tap water and placed in the appropriate space behind the filters. A sample of jelly is placed in a fifth bottle, and this bottle is moved from space to space between the filters until a match is obtained. The color designations are water white, extra white, white, extra light amber, light amber, and amber.

B. Calculation of Unknown Ratio of Fruit to Sugar

Much fruit comes to the manufacturer coated with sugar. Compliance with product specifications for fruit content relative to sugar requires knowledge of the blend ratio of lb of fruit (F) to lb of sugar (S) in any such presweetened fruit stock. This ratio (F:S) is derived from the ss assay of a uniformly blended sample of the fruit–sugar stock and the known or estimated ss of the fruit. Assume the following data are given:

Weight of blended fruit-sugar stock (W)	100 lb
Ss in 100 lb of W (S_w)	26.5 lb
Known ss in 100 lb fruit (F_{ss})	8 lb
Nonsugar solids in 100 lb fruit (F_{nss})	92.0 lb
Nonsugar solids in W (F_{nsw})	73.5 lb.

If 92 lb nonsugar solids in the fruit (F_{nss}) accompanies 8 lb fruit sugar (F_{ss}), then 73.5 lb F_{nsw} accompanies $(8)(73.5)/92$ lb or 6.39 lb F_{ss}. If 8 lb F_{ss} is obtained from 100 lb fruit, then 6.39 lb F_{ss} is obtained from $(100/8)[8(73.5)/92]$ lb or 79.9 lb fruit. The quantity of sugar added to the fruit to make W is 26.5 − 6.39 lb or 20.1 lb (i.e., $26.5 - [8(100-26.5)/92]$), and F:S = 79.9/20.1 or 4.0 or 4 parts fruit to 1 part sugar. The generalized derivation is $F/S = (100 \cdot F_{ss} \cdot F_{nsw})/[F_{ss} \cdot F_{nss} \cdot (S_w - F_{ss} \cdot F_{nsw}/F_{nss})]$, which may be approximated to $F/S = F_{nsw}/S_w - F_{ss}$. These equations presuppose that no water was added to the fruit–sugar mixture.

C. Calculation of Jelly Yield

Commercially, it is important to know the yield of product from the starting materials. The following exercise is a materials balance of input and output. Using the following ingredients,

100 gal grape juice (15% sugar, 0.96% acid as citric acid),
corn syrup (42 d.e., 80.0% total solid),
150-grade pectin,

and the following product specifications:

65.2% ss,
0.55% acid as citric acid, and
45:55 fruit juice:sugar ratio,

a typical calculation of the quantity of grape jelly obtainable is outlined as follows:

Weight of fruit juice (100 gal × 8.83 lb/gal)	883.0 lb
Weight of sugar in fruit juice (883 lb)(0.15)	132.5 lb
Weight of corn syrup sugar [55(883 lb/45)]	1079.2 lb
Total weight of sugar (132.5 lb + 1079.2 lb)	1211.7 lb
Weight of dry pectin (1211.7 lb/150)	8.1 lb
Total ss less acid (1211.7 lb + 8.1 lb)	1219.8 lb
Total jelly less added acid (1219.8 lb/0.652)	1870.9 lb
Total acid [0.0055(1870.9 lb/0.9945)]	10.34 lb
Yield (1870.9 lb + 10.34 lb)	1881.2 lb.

It is important to note that the fruit-juice sugar was not included in the 55 parts of sugar in the 45:55 ratio. The 55 parts sugar came exclusively from the corn syrup.

A manufacturer who may prefer to work with the Brix value instead of the sugar content of a juice should proceed as follows. Assuming the grape juice is 15°Brix, the combined sugar and acid content is 132.5 lb, and the calculation is modified accordingly:

Weight of ss in fruit juice	132.5 lb
Total ss (132.5 lb + 1079.2 lb + 8.1 lb)	1219.8 lb
Yield (weight of jelly) [100 (1219.8/65.2)]	1870.9 lb.

Based on these assumptions, the amount of added acid was as follows:

Total weight of acid (0.0055)(1870.9 lb)	10.29 lb
Weight of citric acid from fruit juice (0.0096)(883 lb)	8.48 lb
Weight of added acid (10.29 lb − 8.48 lb)	1.81 lb.

D. Calculation of Acid

On a factory scale, accuracy to more than one decimal place is seldom necessary. The acid, however, is a critical component that should be measured to two decimal places. The acid is almost always in a water solution, for convenience. The amount to be added may be calculated, as follows:

Weight of acid from juice [0.0096 (883 lb)]	8.48 lb
Weight of acid to be added (10.34 − 8.48 lb)	1.86 lb.

E. Calculation of Liquid Additions

If the pectin were in a 5% wt/wt aqueous dispersion, the weight of dispersion giving 8.1 lb would be (8.1/0.05) 162 lb. Thus, the weight of pectin to be added from an aqueous dispersion is 100 times the calculated weight of pectin required, divided by the wt/wt% of dispersed pectin. The weight of acid to be added from an aqueous solution may be similarly calculated. During the cooking, the excess water contributed by these additions is evaporated to a refractometer reading of 65.2% ss. The hydraulic load may also be lessened by dispersing and dissolving the dry pectin and acid in an aliquot of juice.

References

Ahmed, G. E. (1981). "High-Methoxyl Pectins and Their Uses in Jam Manufacture—A Literature Survey." Scientific & Technical Surveys No. 127. British Food Manufacturing Industries Research Association, Leatherhead, Surrey, England.

Angalet, S. A. (1986). Evaluation of the Voland-Stevens LFRA Texture Analyzer for measuring the strength of pectin–sugar jellies. *J. Texture Studies* **17**, 87–96.

Baker, G. L., and Goodwin, M. W. (1944). "Fruit Jellies: XII. Effect of Methyl Ester Content of Pectinates upon Gel Characteristics at Different Concentrations of Sugar." Delaware Agr. Exp. Sta., Bull 246, Tech. no. 31.

Beach, P., Davis, E., Ikkala, P., and Lundbye, M. (1986). Characterization of pectins. *In* "Chemistry and Function of Pectins" (M. L. Fishman and J. J. Jen, eds.), pp. 103–116. Am. Chem. Soc., Washington, D.C.

B.F.M.I.R.A. (1951). Determination of the grade strength of pectins. Report of the Pectin Subcommittee of the Jam Panel, British Food Manufacturing Industries Research Association. *The Analyst* **76**, 536–540.

Charley, H. (1970). "Food Science." Ronald Press, New York.

Cole, G. M., Cox, R. E., and Joseph, G. H. (1930). Does sugar inversion affect pectin jelly formation? *Food Industries* **2**, 219–221.

Cox, R. E., and Higby, R. H. (1944). A better way to determine the jellying power of pectins. *Food Industries* **16**, 441ff.

Crandall, P. G., and Wicker, L. (1986). Pectin internal gel strength: Theory, measurement, and methodology. *In* "Characterization of Pectins" (M. L. Fishman and J. J. Jen, eds.), pp. 88–102. Am. Chem. Soc., Washington, D.C.

Deuel, H., and Eggenberger, W. (1956). Formation of pectin by a cold process. *Kolloid-Z.* **117**, pp. 97–102.

Doesburg, J. J., and Grevers, G. (1960). Setting time and setting temperature of pectin jellies. *Food Research* **25**, 634ff.

Ehrlich, R. M. (1978). Personal communication. Sunkist Growers, Inc., Ontario, California.

Food & Nutrition Encyclopedia. (1983). 1st, Vol. 2. California: Pegus Press, Clovis, p. 1739.

Hang, Y. D., and Walter, R. H. (1989). Treatment and utilization of apple-processing wastes. *In* "Processed Apple Products" (D. L. Downing, ed.), p. 370. AVI, Van Nostrand Reinhold, New York.

Hercules, Inc. (1985). "Handbook for the Fruit Processing Industry." The Copenhagen Pectin Factory Ltd., Lille Skensved, Denmark.

Hinton, C. L. (1950). The setting temperature of pectin jellies. *J. Sci. Food Agric.* **1**, 300–307.

IFT. (1950). Pectin standardization. Final Report of the IFT Committee, Institute of Food Technologists. *Food Technol.* **13**, 496–500.

Johnson, R. M., and Breene, W. M. (1988). Pectin gel strength measurement. *Food Technol.* **42** (2), 87–93.

Joseph, G. H. (1953). Better pectins. *Food Eng.* **25**, 71–73.

Joseph, G. H., and Baier, W. E. (1949). Methods of determining the firmness and setting time of pectin test jellies. *Food Technol.* **3**, (1), 18–22.

Joseph, G. H., and Havighorst, C. R. (1952). Engineering quality pectins. *Food Engin.* **24**, 87ff.

Joseph, G. H., Kieser, A. H., and Bryant, E. F. (1949). High-polymer, ammonia-demethylated pectinates and their gelation. *Food Technol.* **3**, 85–90.

Kawabata, A., Sawayama, S., and Kotobuki, S. (1976). Effect of sugars and sugar-alcohols on the texture of pectin jelly. *Eiyogaku Zasshi* **34**, 3–10.

Keller, J. (1983). Pectin. *In* "Gum and Starch Technology." 18th Annual Symposium, Special Report No. 53. Cornell University, Geneva Campus, New York.

Keller, J. (1984). Commercially important pectin substances. *In* "Food Hydrocolloids" (H. D. Graham, ed.) pp. 418–437. AVI Westport, Connecticut.

Kertesz, Z. I. (1951). "The Pectic Substances." Interscience, New York.

May, C. D., and Stainsby, G. (1986). Factors affecting pectin gelation. *In* "Gums and Stabilizers for the Food Industry 3" (G. O. Phillips, D. A. Wedlock, and P. A. Williams, eds.), pp. 515–523. Elsevier, London.

Michel, F., Doublier, J. L., and Thiboult, J. F. (1985). Etude viscométrique de la première phase de gélification des pectines hautement méthylées. *Sci. Aliments* **5**, 305–319.

Morris, E. R, Gidley, M. J., Murray, E. J. Powell, D. A., and Rees, D. A. (1980). Characterisation of pectin gelation under conditions of low water activity, by circular dichroism, competitive inhibition, and mechanical properties. *Int. J. Biol. Macromol.* **2**, 327–330.

Owens, H. S., McReady, R. M., and Maclay, W. D. (1949). Gelation characteristics of acid-precipitated pectinates. *Food Technol.* **3**, 77–82.

Owens, H. S., Porter, O., and Maclay, W. D. (1947). New device for grading pectins. *Food Industries* **19**, 606*ff*.

Qiu, G.-G., Rao, M. A., and Walter, R. H. (1990). Creep-compliance and yield behavior of food grade pectin jellies. Unpublished data.

Sunkist Growers. (1964). "Preservers (Exchange Citrus Pectin) Handbook", 7th Ed. Ontario, California.

Swajkajzer, A. (1962). British Food Manuf. Industrial Research Association, Annual Report, *In* Doesburg, J. J. (1965). "Pectic Substances in Fresh and Preserved Fruits and Vegetables," p. 41. Institute for Research on Storage and Processing of Horticultural Produce, I.B.V.T.-Communication No. 25, Wageningen, The Netherlands.

Tressler, D. K., and Woodroof, J. G. (1976). "Food Products Formulary. Vol. 3, Fruit, Vegetable, and Nut Products." AVI Publishing, Westport, Connecticut.

U.S. Department of Commerce, National Bureau of Standards (1946). "Polarimetry, Saccharimetry, and the Sugars." No. 457. The U.S. Department of Commerce.

Van Buren, J. P. (1974). Heat treatments and the texture and pectins of red tart cherries. *J. Food Sci.* **39**, 1203–1205.

Walter, R. H., and Sherman, R. M. (1986). Rheology of high-methoxyl pectin jelly sols prepared above and below the gelation temperature. *Lebensm.-Wiss. u.-Technol.* **19**, 95–100.

CHAPTER 3

Other Pectin Food Products

A. C. Hoefler

Hercules, Incorporated
Middletown, New York

I. INTRODUCTION

Pectin is best known as a gelling agent, and less importantly, as a texturizer, emulsifier, thickener, and stabilizers. Why this biopolymeric extract, unlike others, develops a jelly in the presence of water, sugar, and acid is still a mystery.

The multifunctionality of pectin originates from the nature of its molecules, in which there are polar and nonpolar regions that enable it to be incorporated in food systems as diverse as dietetic soft drinks, chocolate milk, mayonnaise, and yogurt.

II. PRINCIPLES OF PECTIN SELECTION

Some questions to be asked when selecting a pectin for use in food are

1. What physical characteristics (shear thinning, rigidity, *etc.*) should the product display;
2. What is the nature, size, and quantity of the particulate matter;

51

3. Is protein present;
4. What are the pH and temperature ramifications;
5. What is the expected shelf life of the final product; and
6. Will the product be pumped?

Different applications might have different requirements, and therefore might call for different pectins.

The pectin trade classes based on the degree of esterification are the high-methoxyl (HM) pectins, and the low-methoxyl (LM) pectins that are either the conventionally demethylated (LMC) or the amidated (LMA) molecule.

A. HM Pectin

The physical state of a dispersion containing HM pectin will depend on many external factors, *e.g.,* ingredient composition, temperature, and treatment. The gelation mechanism is a physical association of macromolecules in the presence of water, sugar (65%), and acid (pH = 2.9–3.2). The important features about HM pectin gels are their temperature and shear irreversibility under normal conditions encountered in food. An HM pectin gel will soften but will not melt with heating. The surfaces of a broken or cut HM pectin jelly will not reunite, and will release liquid with the passage of time (syneresis).

B. LM Pectin

The setting temperature of an LM pectin jelly is controlled by the calcium reactivity of the pectin, and the amount of pectin and calcium in the product formula. By calcium reactivity is meant the calcium equivalence necessary to promote bonding with the pectin. Typical calcium requirements for LMA and LMC pectins at 30% soluble solids (ss) are seen in Fig. 1.

LMC pectins tend to gel at a higher temperature than do LMA pectins with equal gel strength. LMC pectins would therefore be preferable for hot-filled products with fruit pieces that must remain suspended. The pH is not a critical factor in the development of an LM pectin gel. Because of their shear reversibility, LM pectin gel products can be pumped without fear of damaging the gel. The choice of either LM pectin is made on the basis of the preferred product texture. Neither of these subtypes alone yields a gel similar to HM gels containing 65% ss, but a combination will do so, in the presence of as little as 30–35% ss. Texture in an LM pectin jelly is directly a property of the pH, ss content, calcium concentration, and calcium reactivity. Texture can also be modified by a changing formula ratio of LMA to LMC pectin. The main properties of the pectins are listed in Table I.

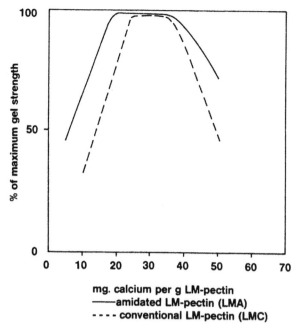

Figure 1 Calcium requirements for LMA and LMC pectins at 30% soluble solids content.

III. DISPERSION OF PECTIN

The granules in a dry sample of pectin either are in intimate contact with each other, or are separated by a barrier of air. Their cohesiveness does not facilitate dispersion in water, but instead, they congeal into a lump, when wetted. Diffusion

Table I General properties of high-methoxyl, conventional low-methoxyl, and amidated low-methoxyl pectin and their gels

Property	HM[b]	LMC	LMA
De[a]	≥50%	≤50%	≤50%
Gelation conditions	55% ss	0–80% ss	0–80% ss
	pH <3.5	1 < pH < 5.0	1 < pH < 5.0
Thermal stability	Irreversible	Reversible	Reversible
Shear stability	Irreversible	Reversible	Reversible
Gel temperature	35–90°C	40–100°C	30–70°C
Gel state pH < 3.4	Elastic	Soft gel	Rubbery gel
pH > 3.4	Sol	Thixotropic gel	Thixotropic gel
Calcium sensitivity	Negative	Positive	Positive
pH sensitivity	Large	Small	Small

[a]Degree of esterification.
[b]HM, high methoxyl; LMC, conventional methoxyl; LMA, amidated low methoxyl.

is slow across the air–air or air–water interface, and the pectin–water mixture must therefore be agitated, in order for the pectin to be provided with sufficient energy to be dispersed in a practicable period. A variety of techniques may be employed to effect dispersion. First, a soluble diluent, *e.g.,* sugar, may be interspersed between the granules before adding water. Pectin hydration takes place concurrent with sugar dissolution. Similarly, pectin may be dispersed by first wetting it thoroughly with ethanol before adding water. Ethanol addition to food is seldom an acceptable practice, but ethanol as a solvent for food additives, *e.g.,* essences and flavor extracts, dissolved in it may be used as the wetting agent.

Dry pectin may be dispersed in a *crowded* (high concentration of total solids) aqueous system like corn syrup, and the mixture, poured into water and stirred. The high concentration of glucose lowers the water activity to the extent that the rate of dispersion of pectin exceeds the rate of hydration, and lumps do not form.

Dry pectin may be mixed with the aid of a mixing device developed by Hercules, Inc. (1980). The pectin is fed into a jet eductor and is rapidly dispersed by a high-velocity stream of water. The pectin is essentially atomized before it becomes hydrated. In practice, the eductor is connected to a water line that feeds into a tank at a rate of approximately 20 gal per min.

A Waring blender may be used to disperse pectin in water without any kind of pretreatment. The granules are trickled slowly into the vortex created by the rapid swirling. This technique makes initial hydration and lumping irrelevant, because the high shearing rate causes the lumps to disintegrate. The Breddo Likwifier is in effect a large Waring blender (25–50 gal capacity), capable of making a 7% concentration of pectin in water. Pectin is among the most shear-stable group of biopolymers. It does not depolymerize with a loss of viscosity under high-shear stress.

IV. STANDARD PECTIN

Climate, weather, fruit-processing, and waste-processing practices have a great influence on the quality and properties of extracted pectin. Manufacturers, not wishing to make adjustments to their formulas with every new batch of pectin delivered, have relied on standardized pectin. Until recently, this has meant a constant gel strength as a property of pectin grade. For example, a 150-grade pectin implied that 1 lb pectin will gel 150 lb sugar to a certain sag percentage, under specified conditions (IFT, 1959). As pectins have come into use for sour-milk drinks, yogurt, dietetic beverages, and other applications that do not require gelling, conventional standardization has become relatively mean-ingless. It must now be understood that the conventionally defined gel strength is irrelevant to a pectin used in yogurt, for example. Conversely, a pectin standardized for yogurt might have an unknown ability to get with sugar and acid or calcium.

V. PECTIN FORMULATIONS

Food uses for pectin other than jams, jellies, and preserves include low-calorie syrups and beverages, flavor emulsions, salad dressings, cream-whipping aids, bakers' glazes, malted-milk thickener, milk gels and puddings (Joseph, 1953), and a wide assortment of fruit spreads, ripples, sauces, compotes, ketchup, and drinks (Hercules, Inc., 1985). Pectin is a good stabilizer of oil-in-water emulsions. Examples of the numerous recipes that utilize one or more of the properties of pectin follow.

A. Reduced-Calorie Jams, Jellies, and Preserves

Reduced-calorie foods cannot be labeled as traditional jams, jellies, preserves, *etc.* Such a claim, by federal regulation, must be substantiated by the products' having at least one third less caloric value than their regular counterparts. Many manufacturers have lowered the percentage from 43% (*i.e.*, two thirds of 65%) to 30 to 35%. In this ss range, LM pectins are the efficacious pectins. Artificially sweetened fruit preserves and jams, also by federal regulation, must contain at least 55% fruit by weight of the finished item. For all these products, texture as well as flavor must be competitive, from a consumer standpoint, with their standard counterparts. A typical formula for a low-calorie jelly is outlined in Table II. The step-by-step directions for its preparation are as follows:

1. Combine the A ingredients in a suitable vessel and heat to 80°C.
2. Prepare the pectin dispersion (B) with hot water (at least 60°C).
3. Pour B into A, then add C with stirring.
4. Pasteurize, cool to 65°C and hot fill (preferably in glass).

Table II Typical formula of a low-calorie jelly

Ingredients	g/kg	ss[a]	Code
Fruit or juice (10% ss)	500.0	50.0	A
$CaCl_2 \cdot 2 H_2O$	0.37	0.37	A
Citric acid	To taste	—	A
Sodium citrate dihydrate	To taste	—	A
Water	250.0	0	B
LM pectin	10.0	10.0	B.
Sucrose	260.0	260.0	C

[a]Soluble solids (ss) content in grams/kg

B. Conserves

Conserves are a relative newcomer to the jam and jelly industry. They do not meet the jam and jelly standards of identity, because they do not contain a sweetener other than fruit juice or fruit juice concentrate; hence the ss content is slightly lower. They are perceived by the consumer as being of a higher quality than regular jams and jellies, and to be more *natural,* because they contain only fruit and/or fruit juice. The ss content of commercial conserves is 55–62%, and consequently, they do not fall into the classification of a reduced-calorie or a dietetic product. At the upper ss limit, a rapid-set HM pectin may be used. At the lower limit, an LM pectin is required. The inherent calcium concentration in the large amount of fruit in the formula usually makes additional calcium unnecessary.

Fructose (fruit sugar) is the major sweetening compound in fruits. It has the same caloric value as sucrose, but conserves seem, nevertheless, to be accepted more by diabetics, probably because much less fructose is needed to arrive at the same sweetness intensity as given by sucrose and glucose.

LM pectin gels with fructose are softer than LM pectin gels with sucrose. As a result, more pectin is required for an LM pectin conserve with fructose to have the same textural characteristics as a sucrose–pectin jam. A typical formula for a conserve is outlined in Table III. The step-by-step directions are as follows:

1. Mix the fruit and juice concentrates (A) in a suitable vessel, add water, and heat to 80°C.
2. Prepare B in 60°C water.
3. Add B to A, and evaporate the excess water under vacuum to the desired ss content.
4. Pasteurize the product and fill containers.

C. Bakery Jams and Jellies

Some bakery products contain jelly centers, whose moisture and flavor must remain within the confines of the dough surrounding the jelly. So that the bound-

Table III Typical formula of a strawberry conserve

Ingredients	g/kg	Code
Strawberries (10% ss)	400	A
Apple juice concentrate (72% ss)	700	A
Water	93	A/B
LM pectin	7	B

Table IV Typical formula of a bakery jelly

Ingredients	g/kg
Fruit (10° Brix)	300
Sucrose	350
Corn syrup (42 DE[a])	350
Water	190
HM pectin	10
Citric acid (50% solution)	2
Sodium citrate dihydrate	1

[a]Dextrose equivalent.

ary is maintained, the heated gel should not revert to a sol. HM pectin gels, being thermally irreversible, are suited to this restriction, and a baker may place a previously prepared HM gel in a dough or batter, and bake it without fear of having it fluidized. If the fiber content of the jelly formula is high, fiber entanglements will further reinforce the gel structure, making it yet more heat stable.

Jelly is pumped into some doughnuts and other pastries after the pastries have been baked. HM pectin in this jelly should amount of 0.75 to 1.0% by weight of the jelly. This level—two or three times the level in regular jams and jellies—is necessary to maintain the integrity of the centers while the jelly is being pumped. The higher level of pectin can accommodate pH 3.3–3.6, as opposed to pH 3.0–3.2 in regular jellies. A typical formula for a bakery jelly is given in Table IV.

LM pectins offer bakery jams and jellies a wider applicable ss range (as low as 45°Brix) and acidity (pH 3.4–3.7), in addition to thermal reversibility. An LM pectin jelly formula should contain 10% more pectin than an HM pectin jelly formula to approximate the same firmness. Too much calcium will induce premature gelation (pregelation), *i.e.*, gelation at an inopportune time, *e.g.*, before completion of decanting or filling retail containers. Too little calcium will accelerate liquid flow from the collapsible structure while the product is in the oven. The optimal calcium concentration for a specific application and a specific pectin is usually available from the pectin manufacturer. Any recommendation is merely an estimate, and therefore some trial-and-error experimentation should be done beforehand. A formula of a typical LM pectin, 60°Brix bakery jelly, capable of being pumped, is given in Table V. The step-by-step directions are as follows:

1. Prepare the solution of A ingredients, and heat to 85°C.
2. Disperse B in 85°C water.
3. Add B to A, mix well, and evaporate excess water.
4. Add the combined C slowly with stirring.
5. Cool to fill temperature (70–30°C).

Table V Typical formula of a bakery filling, capable of being pumped

Ingredients	g/kg	g/kg	Code
Fruit (10°Brix)	300	30	A
Sucrose	370	370	A
42 DE[a] corn syrup (80°Brix)	235	88	A
Water	150	0	A/B
LM pectin	9.8	9.8	B
50% citric acid	—	—	C
Sodium citrate dihydrate	—	—	C
Calcium citrate	0.55	0.55	C

[a]Dextrose equivalent.

D. Cold-Setting Flan Jelly

This application, a common practice in Europe, but less known in the United States, is based on the principle that HM pectin can be made to gel when cold. A baker may purchase customized syrup containing pectin, to which is added a prescribed quantity of citric acid. If a thin film of this syrup is layered over a pastry, for example, within minutes, the layer will gel in the form of a glaze. A typical formula for this application is outlined in Table VI. The directions are as follows:

1. Blend the A ingredients.
2. Prepare B.
3. Combine A and B, heat to boiling, and continue boiling.
4. Add C, and continue boiling to a refractometer reading of 61% solids.
5. Add D, stir, and fill containers.

Table VI Typical formula of a cold-setting flan jelly

Ingredients	g/kg	Code
Sugar	35	A
Rapid-set pectin	7	A
Water	420	B
Sodium citrate dihydrate	0.8	B
50% citric acid	2.5	B
Sugar	477	C
42 DE[a] corn syrup (80°Brix)	110	C
Flavor, color, *etc*.	—	D

[a]Dextrose equivalent.

Table VII Gelation interval as a function of citric
acid concentration per kilogram of flan jelly

Interval	CA[a]	Final pH
4–5 min	10 ml	2.8
1–2 min	15 ml	2.6
50 sec	20 ml	2.5

[a]Citric acid concentration in ml of a 50% wt/vol solution.

The amount of citric acid should be adjusted to the length of time available
between cooking and layering (Table VII).

E. Neutral-Flavored Confectionery Pectin

This section mentions a *neutral-flavored* pectin, to indicate that no fruit flavor
is carried over by the pectin into the product. The extraneous flavor of choice
(vanilla, spearmint, licorice, etc.) is conferred on it. One formula is designed
around an LM pectin and phosphoric acid salts. When the product pH approximates 4.0, perception of the acid is at a minimum. With this simple prescription,
there is only one weighing step. The directions for preparing a neutral-flavored
pectin confectionery (Table VIII) are as follows:

1. Measure A into a kettle suitable for boiling the mixture.
2. Dry-blend the B ingredients, and add to A with vigorous stirring.
3. Heat A plus B slowly, and boil for 2 min with stirring (to ensure complete
   dispersion of pectin).
4. Add the C ingredients, and stir until dissolution is complete.
5. Boil to net weight, add D, and deposit the hot sol into starch or metal molds.

Table VIII Typical formula of a neutral-flavored pectin candy

Ingredients	g/kg	Code
Water	300	A
LM pectin	25	B
Sugar	75	B
Sugar	445	C
42 DE[a] corn syrup	300	C
Flavor and color	—	D

[a]Dextrose equivalent.

Table IX Typical formula for an ice cream fruit syrup, ripple, and related items

Ingredients	g/kg	Code
Fruit puree (10°Brix)	25	A
Sugar	61	A
Citric acid	0.25	A
LM pectin	0.30	B
Water	15	B

F. Ice Cream Fruit Syrups, Ripples, and Related Items

These applications call for a shear-reversible fluid, containing fruit pieces in suspension, that is capable of being pumped. A typical formula is outlined in Table IX. The directions are as follows:

1. Mix the A ingredients in a cooking utensil, and begin the heating.
2. Blend the B ingredients, then pour into A.
3. Boil to the net weight, and pack.

G. Barbecue Sauce

In the United States, some retail brands of barbecue sauce contain LM pectin to take advantage of its excellent adhesive and flavor-release attributes, and the pulpy texture attributed to it. The pectin system should have a high melting temperature to ensure that the sauce will not flow during barbecuing. The LM pectin and calcium contents in the formula determine the product's final con-

Table X Typical formula for a commercial barbecue sauce

Ingredients	g/kg	ss[a]	Code
Tomato paste (30% ss)	250	75	A
Corn syrup (42 DE[b])	100	80	A
Water	155	0	A
LM pectin	4	4	B
Water	100	0	B
Sucrose	160.6	160.6	C
NaCl	30	300	C
$CaCl_2 \cdot 2H_2O$	0.4	0.4	C
Spices and flavoring	—	—	C
Vinegar (50 grain)	200	0	D

[a]Soluble solids (ss) content in grams/kg.
[b]Dextrose equivalent.

sistency and texture. A smoother sauce will result from more pectin and less calcium, and *vice versa* for a pulpier product. These sauces almost always contain tomato paste, in which the calcium content is variable, thereby necessitating small adjustments (with calcium chloride) from time to time. Table X is a typical, commercial formula for barbecue sauce containing LM pectin. The directions are as follows:

1. Mix the A ingredients in a stainless steel pot, and heat to 80°C.
2. Prepare B, heat to 80°C, and add to A.
3. Dry-blend C, add to A + B, and stir.
4. Add D, pasteurize, and hot-fill suitable containers.

H. Dietetic and Fruit Juice Beverages

Dietetic soft drinks enjoy widespread popularity, and the market is increasing at a greater rate than that for their traditional sugar-containing alternatives. A regular soda contains approximately 10–12% sweetener (sucrose, high-fructose corn syrup, or a combination of the two), flavor, color, and other additives, but the major ingredient is water. The Newtonian flow characteristics of these conventional beverages gives them a certain mouthfeel or *body*. Replacing a large quantity of sugar with a very small quantity of artificial sweetener deprives the beverage of this expected mouthfeel. However, in the case of dietetic beverages, this property has been restored with 0.05 to 0.10% HM pectin.

The pulp present in the fruit juice component, if any, of a dietetic fruit-juice beverage is prone to deposition into a hard mass that is difficult to redisperse. This deposition is known as *hardpacking*. Addition of pectin to the formula has been found to mitigate hardpacking.

A dietetic beverage base (Table XI) is manufactured by dispersing HM pectin in a small volume of water (A) with the aid of a high-speed mixer, adding the

Table XI Typical formula of a dietetic soft drink base

Ingredients	Amount (wt/wt%)	ss[a] (wt/wt%)	Code
Water	5.5	0	A
HM Pectin	0.29	0.29	A
Water	91.8	0	B
Sodium citrate dihydrate	0.43	0.43	B
Saccharin (sodium form)	0.070	0.070	B
Aspartame	0.070	0.070	B
Sodium benzoate	0.044	0.044	B
Flavor and color	—	—	B
50% citric acid solution	1.8	0.9	C

[a]Soluble solids content.

Table XII Typical formula of a syrup for a pectin pudding dessert

Ingredients	g/kg
Fruit or juice	150
Water	200
Sugar	250
LMA pectin	16
Sodium citrate dihydrate	10
Water	400
50% citric acid solution	13

dispersion to a solution of the B ingredients, then adding C, and pasteurizing the formula. It is seen from Table XI that the ss content of a dietetic-beverage base is approximately 2%. To reach beverage strength, the base is diluted with water in the proportion of 35 parts base to 165 parts water.

I. Gelled Pudding Dessert

The advantages of a gelled puddling-like dessert are convenience and an economy of time. The consumer buys a container of fruit syrup that is easily mixed with an equal volume of cold milk, to obtain in a few minutes without refrigeration a dessert with the consistency of a pudding. An LMA pectin standardized for this purpose is used (Table XII). Calcium sensitivity is a critical property of the LMA pectin for this application. For best results, pH should be 4.0–4.2. Above this range, the product either will not set or will set very slowly.

J. Frozen Novelties

This group, in which controlling the rate of crystal growth is the primary consideration, includes Italian ices, sorbets, and sherberts. Too-small crystals will

Table XIII Typical formula of an Italian ice

| Ingredients | Texture (%) | | |
	A	B	C
Dextrose or sucrose	13	16	22
Corn syrup (32 or 42 DE[a])	0	0	7
HM rapid-set pectin	0.2	0.2	0.2
Citric acid (50% solution)	1.0	0.7	0.8
Fruit juice or puree	15	15	15
Water	43	56	55

[a]Dextrose equivalent.

impart a smooth albeit heavy texture (Table XIII, column A). If the crystals are too large, the product will have a light but coarse, sandy, or grainy texture (Table XIII, column C). For those persons who like neither extreme, column B is a safe compromise. Pectin may be used alone or in combination with other gums to control crystal growth. The pectin is mixed with five parts of sugar and dispersed in water. All the ingredients in Table XIII are then combined.

VI. PECTIN–PROTEIN COMPLEXES

Pectin is a stabilizer of protein in a low-pH fluid milk system. The isoelectric point of casein is pH 4.7, below the positively charged species becomes attached to the negatively charged, linear pectin molecules at the latter's carboxyl sites, suggesting the model of a golf ball coated with a loose fuzz (Fig. 2). Some degree of stabilization of the dispersed phase results from repulsion of the net negative charge on the fuzz.

In milk–protein complexes, the initial pH is an important factor to consider. If milk and a pectin are combined at pH 6.8, and the combination is refrigerated for an extended period, a precipitate forms. Casein, in the midst of 0.3% pectin, precipitated this way is claimed to be easily redispersed, to be more digestible than that separated by other processes, and is therefore used to prepare some types of cheeses (Joseph, 1953). If, immediately after the addition of pectin to the milk, the pH is gradually lowered to 4.2, a soluble complex forms, based on mutual attraction of the positively charged casein and negatively charged pectin. Lowering the milk pH to 4.2 before adding the pectin will precipitate casein only.

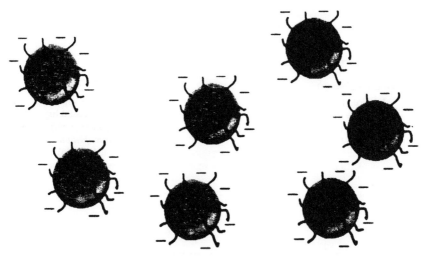

Figure 2   The fuzzy golf ball model of casein stabilized by pectin at pH 4.2.

Table XIV Typical formula of a yogurt–fruit preparation

Ingredients	g/kg	Code
Fruit (10°Brix)	400	A
Sucrose	500	A
$CaCl_2 \cdot 2H_2O$	As required	A
Citric acid	As required	A
Sodium citrate dihydrate	As required	A
LM pectin	7.5	B
Water	150	B

A. Yogurt

Yogurt is a milk- and fruit-based product of lactic acid fermentation that fits into currently fashionable, weight-reducing regimens, because of its naturalness. The equilibrium pH is about 4.2 to 4.4, where much of a fruit's flavor and color are lost. Below, curdled casein will give the product an unwholesome texture. The fruit and surrounding gel should be at the same pH of 3.5 to 3.9, a safe compromise between flavor and colorretention and casein stability.

Yogurt containing fruit is divided into two kinds, *viz.*, *stirred* or *Swiss-style*

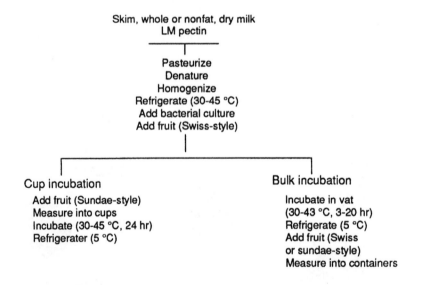

Skim, whole or nonfat, dry milk
LM pectin

Pasteurize
Denature
Homogenize
Refrigerate (30-45 °C)
Add bacterial culture
Add fruit (Swiss-style)

Cup incubation

Add fruit (Sundae-style)
Measure into cups
Incubate (30-45 °C, 24 hr)
Refrigerater (5 °C)

Bulk incubation

Incubate in vat
(30-43 °C, 3-20 hr)
Refrigerate (5 °C)
Add fruit (Swiss
or sundae-style)
Measure into containers

Figure 3 Flow diagram for the manufacture of yogurt.

and *sundae-style*. The choice of either style is a matter of personal preference. In the former, the fruit is uniformly mixed with the other ingredients; in the latter, the fruit is deposited as a layer at the bottom of the container. Flotation and uneven distribution of the fruit pieces are prevented by including a stabilizer. Compared with starch and gum, a pectin-stabilized yogurt–fruit preparation is believed to have superior flavor-release characteristics. In practice, LMC and LMA pectins are used in both types of yogurt. However, this author's preference is LMC pectin in cup-incubated yogurt, and LMA pectin, in bulk-incubated (stirred) yogurt.

Ordinarily, yogurt contains approximately 1000 parts per million of calcium. Over its storage life (4–6 weeks), water and calcium will migrate from the surrounding fluid into the fruit layer, and the latter might harden from uptake of extra calcium. The solution to this hardening is to saturate the fruit preparation with calcium before adding to the yogurt. For fruit on the bottom (sundae-style), the fruit layer must be soft enough to be easily cut with a spoon. A typical formula for a yogurt–fruit preparation is listed in Table XIV. The A and B ingredients are mixed separately, then combined. The pectin (B) must be dispersed in water (B). The formula contains approximately 40–60% ss by weight. In an *all natural* yogurt, lemon juice concentrate is substituted for citric acid. The manufacture of yogurt is outlined in Fig. 3.

B. Juice–Milk Beverages

Juice–milk (and juice–yogurt) combinations must ideally be minimally sensitive to acid for the reasons previously stated. HM pectins are calcium insensitive, and LM pectins are not. HM pectins are preferred, presumably because calcium sensitivity is deleterious to the proper physical characteristics of these products. Unlike LM pectin, there is no upper limit, except an economic one, on the use of HM pectin in low pH, milk-based formulations, because HM pectins are insensitive to calcium. The absence of a reaction of HM pectin with calcium ensures a creamier, smoother texture. The HM pectin should be limited to the level that contributes the ideal texture. Here, the concentration is 0.2–1.0%. In the fabrication of this beverage, juice is gradually added to milk fortified with pectin, lowering the milk pH to below pH 4.7. Citric acid may provide the final adjustment. The beverage is pasteurized at 90°C for 30 sec, and homogenized at 2500 psi before being hot-filled (70–80°C). Acidity and pasteurization ensure biological stability. Alternatively, the beverage may be cooled to 15°C or lower, and aseptically poured into containers.

VII. CONCLUSION

Numerous uses exist for pectin in food manufacture other than in jellies, jams, and preserves. Gelation in varying degrees remains the outstanding application, but there are other possibilities still to be explored. HM and LM pectins are largely interchangeable in a food formula, yet one may be preferred over the other, on the basis of a narrow albeit important criterion. New trends in microwave heating and convenience-food preparation should begin to take advantage of the versatility of pectin, that, in a number of instances, is more conducive to food fabrication than are many other edible hydrocolloids.

References

Hercules, Inc. (1980). "Cellulose Gum," 250-10 Rev. 5-80. Wilmington, Delaware.

Hercules, Inc. (1985). "GENU Handbook for the Fruit Processing Industry." Middletown, New York.

Institute of Food Technologists (IFT). (1959). Pectin standardization. Final Report of the IFT Committee. *Food Technol.* **13,** 496–500.

Joseph, G. H. (1953). Better pectins. *Food Eng.* **25,** 71–73.

CHAPTER 4

Tropical Fruit Products

A. S. Hodgson
University of Hawaii at Manoa
Honolulu, Hawaii

L. H. Kerr
Food Technology Institute
Kingston, Jamaica

I. INTRODUCTION

Tropical fruits are finding their way to supermarket shelves, worldwide, as juice blends, natural flavors, dried fruit mixes, marmalades, jams, jellies, and other confections. These items are sometimes interchangeable in the vernacular. Generally, a jelly is considered to be made from clarified or filtered fruit juice, with or without added water; jams contain fruit pulp or unclarified juice; marmalades have fruit pieces suspended in the pulp; confections are sugar–fruit preparations.

Tropical fruits have not been widely exploited for their pectin content, except as jellying agents in indigenous manufacture. One of the reasons for this is the comparative advantage afforded in the temperate-climate regions to large-scale industrial factories that extract pectin from citrus and apple byproducts, and export it to the tropics more economically than the local costs of production would allow.

II. PECTIN CONTENT OF TROPICAL FRUITS

The pectin content of some tropical fruits is listed in Table I. The overwhelming tonnage of citrus is harvested in the subtropical regions, making pectin from this source also a subtropical commodity. Those commonly used to make jams and jellies, and their essential properties, are listed in Table II. They mostly suffer the disadvantages of low acidity and low soluble-solids content. The data also show that papaya, guava (strawberry and common), soursop, tamarind, natal plum, and pomelo all contain considerable amounts of pectin. Nagy and Shaw (1980) describe most of the available information on composition, methoxyl content, and properties of tropical fruits.

Table I Pectin content of some tropical fruits

Fruit	% w/w
Strawberry guava (*Psidium cattleianum* S.)	0.81[a]
Papaya (*Carica papaya*)	1.00[a]
	0.66[b]
Banana (*Musa acuminata* C.)	0.62[b]
Plantain (*Musa paradisiaca* L.)	0.60[b]
Guava (*Psidium guajava* L.)	0.99[c]
	0.71[b]
Lychee (*Litchi chinensis* S.)	0.424[d]
Mango (*Mangifera indica* L.)	0.38[b]
Alphonso cultivar (from mango marc)	0.265[e]
Kitchener cultivar (from mango marc)	0.298[e]
Abu Samaka cultivar (from mango marc)	0.423[e]
Mountain apple (*Eugenia malaccensis*)	0.47[a]
Pineapple (*Ananas comosus* L.)	0.13[a]
	0.04[b]
Surinam cherry (*Eugenia uniflora*)	0.61[a]
Soursop (*Annon muricata* L.)	2.07[a]
	0.36[b]
Carambola (*Averrhoa carambola*)	0.66[a]
Tamarind (*Tamarindus indica* L.)	1.71[a]
Thimbleberry (*Rubus rosalfolius*)	0.72[a]
Natal plum (*Carissa grandiflora*)	1.81[a]
Pomelo (*Citrus grandis*)	1.52[a]
Giant granadilla (*Passiflora quandrangularis* L.)	0.40[b]
Passion fruit (*Passiflora edulis* Sims)	0.05[b]
Passion fruit rind	2.1–3.0[f]

[a]Sherman and Kanshiro (1947).
[b]Garces Medina (1968).
[c]Wilson (1980).
[d]Cavaletto (1980).
[e]Saeed *et al.* (1975).
[f]Sherman *et al.* (1953).

4. Tropical Fruit Products

Table II   Soluble-solids content (°Brix) and pH of a selection of tropical fruits

Commodity	°Brix	pH
Ripe banana	21.4–26.0	5.0–5.5
Grapefruit	4.6–8.8	2.8–3.2
Guava	7.6	2.5
Mango	21.0	4.1
Orange	10.0	3.4
Otaheite apple	5.5–6.3	3.7–5.5
Papaya	8.0–11.8	5.0–5.5
Pineapple	12.4–16.0	3.3–3.7

III. PECTIN EXTRACTION

Technical aspects of pectin extraction from tropical fruits were examined by Muroki and Saint-Hilaire (1977), who focused on guava, and Simpson *et al.* (1984), who focused on grapefruit rinds. Commercial pectin extraction is conducted at pH 1.5–3.0 and 60–100°C. The extract may be isolated by precipitation as pectin salts or with isopropanol or ethanol. In the first instance, the salts are washed with acidified alcohol to remove excess ions. The isolate may be concentrated in liquid or solid form. In international commerce, the tropical *gourmet* and *exotic* fruit packs (Fig. 1), in which pectin is the major functional ingredient, contain one or more of the products popularized locally. The better-known pectins are discussed briefly below.

A. Guava Pectin

The total pectin content of guavas was found to vary from 346 mg to 396 mg/ 100 g for unripe fruit, and 705–804 mg/100 g for fully ripened guavas (Jagtiani *et al.*, 1988). Pal and Selvaraj (1979) reported earlier that the pectin content of the Beaumont, Hybrid I (Allahabad Safeda × Banaras), and Hybrid II (Allahabad Safeda × Red Flesh) varieties, grown in India, were highest when the fruit was immature, and decreased as it passed maturity. In contrast, the Allahabad Safeda and Red Flesh varieties each had a maximal pectin content when the fruit was ripe, which decreased at the overripe stage.

The Muroki and Saint-Hilaire (1977) method of extracting pectin from guava yielded 11% relatively pure pectin containing a degree of methylation of 55%. The authors suggested that the extracted pectin was suitable for low-solids, sugar–acid gels.

Figure 1 Hawaii fruit products packaged in glass containers ranging in size from 1 oz to 18 oz. Bottom group is made from fruits grown only in the Hawaiian Islands.

Guava pectin was extracted by boiling equal parts of guava and water for 20 to 30 min (Kanehiro and Sherman, 1946). The hot mixture was filtered successively until 8.5% total soluble solids and 1.2% pectin were obtained. The resulting extract was concentrated under vacuum to a total soluble-solids content of 20%. To each part of the concentrated extract was added 2 parts sugar, and the mixture was further vacuum dried without heat. This method produced a cream-colored, crystalline, dehydrated guava pectin with excellent aroma, flavor, and gelling characteristics, and retaining approximately 60% of the ascorbic acid content of the original fruit. The pectin contained 76.65% uronic acid and 8.25% methyl esters, similar to the composition of commercial pectins. This guava pectin may be used as a flavoring as well as a gelling agent.

B. Papaya Pectin

Papaya contains sufficient pectin to structure a gel, but does not contain enough acid to impart firmness to it. Incidentally, papaya is an excellent source of provitamin A, vitamin C, and papain, a proteolytic enzyme.

Washed papayas were cut in a mincing machine with an 18-mesh screen and allowed to soak in cold water overnight (Bhatia et al., 1959). To prevent the discoloration of the fruit slices, 100 ppm SO_2 was added. Pulping was effected by a hydraulic press, and pectin was extracted at pH 2.8–3.0 with 0.2 N boiling HCl. Four successive extractions were made, each lasting 30 min. Solid and semisolid materials were separated from the mixtures and cooled immediately. All extracts were combined and adjusted to pH 4.0 with ammonia. After an overnight soak with 200 ppm SO_2, the combined extract was decanted and centrifuged. Pectin was isolated as the aluminum salt by successive, 30-min immersions of the extract in 1 M aluminum chloride at pH 4.0–4.5 and separated with 50 to 60 mesh vibrating sieves and a cloth-and-basket press. The precipitates were combined, cut and sized with a mincing machine equipped with an 18-mesh sieve, and dried at 55 to 56°C for 5 to 6 hr in a cross-flow, hot-air drier. The isolated pectin was washed in two stages with acidified (HCl) 95% alcohol, followed by three rinsings with 95% alcohol. Purified pectin was separated from the washings with a basket press, dried at 55°C for 2 hr in a cross-flow drier, packed in airtight containers, and stored in a dry, cool place. The purified, dried pectin may be mixed with sugar to a standardized jelly grade before packaging and storage.

C. Passion Fruit Pectin

Passion fruit juice is obtained from only one third of the fruit. Approximately 90% of the remaining two thirds are the rinds. Pectin content of the juice is expectedly small, but it is considerable in the rind. Isolation of pectin from this source was hailed as one solution to a waste problem (Sherman et al., 1953).

Analysis of passion fruit rind indicated approximately 3% pectin containing 76.6–78.0% galacturonic acid and 8.9–9.2% methyl esters. The pectin was capable of giving a jelly grade above 200, i.e., 200 kg sugar cooked to 65°Brix with 1 kg of that pectin yielded a satisfactory gel.

Fresh passion fruit rinds from a processing plant were cut with a meat chopper, and six times their weight of water was added. The mixture was acidified with sulfurous acid to pH 2.5. Pectin was extracted by heating the rind–water mixture for 6 hr at 80°C. The suspension was filtered through a muslin cloth followed by filtration through Whatman No. 4 filter paper. Previously added sulfur dioxide was removed by aerating the extract. Pectin was precipitated with an equal volume of 95% alcohol, filtered, and washed several times with ether. The precipitate was dried in a force-draft oven at 45°C, ground and sized to pass a 60-mesh screen. Three pectic enzymes, viz., protopectinase, pectinesterase, and polygalacturonase, were identified. Pectin yield decreased during storage, when pectinesterase was left inactivated in the rind. The deterioration with time was arrested by blanching the rinds with steam for 5 min (Sherman et al., 1953).

D. Mango Pectin

Saeed et al. (1975) extracted pectic substances from the marc (insoluble residue) of three mango cultivars, viz., Alphonso, Kitchener, and Abu Samaka. The mangoes were harvested during peak season and allowed to ripen at 28 to 30°C and 45% relative humidity. The pectin content of Abu Samaka was higher than that of Kitchener, which in turn was higher than that of Alphonso (Table I). Abu Samaka and Kitchener had degrees of esterification of 75.5% and 70.3%, respectively, while Alphonso had only 58.2%. The authors explained that since Abu Samaka and Kitchener had degrees of esterification greater than 70%, pectins from these cultivars would form a firm gel rapidly and at a higher temperature than would pectins from Alphonso, which had a low degree of esterification (50–70%). These data indicate that Abu Samaka and Kitchener pectins, extracted by the method outlined, would not be sensitive to calcium, whereas Alphonso pectin would be.

Pectin preparation from mango marc involved inactivation of enzymes with 95% ethanol. Concurrently, residues of sugar, soluble organic acids, and salts were dissolved. The alcohol-insoluble solids were blended, filtered, washed first with 70% ethanol then twice with acetone, air-dried at room temperature, and ground to pass a 40-mesh screen (McCready and McComb, 1952; Gee et al., 1958). McCready and McComb (1952) added a 0.5% Versene solution (sodium salt of ethylenediaminetetraacetic acid EDTA in water) at pH 11.5 to the marc to sequester cations and chemically deesterify the pectic substances, through reactions that were allowed to proceed at 25°C for 30 min, before treatment with pectinase (0.05% by weight) and glacial acetic acid (to pH 5.0–5.5).

E. Banana Pectin

A laboratory method for extracting pectin from banana peel was developed by von Loesecke (1950). The peel was ground and mixed with an equal weight of 0.2% citric acid solution. The suspension was allowed to stand at 98 to 100°C for 1 h before being filtered. The extraction was repeated, and the extracts were combined and incubated with *Aspergillus oryzae* at 37°C for 3 to 4 days to remove starch. The starch-free extract was concentrated under vacuum initially, then dried at 50°C, which resulted in a dark-brown, hygroscopic material containing approximately 10.19–25.79% calcium pectate.

IV. REGULATIONS AND SPECIFICATIONS

A. The United States

Given the relative paucity of technical information, the general lack of uniform standards in the tropics, and the burgeoning, international trading in tropical fruit products, this section summarizes the United States Standards of Identity, as they pertain to tropical fruit products for export into that country. The standards are found in the Code of Federal Regulations, Title 21, Part 150, Sections 150.140 (fruit jelly), 150.141 (artificially sweetened fruit jelly), 150.160 (fruit preserves and jams), and 150.161 (artificially sweetened fruit preserves and jams) (FDA, 1989).

1. Fruit Jelly

Of the fruits listed in the federal standards for jams and jellies, only guava, orange, grapefruit, and pineapple are considered tropical. Fruit jams and jellies containing a mixture of fruits not listed may be considered to be nonstandardized, a category recommended to contain 45 parts of fruit to 55 parts of sugar, and have a final soluble-solids content of 65%. An exception to the rule is blackcurrant jelly, which is permitted to have a fruit:sugar ratio of 27:55, as a result of its high pectin content (FDA, 1980).

Fruit jelly is the jelled food prepared from a fruit juice or up to a combination of five fruit juices in equal amounts, and one or a combination of permitted optional ingredients (nutritive sweetener, saccharin and its salts, pectin, spice, flavor, acid, etc.). It may not contain less than 45 parts by weight of fruit juice ingredients to 55 parts by weight of sugar, and 65% soluble solids. If it is made with a single fruit juice, the label is to read *Jelly* preceded or followed by the name or synonym of the corresponding fruit. If it is made from more than one fruit juice, the label should read *Jelly,* preceded or followed by *Mixed fruit* or by the corresponding fruit names or synonyms in decreasing order of their predominance by weight.

The weight of fruit or fruit juice (w) required for a particular jelly may be calculated from the equation:

$$w = \{[\text{weight of jelly (W)} \times \text{\% jelly soluble solids}/100]$$
$$- \text{weight of sugar}\} \times \text{factor} \quad (1)$$

The factor is the reciprocal of the average soluble-solids content of the fruit or juice involved, multiplied by 100. For example, if the soluble-solids content of pineapple juice is an average 14%, then the factor for calculating the weight of fruit juice in a predetermined quantity of jelly (W) is 100/14, or approximately 7 (FDA, 1989). Similarly, the listed factor 13 (i.e., 100/8), for guava assumes an average soluble-solids content in the juice of 8%. Tables showing the amounts of various fruit juices, some of them tropical, used in making pure jellies according to the federal standards are included in the Preservers Handbook (Sunkist Growers, 1964).

2. Artificially Sweetened Fruit Jelly

FDA Standards of Identity are also prescribed for artificially sweetened fruit jelly, otherwise known as dietetic jelly, whose juice composition is defined as in the standards of identity for fruit jelly. However, water may be added. In this category, no nutritive sweetener is permitted, except that amount used separately to standardize the gelling agent. These jellies rely on chemical bonding between calcium (Ca) and low-methoxyl pectins for their gel structure. They contain about 1.25 to 1.5% by weight of pectin and 15–20 mg (Ca)/g pectin (Sunkist Growers, 1964). The nutritive sweetener cannot amount to more than 44% of the pectin, and the pectin cannot amount to more than 3% of the jelly. The fruit juice ingredient cannot be less than 55% by weight of the jelly, thereby necessitating a heat process (hot-filling) to prevent spoilage, and inclusion of sodium benzoate or benzoic acid (0.05–0.10%) and potassium sorbate, not to exceed 0.10% of the finished product (Sunkist Growers, 1964). The label must include *Artificially sweetened*, prominently and conspicuously displayed, immediately followed by the name or synonym of the fruit from which the juice that describes the jelly was extracted.

3. Fruit Preserves and Jams

Fruit preserves and jams are defined as viscous or semisolid foods in which the fruit ingredients are substantially governed by the standards of identity of the previous products (Sections 1 and 2). They are prepared with or without added water by concentration with or without heat. Pectin may be added to compensate for a deficiency, if any, in the natural pectin content of the fruit.

Pieces of fruit in tropical preserves are as likely to float occasionally as in temperate-climate preserves.

The standards of identity for this category distinguish between two groups (I and II) of fruit ingredients in fresh, concentrated, frozen, and/or canned form. A Group I preserve or jam must have the ratio of fruit ingredient to sugar at 47:55. For all others, including Group II, the ratio must be 45:55. Pineapple, orange, and grapefruit are in Group I; guava is in Group II. The reason for the group distinction, I and II, is not known, but it is widely believed to be simply a contemporaneous rule of thumb at the time the regulation was instituted. The soluble-solids content of the finished product should not be less than 65% by weight. If the fruit ingredient is a single fruit, the label should read *Preserve* or *Jam,* preceded or followed by the name or synonym of the fruit. If the fruit ingredient is a combination of up to five fruits, these designations should be preceded or followed by *Mixed fruit,* or by the fruit names or synonyms in order of their predominance on a weight basis.

4. Artificially Sweetened Fruit Preserves and Jams

This category is similar to that for artificially sweetened fruit jelly, with the same limitations on nutritive sweeteners. The fruit rather than the fruit juice is the source of the fruit ingredients. A low-methoxyl pectin concentration of about 1.0% is usually satisfactory (Sunkist Growers, 1964). Naming and labeling this category follow the same convention previously described.

5. Imitation Products

It is noteworthy that the Supreme Court of the United States has ruled that a jam, for example, labeled *Imitation* is deemed not to be misbranded, even though its fruit content is less than what the standard of identity for regular jam requires, since it did not purport to be a standardized food.

B. Jamaica

In recognition of the sometimes special conditions prevailing in the tropics, indigenous standards have from time to time been promulgated by local governmental organizations. For example, in 1964, the Jamaica Bureau of Standards, Jamaica, West Indies, instituted styles, standards, and grades for the local manufacture of jelly.

1. Style I

Style I jellies must bear the name of the fruit, and must be made by boiling the juice of the named fruit, free from seeds and pulp, or a juice concentrate to which water and a sweetening agent have been added. It contains not less than 60% of the water-soluble solids, estimated by refractometer at 30°C (86°F) uncorrected for insoluble solids. It may contain added pectin and acid to compensate for any deficiency in the natural pectin content or acidity of the fruit.

2. Style II

Style II jellies are labeled and prepared similarly to Style I jellies, with the additional stipulation that they contain not less than 32% of the juice of the named fruit and not less than 60% of water-soluble solids. The juice of another fruit is permitted. Sulfur dioxide not exceeding 40 parts per million may be present. When the product is to be exported, sulfur dioxide may not exceed the maximum allowed in the regulations of the importing country.

3. Objective Testing

The routine physical measurements of the two styles of jelly are soluble-solids content, pH, and gel point. The first two are performed conventionally (see Chapter 2). The gel point is ascertained by an alcohol or ice-water test. If a firm, single clot forms in the bottom of a beaker containing alcohol or ice-water, when a few drops of the hot sol are immersed in it, the gel point has been reached. Occasionally, the sol fails to gel, although quality-control monitoring indicates that the gel point has been reached. The better-equipped laboratories measure texture (gel strength) by the ridgelimeter.

4. Jelly Grades

The grades enunciated by the Jamaica Bureau of Standards are *Fancy* and *Choice*, subjectively based on consistency, color, freedom from defects, and flavor.

Texture and acceptability are evaluated principally by taste panels. Irrespective of the style or grade, the ideal jelly should be transparent, should set to an elastic body, and should have an aroma reminiscent of the fruit from which the juice was extracted.

V. MANUFACTURE

Many of the tropical fruits for export are grown in commercial orchards. Many more are not so grown, but are harvested from diverse locations, often in different stages of maturity. In the majority of instances, quality-assurance programs in the field are ineffective. Special precautions must therefore be taken to ensure raw materials of the highest quality.

The tropical pectins are comparable with temperate-climate pectins and are similarly graded. However, when this is not the case, jelly grades may be interchanged through the equation (Sunkist Growers, 1964)

$$P = \frac{pg}{G} \tag{2}$$

where P = amount of pectin to be used, G = grade of pectin to be used, p = weight of pectin being used, and g = grade of pectin being used. Low-methoxyl products shipped to and stored in warm climates should be prepared with more pectin and/or calcium than that used in temperate climates, because low-methoxyl gels soften at even moderately elevated temperatures. Since high-methoxyl pectin gels are relatively indifferent to temperature increases, temperate-climate formulae for HM pectin jellies are adequate for tropical climates.

The failure of some jellies to gel may be the consequence of a number of factors, viz., improper dispersion of the pectin, insufficient pectin, too low soluble-solids content, and too high pH.

The problem of floating fruit can be solved by using a rapid-set pectin that initiates gelation before the fruit pieces have an opportunity to rise to the surface of the sol. Vacuum cooking accelerates the diffusion of sugar into the interior of fruit tissue by exhausting the interior air, causing the tissue to approach the density of the liquid medium, hence making it less likely to float.

A. Fruit Jellies

In Hawaii, most fruit jellies are prepared as follows: fruits are washed, cut, and boiled in just sufficient water to cover them, until the pieces have been softened, usually about 20 min per 10 kg. The cooked mixture is then poured into a jelly bag of filtered through several thicknesses of cheesecloth to obtain a clear juice. Alternatively, pulpers with 0.030-in and 0.014-in screens may be used. If a pulpy jelly is desired, a coarser screen is substituted. The quantity of sugar may be determined with a jelmeter. This device is merely a glass tube leading into a glass capillary through which an aqueous pectin dispersion is permitted to flow. The tube is calibrated in cups of sugar per cup of pectin or fruit-pectin extract. The slower the flow, *i.e.,* the more viscous the sol, the higher is the alleged sugar-binding capacity of the pectin, and *vice versa* (Charley, 1970).

The jelly ingredients other than pectin are weighed and placed in a steam kettle. This jelly stock is then heated to 71°C. Pectin is mixed with about five to ten times its weight of sugar, and the sugar–pectin mixture is added to the kettle. The suspension is brought to a vigorous boil, and any sugar remaining is added. Boiling is then reduced to a simmer until the temperature of the sol reaches 105–106°C. The hot sol is allowed to stand for a few minutes. At this point, the acid solution is added to give a pH of approximately 3.2, and the soluble-solids content is measured. Boiling is continued if the value is less than 65°Brix. Foam is skimmed off the surface of the sol before the jelly is hot-filled into presterilized containers. A 1.27 cm headspace that may later be covered with a layer of paraffin is left in the container. The containers are closed at a temperature of at least 82°C and inverted (if paraffin is not used) for 2 min to sterilize the cover. The containers are then allowed to air-cool, and the jelly is allowed to set in the containers in an upright position.

1. Guava Jelly

Most guava jelly formulations recommend the use of firm or underripe fruits, because in this condition, the pectin content is at a maximum. A typical formula is as follows:

Guava juice	4 parts;
Sugar	4–4.5 parts.

When ripe guavas are used, the jelly has a more pronounced guava flavor, but the pectin and acid may be insufficient to give a satisfactory gel strength. In this circumstance, a commercial pectin is added (Lynch *et al.,* 1959), and the formula is revised to

Guava jelly stock	45–50 parts;
Sugar	50–55 parts;
Pectin (rapid set, 150 grade)	0.42 part.

2. Passion Fruit and Passion Fruit–Papaya Jelly

The yellow passion fruit is reported to be more acidic (3.9% as citric acid) than the purple passion fruit (2.3% as citric acid) (Miller *et al.,* 1981). The two types contain insufficient pectin to produce a firm gel; hence, the addition of pectin is necessary. A typical formula is as follows:

Passion fruit juice (fresh)   50 kg;
Papaya puree   15 kg (optional);
Sugar   60 kg;
Pectin (slow set, 208 grade)   0.41 kg.

The juice stock may be modified by pulping the passion fruit through 0.033-inch screens and finishing with a 60-mesh screen. Papaya puree, if included, may be prepared by the method of Brekke *et al.* (1973), whereby papayas are steamed for 2 min, spray-cooled with water, and sliced with a crusher–scraper device. The seeds are separated from the pulverized flesh by a centrifugal separator, and the pulp is screened (0.033-inch), acidified with citric acid to pH 3.5, and finished with 0.020-inch screen. The alternative formula with papaya puree produces a jelly that is less tart and pulpier.

3. Natal Plum (*Carissa*) Jelly

Natal plum contains sufficient pectin and acid to prepare a firm gel. The resulting jelly is a bright red color with a slight raspberry flavor (Miller *et al.*, 1981). The formula consists of Natal plum juice and sugar in equal amounts.

4. Acerola Cherry Jelly

The acerola cherry pH range was reported to be 3.2–3.5 (Miller *et al.*, 1981). Although this cherry is adequately acidic, it does not contain sufficient pectin to prepare a firm gel. It is therefore advisable to combine Acerola fruit or juice with fruit or juice containing more pectin, or alternatively to add commercial pectin. The following formula is a modified version of the formula of Miller *et al.* (1981)

Acerola cherry juice   3 parts;
Sugar   3 parts;
Pectin (150 grade)   0.45 part.

The juice is boiled for 2 to 3 min before the addition of the pectin–sugar mixture. The balance of sugar is then added, and the final boiling temperature is reduced to 103 to 104°C.

B. Jams and Preserves

In Hawaii, most fruit jams are prepared as follows: ripe, whole fruits are washed, peeled (when necessary), sliced, and pulped through 0.030-inch screens or pulverized by other appropriate methods. The puree is heated in a steam-jacketed kettle to 71°C. Pectin, dispersed in five to ten times its weight of sugar, and an acidulant (if necessary) are added to the hot puree, and the mixture is brought to a vigorous boil. The remainder of the sugar is added, and moderate boiling is continued at 105°C. Constant stirring prevents the product from burning. The jam or preserve is filled hot into presterilized containers without a headspace and closed at a minimum temperature of 82°C. The filled containers are inverted for 2 min to sterilize the cover and allowed to air-cool or, with cans, are water-cooled. The sol is allowed to set with the containers in an upright position.

1. Guava Preserve and Jam

A formula for guava jam is given by Lynch *et al.* (1959), as follows:

Guava puree	45–50 parts;
Sugar	50–55 parts;
Pectin (rapid set, 150 grade)	0.45 part.

2. Mango Preserve and Jam

A typical mango jam formula is given by Jagtiani *et al.* (1988), as follows:

Mango puree	45 kg;
Sugar	55 kg;
Pectin	0.4 kg;
Citric acid	0.4–0.5 kg (to pH 3.2).

As discussed earlier, preparation of puree may differ somewhat owing to fruit characteristics. Jagtiani *et al.* (1988) described the methods of fruit preparation, peeling, and pureeing of mangoes. The mangoes were washed, steam-blanched (sometimes necessary), manually peeled, and pulped with paddle pulpers equipped with 0.033-inch screens. Seeds were separated by a 60-mesh screen. The resulting puree was heated in a steam kettle to 103°C. The rest of the procedure was as previously described.

3. Papaya–Pineapple Preserve and Jam

The following formula may be prepared from commercially available papaya puree:

Papaya puree	25 kg;
Pineapple (crushed)	20–25 kg;
Sugar	55 kg;

Pectin (rapid set, 150 grade)	0.31 kg;
Citric acid	0.25 kg.

Papaya puree may be obtained by the method of Brekke *et al.* (1973). Fresh pineapple is peeled, cored, and cut into 1/4-inch dices. For this product, the conventional jam procedure is slightly modified, in that the balance of the sugar is added in three equal portions, using an initial boiling temperature that may be lower than the final boiling temperature (105°C).

4. Banana Jam

The pH of the banana is one of the highest among the tropical fruits, posing problems for banana puree in storage. It was therefore necessary to acidify banana puree to pH 4.2, but this resulted in an unappealing flavor (Forsyth, 1980). Banana jam, utilizing sugar as a preservative, was produced commercially by the United Fruit Company (von Loesecke, 1950). The formula was as follows:

Banana pulp	200 lb;
Sugar	200 lb;
Water	10 gal;
Cream of tartar	12 oz;
Lemon juice (or other acid)	2.5 gal.

The product was prepared by first mixing weighed amounts of sugar, water, and cream of tartar. This syrup was boiled at 110°C. Subsequently, an acidic fruit juice, e.g., pineapple, cranberry, or lemon, was added, and the prescribed quantity of ripe, peeled bananas (as pulp) was added to the acidified syrup. The matrix was boiled at 106.7°C until the desired consistency was reached. The product was shipped in bulk in wooden candy tubs. Until the advent of high-temperature–short-time sterilization combined with aseptic packaging, maintaining flavor and acceptable physical qualities was a problem with this product.

C. Low-Sugar Fruit Spread

The jellying of these products is achieved by the reaction of low-methoxyl pectin with calcium (chloride). Inasmuch as the soluble-solids content is low, the resulting confection does not conform to the U.S. Standards of Identity for jams and jellies, and is therefore labeled a *fruit spread* in some locations. Boyle *et al.* (1957) gave a formula and described the preparation of a low-sugar guava spread, as follows:

Guava puree	100	lb;
Sugar	67	lb;
Low-methoxyl pectin	1.8	lb;
Calcium chloride (anhydrous)	0.1	lb.

The fruit puree is prepared as before. Pectin is weighed in a dry container and mixed with 10 times its weight of sugar. The puree is divided into two equal batches. The pectin–sugar mixture is added to the first batch. This combination is allowed to boil for 1 min before heating is discontinued. The remaining sugar is added to this batch, which is again boiled for 1 min before heating is discontinued a second time. Calcium chloride, previously dissolved in water, is added to the second batch. This second batch is allowed to boil before the heating is discontinued. The second batch is added to the first batch with constant stirring, and the final mixture is boiled for 1 min. Presterilized containers are filled with the fruit spread at a minimum temperature of 87°C. The containers are closed, and cooled in water.

D. Frozen Fruit Spread

The flavor and aroma of the fruit spread described in the previous section may be conserved in a condition similar to that of the original fruit, if the confection is made without heat. To this end, Boyle *et al* (1957) described a method for preparing frozen guava spread containing 56.5% soluble solids, calculated from the weights of sugar and guava puree in Eq. 1. The pH of the guava puree is adjusted to 3.0 with citric acid. Pectin (at least 150 jelly grade) is weighed so that it is 0.45–0.50% of the combined weight of the jelly stock. It is mixed with eight times its weight of sugar to facilitate its dispersion in the fruit spread. The puree is divided into three portions—9, 50, and 41% of the total spread weight. The pectin–sugar mixture is added to the 9% lot and stirred until thoroughly mixed, care being taken not to incorporate air. The 50% lot is added to the puree–pectin–sugar mixture and is stirred for at least 20 min. The resulting mixture should have at least 25°Brix to prevent the precipitation of pectin. The balance of sugar is dissolved in the 41% lot that is carefully added to the first mixture. The spread is allowed to set undisturbed overnight in open containers. Afterwards, the containers are sealed and frozen at a temperature of at least −18°C. The frozen fruit spread is thawed before consumption.

VI. TRENDS

Planners of third-world economies anticipate that tropical fruits and fruit products will continue to expand as items of international commerce into temperate-climate markets, where they will be incorporated in various forms of beverages, baked

Figure 2   The Poha, also called the Cape Gooseberry (dark lobes are the bare fruits; the others are partially husked fruits). Courtesy of Chian Leng Chia, University of Hawaii at Manoa.

goods, entrees, snacks, etc. With the exception of the intact fruits, jams and jellies are currently the more favored products, because of their long history of use. They have the advantage of being inherently stable, and can therefore be transported internationally at nominal cost. Promotional strategies for the non-traditional items will have to be developed.

In the tropics, manufacturers are beginning to convert to vacuum cooking, because they realize the vast improvement in color and flavor that can be achieved by this method over atmospheric cooking. Essence recovery is an additional benefit accruing to vacuum cooking, and it is expected that tropical fruit essences will grow into a premium item of international commerce.

The tropics are a vacation and recreational venue, attracting an ever-increasing number of tourists from temperate-climate countries. This transient population

Figure 3   The Ohelo, a fruit uniquely indigenous to Hawaii. Courtesy of Chian Leng Chia, University of Hawaii at Manoa.

Figure 4   The Otaheite (Malacca) Apple, a purple-red fruit indigenous to Jamaica. This fruit has been immersed in syrup and sold as a confection.

is contemplated as a promotional mechanism for transporting to the temperate regions the novel products available in small container packs as gift and souvenir purchases (Fig. 1). Thus, fruits unknown to temperate-climate consumers but indigenous to the tropics, such as the Poha (*Physalis peruviana*), also called Cape gooseberry (Fig. 2), the wild Ohelo berry (*Vaccinium reticulatum*) of Hawaii (Fig. 3), and the Otaheite (Malacca) apple (*Syzygium malaccense*) of Jamaica (Fig. 4), for example, can be introduced initially into the international market-place without inordinate expenditures. It is noteworthy that 1- to 6-oz containers of the novel Hawaii preserves have already been offered for sale through mail-order channels.

International food fairs are another mechanism for bringing tropical *exotic* and *gourmet* fruit products to the attention of targetted, external populations. Tamarind, hot-pepper, and pimento are confections of the new commerce in the temperate-climate regions. The possibility exists that these and an ever greater assortment of products will become available, as technologies of fruit and vegetable processing and preservation are developed or transferred to the region, and as quality standards for them are established. Unfortunately, low production volumes and high production costs, aggravated by raw materials imports (sugar, acid, containers, etc.) in many tropical climates adversely affect the extent to which any new, external market can be penetrated. Nevertheless, development strategies should call for promoting the novel fruit sauces, dressings, glazes, and condiments on a global scale, since these value-added products command higher prices than do their fresh fruit and vegetable counterparts.

References

Bhatia, B. S., Krishnamurthi, G. V., and Lal, G. (1959). Preparation of pectin from raw papaya (*Carcia papaya*) by an aluminum chloride precipitation method. *Food Sci.* **8,** 314.

Boyle, F. P., Seagrave-Smith, H., Sakata, S., and Shermand, G. D. (1957). "Commercial Guava Processing in Hawaii." Bull. 111, Hawaii Agric. Exp. Sta., Univ. of Hawaii, Honolulu.

Brekke, J. E., Chan, H. T., Jr., and Cavaletto, C. G. (1973). "Papaya puree and nectar." Res. Bull. 170. Hawaii Agric. Exp. Sta., Univ. of Hawaii, Honolulu.

Cavaletto, C. G. (1980). Lychee. *In* "Tropical and Subtropical Fruits. Composition, Properties, and Uses." (S. Nagy and P. E. Shaw, eds.), pp. 469–478. AVI Publishing, Westport, Connecticut.

Charley, H. (1970). Food Science. Ronald Press, New York,

FDA. (1980). Fruit. Compliance Policy Guide 7110.14, p. 10. Food and Drug Administration. Washington, D.C.

FDA. (1989). Code of Federal Regulations, pp. 334–341. U.S. Government Printing Office, Washington, D.C.

Forsyth, W. G. C. (1980). Banana and Plantain. *In* "Tropical and Subtropical Fruits. Composition, Properties and Uses." (S. Nagy and P. E. Shaw, eds.), pp. 258–278. AVI Publishing, Westport, Connecticut.

Garces Medina, M. (1968). Pectin, pectin esterase, and ascorbic acid in tropical fruit pulps. *Arch. Latinoam. Nutr.* **18** (4), 401–412.

Gee, M., McComb, E. A., and McCready, R. M. (1958). A method for the characterization of pectic substances in some fruit and sugar-beet marcs. *Food Res.* **23,** 72–75.

Jagtiani, J., Chan, H. T., Jr., and Sakai, W. S. (eds.) (1988). "Tropical Fruit Processing." Academic Press, San Diego, California.

Kanehiro, Y., and Sherman, G. D. (1946). "Guava-Flavored Pectin Powder Is Rich in Ascorbic Acid." Hawaii Agric. Exp. Stn., Tech. Paper 142. University of Hawaii, Honolulu.

Lynch, L. J., Chang, A. T., Lum, J. C. N., Sherman, G. D., and Seale, P. E. (1959). "Hawaii Food Processors Handbook." Hawaii Agric. Exp. Stn., Circ. 55. University of Hawaii, Honolulu.

McCready, R. M., and McComb, E. A. (1952). Extraction and determination of total pectic materials in fruits. *Anal. Chem.* **24,** 1986–1988.

Miller, C. D., Bazore, K., and Bartow, M. (1981). "Fruits of Hawaii." The University Press of Hawaii, Honolulu, Hawaii.

Muroki, N. M., and Saint-Hilaire, P. (1977). Pectin from guava fruit (*Psidium guajava* L.). *Lebensm-Wiss. u. Technol.* **10,** 314–315.

Nagy, S., and Shaw, P. E. (1980). "Tropical and Subtropical Fruits." AVI Publishing, Westport, Connecticut.

Pal, D. K., and Selvaraj, Y. (1979). Changes in pectin and pectinesterase activity in developing guava fruits. *J. Food Sci. Technol.* **16**(3), 115–116.

Saeed, A. R., El Tinay, A. H., and Khattab, A. H. (1975). Characterization of pectic substances in mango marc. *J. Food Sci.* **40,** 205–206.

Sherman, G. D., Cook, C. K., and Nichols, E. (1953). "Pectin from Passion Fruit Rinds." Hawaii Agric. Expt. Stn., Prog. Notes 92. University of Hawaii, Honolulu.

Sherman, G. D., and Kanehiro, Y. (1947). Tropical plants as a source of pectin. *The Chemurgic Digest* **6**(4), 65–68.

Simpson, B. K., Egyankor, K. B., and Martin, A. M. (1984). Extraction, purification, and determination of pectin in tropical fruits. *J. Food Process. Preserv.* **8,** 63–72.

Sunkist Growers. (1964). "Preservers Handbook." Ontario, California.

The Almanac. (1989). "74th Annual Compilation of Basic References for the Canning, Freezing, Preserving, and Allied Industries." Edward E. Judge, Westminster, Maryland.

von Loesecke, H. W. (1950). "Bananas." Interscience, New York.

Wilson, C. W., III. (1980). Guava. *In* "Tropical and Subtropical Fruits. Composition, Properties and Uses." (S. Nagy and P. E. Shaw, eds.), pp. 279–299. AVI Publishing, Westport, Connecticut.

CHAPTER 5

The Chemistry of High-Methoxyl Pectins

D.G. Oakenfull

CSIRO Division of Food Processing
North Ryde, New South Wales
Australia

I. INTRODUCTION

An understanding of the molecular mechanism of pectin gelation and how the gelation process is related to molecular structure would be invaluable in development of new and diverse products. Such an understanding would make it possible to systematically manipulate and control the gel-rheological conditions (pH, temperature cosolutes, etc.), and response to subtle changes in structural detail.

The first part of this chapter describes the chemical composition, physical characteristics, and molecular structure of high-methoxyl (HM) pectins. The second part deals with gelation, relating, where possible, the properties of HM pectins to chemical composition and molecular structure. There have been several excellent reviews in this area (Deuel *et al.*, 1953; Ahmed, 1981), so the emphasis in this chapter is on recent advances.

II. CHEMICAL COMPOSITION AND MOLECULAR STRUCTURE

A. Chemical Composition

The essentially linear polymers of D-galacturonic acid, as shown in Fig. 1, are interrupted occasionally by L-rhamnose, a sugar most clearly involved as part of the main polymer chain (Aspinall *et al.*, 1968), amounting to approximately one monomer in 50, linked 1:2. This sugar appears not to be distributed evenly within the biopolymer, but to be concentrated in zones separating segments rich in galacturonic acid segments. The anhydrogalacturonic acid units are linked by α-1,4-glycosidic bonds, with the linkage *alpha* to the general plane of the pyranose ring. Some of the carboxyl groups are esterified with methanol, some are neutralized with cations, and some of the secondary hydroxyl groups, particularly in pectins extracted from sugar beet, are esterified with acetic acid (Kertesz, 1951). There is a subtle complexity underlying this simple basic structure. The distribution of esterified galacturonic acid groups can be nonrandom along the chain, so pectins from different sources can vary substantially in composition and structure. Anger and Dongowski (1984), for example, have shown that, in the range from 20 to 90% esterification, it is possible to differentiate between random and blockwise arrangement of free carboxyl groups by fractionation on DEAE-Sephacell.

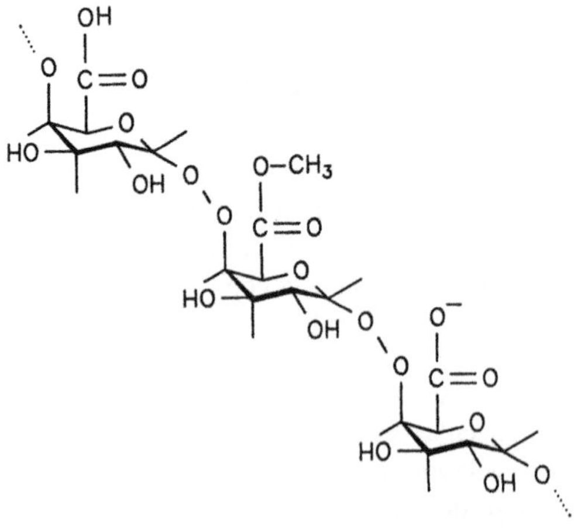

Figure 1   A repeating segment of the pectin molecule.

The precise galacturonic acid content of the polymer depends on the plant source and on the conditions of preparation. Commercial orange and grapefruit pectins generally have over 75%, and lemon pectins, over 85% galacturonic acid (Nelson, 1977). The regularity and frequency of the rhamnose interruptions to the polygalacturonate chain are different in pectins from different plant species.

Fucose, glucuronic acid, xylose, L-arabinose, and D-galactose are minor components of pectin. These sugars, including rhamnose, influence the conformation of the polymer in solution and ultimately its gelling properties.

The commercial pectins are very heterogeneous, and, consequently, the various physical and chemical characteristics measured represent average values.

B. Molecular Structure

Proton nuclear magnetic resonance (NMR) spectroscopy, model-building computations (Rees and Wight, 1971), and X-ray diffraction studies (Walkinshaw and Arnott, 1981a) suggest that pectin molecules in solution adopt a helical configuration with three monosaccharide units per turn and a pitch of 1.33 nm. The structure is stabilized by steric factors with a possible contribution from

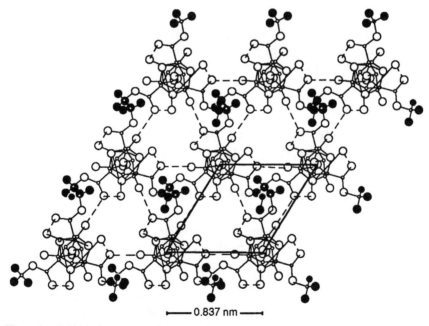

————— 0.837 nm —————

Figure 2 Polygalacturonic acid (DE = 100%) showing parallel chains packed in a hexagonal lattice. Methyl hydrogen atoms are indicated by filled circles. Redrawn from Walkinshaw and Arnott (1981b).

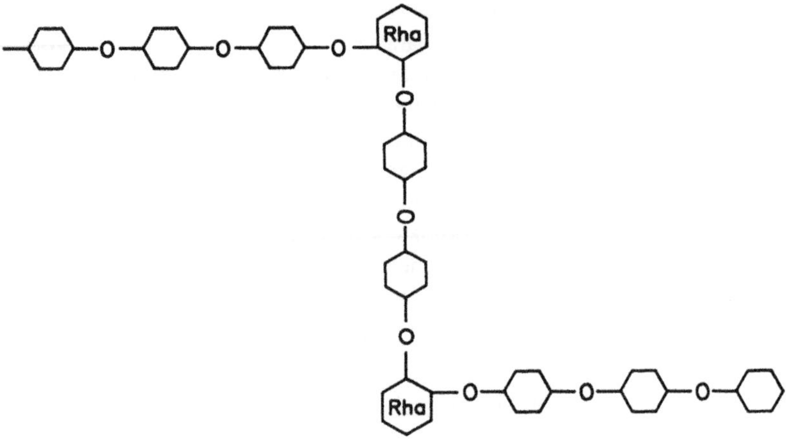

Figure 3. Schematic diagram showing how rhamnose insertions (Rha) cause kinking of the poly-galacturonic acid chain.

intramolecular hydrogen bonding. In the solid state, the molecules appear to form parallel sheets with adjacent helices having opposite symmetry, as shown in Fig. 2. This regular helical structure is disrupted by rhamnose residues (Rees and Wight, 1971). Whether the rhamnose linkage is α or β, the effect of its presence in the chain is to produce a definite "kink" in the otherwise linear structure, as shown in Fig. 3. The presence of kinks must have profound effects on the gelling properties of the pectin, since the occurrence of a kinking residue would interfere with the macromolecular organization required to form a gel network.

III. PHYSICAL PROPERTIES OF
HIGH-METHOXYL PECTINS

A. Effect of Temperature

The effect of temperature on the rate of gelation is analogous to a nucleated crystallization reaction, governed by the equation (Flory and Weaver, 1960)

$$\log (\text{rate}) \propto 1/T(T_m - T) \tag{1}$$

where T is the actual setting temperature and T_m, a thermodynamic quantity, is the sol-gel transition temperature. It has been shown that the rate passes from the boiling temperature ($T_m < T$) through a maximum (Fig. 4). Afterwards, $T_m > T$, and the rate decreases upon further cooling.

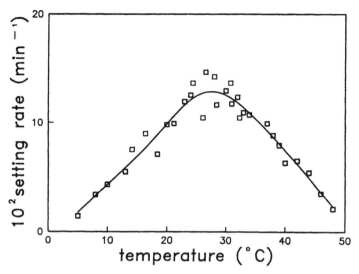

Figure 4   Effect of temperature on the rate of gelation of an HM pectin (DE = 69.7%) at a concentration of 5 g/kg with 550 g/kg sucrose at pH 3.5. Oakenfull and Scott, (1986).

B. Viscosity

The viscosity of aqueous solutions of HM pectins is very dependent on a number of variables, e.g., the degree of esterification (DE), electrolyte concentration, pH and temperature. Different concentrations of a sugar (Fig. 5) and different sugars (Fig. 6) affect the viscosity differently. The viscosity increases markedly as T approaches T_m.

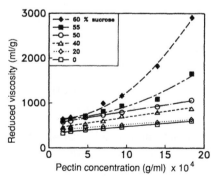

Figure 5   Effect of sucrose on the reduced viscosity of buffered pectin solutions (DE = 72.9%) at pH 3.0 at 25°C. Redrawn from Michel *et al.* (1985).

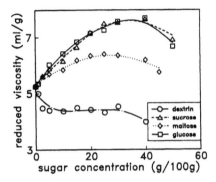

Figure 6   Effect of sugar concentrations on the reduced viscosity of a 0.1% pectin solution. Redrawn from Chen and Joslyn (1967).

Viscosity is conventionally analyzed in terms of the equation

$$\eta_{red.} = [\eta] + \lambda_h[\eta]^2 c \tag{2}$$

where $\eta_{red.}$ is the reduced specific viscosity, $[\eta]$ is the intrinsic viscosity, λ_h is the Huggins constant, and c is the concentration. The observed dependence of viscosity on sucrose concentration was claimed to result more from an increase in λ_h than in $[\eta]$, with the inference that aggregation of the pectin molecules occurs as the gelatin conditions become favorable (Michel *et al.*, 1985).

C. Molecular Weight and Molecular-Weight Distribution

Accurate determination of molecular weights is difficult, partly because of the extreme heterogeneity of commercial pectin samples, and partly because of the tendency of pectin molecules in solution to aggregate, even under conditions not favorable to gelation.

The pectin molecular weights can be expressed either as a weight average ($\overline{M}_w$) or a number average ($\overline{M}_n$) value:

$$\overline{M}_w = \Sigma N_x M^2_x / \Sigma N_x M_x \tag{3}$$

$$\overline{M}_n = \Sigma N_x M_x / \Sigma N_x \tag{4}$$

where N_x is the number of molecules of molecular weight, M_x, present (Richards, 1980). Different experimental techniques give differently averaged molecular weights. Osmometry and quasi-electic light scattering give the number average value; Rayleigh light scattering gives the weight average value. A full description of molecular weights necessitates information on the statistical distribution of molecular sizes about the mean. $\overline{M}_w/\overline{M}_n$ is a convenient measure of the degree of polydispersity. The closer the ratio is to unity the more monodisperse is the sample.

Forty years ago, molecular weights and molecular weight distribution of pectins were carefully and systematically studied by Owens *et al.* (1948) using viscometry and osmometry. They reported molecular weights varying from 20,000 to 300,000, depending on the preparation procedure. More interestingly, they always found a substantial difference between $\overline{M}_n$ (from osmometry) and $\overline{M}_w$ (from viscometry), indicating a high degree of polydispersity. With the aid of light scattering and viscometry, Berth *et al.* (1977) studied a series of pectin preparations made from partial depolymerization by methanolysis of an original pectin (70% DE). Their samples were found to contain a high-molecular-weight fraction that could be removed by ultracentrifugation and fractionation on DEAE-cellulose. The original pectin had an apparent $\overline{M}_w$ of 1.64×10^6, decreasing to 0.2×10^6 after ultracentrifugation and fractionation. For pectin samples with molecular weights within the range from 20,000 to 200,000, they concluded that $\overline{M}_w$ is related to $[\eta]$ by the equation

$$[\eta] = 2.16 \times 10^{-2} \, \overline{M}_w^{0.79} \tag{5}$$

$[\eta]$, in effect, measures the hydrodynamic volume (Richards, 1980). Jordan and Brant (1978) took light-scattering measurements on a pectin fraction in aqueous dimethyl sulfoxide. They found $\overline{M}_w$ between 9×10^5 and 6×10^5, varying with filtration and clarification. In the same study, $\overline{M}_n$ from osmometry was approximately 5×10^4. These results again suggest considerable polydispersity. The authors felt that their results indicated a nonequilibrium association of the pectin molecules whose aggregates (microgel) form very slowly, and gradually reappear after removal by filtration, gel-permeation chromatography and ultracentrifugation (Berth, 1988).

Smith and Stainsby (1977) have used a novel variation on the standard light-scattering procedure to measure $\overline{M}_w$ of a series of pectins. They first removed the microgel particles from the solutions and then carried out the light-scattering measurements before and after treating them with a depolymerizing enzyme. The difference between the intensities of scattered light before and after depolymerization was used to calculate molecular weight. They found values of $\overline{M}_w$ ranging from 2.0×10^5 to 5.5×10^5. Their results confirmed once more that pectins are highly polydisperse. One of the pectins gave $\overline{M}_w/\overline{M}_n \sim 11$. Smith and Stainby's results indicate that Eq. (5) and similar equations relating molecular weight to intrinsic viscosity should be used with caution, because such relationships, when derived from polydisperse pectins differing also in composition, are oversimplified.

In summary, there are enormous practical and interpretational difficulties in determining molecular weights of pectins, largely because of the extreme heterogeneity of the material and because of the tendency of pectin molecules in solution to aggregate.

IV. THE MOLECULAR MECHANISM OF GELATION

In order to understand the conditions required for HM pectin gelation, we need to know in as much detail as possible the molecular mechanism by which gels form. This section outlines the basic concepts whereby molecular properties can be related to the bulk properties of a gel. For more detailed information, the reader is referred to recent reviews by Oakenfull (1987) and Clark and Ross-Murphy (1987).

A. Basic Concepts

Gels consist of polymeric molecules cross-linked to form a tangled, interconnected network (Fig. 7) immersed in a liquid medium (Flory, 1953). In pectin gels and other food systems containing pectin, this liquid is water. The properties of the gel are the net result of complex interactions between solute and solvent. The water solvent influences the nature and magnitude of the intermolecular forces that maintain the integrity of the gel, empowering it with a large water-holding capacity.

In most food gels, the cross-linkages in the network are not point interactions, but involve extensive segments from two or more polymer molecules, usually in well-defined structures called junction zones that are stabilized by a combination of weak intermolecular forces. Individually, these forces are not enough to maintain the structural integrity of the junction zones, but cumulatively, their effect is to impart thermodynamic stability to them.

Figure 7   Schematic diagram of a gel network. The hatched areas represent junction zones.

B. The Intermolecular Forces Stabilizing the Gel Network

Recent evidence has suggested that hydrogen bonding and hydrophobic interactions are important forces in the aggregation of pectin molecules.

1. Hydrogen Bonding

An excellent and readily understandable account of the role of hydrogen bonding in stabilizing macromolecular structures is that given by Jencks (1969). Suffice it to state here that hydrogen bonds are generally weak (Joesten and Schaad, 1974), and only when a number of them cooperate do they confer significant thermodynamic stability to a gel network.

2. Hydrophobic Interactions

Hydrophobic interactions arise from the simple fact that oil and water do not mix. In solution in water, nonpolar molecules, or nonpolar groups attached to macromolecules, tend to coalesce, because this minimizes their energetically unfavorable contact with the water, as shown in Fig. 8. The driving force for this interaction is provided by the unique, three-dimensional hydrogen-bonded structure of water (Némethy and Scheraga, 1962; Ben-Naim, 1980). In this medium, water molecules adjacent to a hydrophobic surface become structured. When two such surfaces approach or contact each other, the unstructured water molecules, in effect, are expressed into the pool of bulk water, and there is a decrease in the entropy at the interface.

Hydrophobic interactions seem to explain more fully Chen and Joslyn's (1967) results. For example, sugars can enhance this phenomenon, depending on the extent to which they influence the structure of water. Glucose is more structure-making than is maltose. A polymer of sucrose prepared similarly to Chen and Joslyn's dextrin appears to be structure breaking (Back et al., 1979).

Hydrophobic interactions are a key distinction in gelation between HM pectins and LM pectins and most of the other gel-forming polysaccharides.

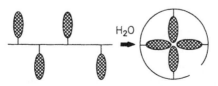

Figure 8   Schematic diagram of how hydrophobic interaction between apolar side-chains (hatched areas) on a polymer could force the molecule into a folded conformation in aqueous solution.

C. Modern Methods of Studying the Gel Network

The more useful, current methods of studying HM pectin gels involve chiroptical, X-ray diffraction, rheological, and kinetic measurements. These are briefly discussed below.

1. Chiroptical Techniques

The molecular asymmetry introduced by the formation of helical structures in polysaccharides provides possibilities for studying structures and conformations by chiroptical techniques. An excellent, recent, general account of the applications of chiroptical techniques in this area is given by Clark and Lee-Tuffnell (1986).

Optical rotation measurements have featured prominently in studies of the gelling of polysaccharides. By extensive studies of the optical rotatory power of polysaccharides in relation to their conformation preferences, Rees (1969) established a semiempirical relationship between the magnitude and sign of optical rotation and the torsion angles specifying the relative orientations of the monosaccharide residues in a polysaccharide chain.

The wavelength dependence of optical rotation in the optical rotatory dispersion spectrum provides information about the molecular environment of the chromophore. This technique and closely related circular dichroism have proved to be powerful means of studying network formation by polysaccharides. They have been particularly fruitful in studies of the mechanism of gelation of the alginates, leading to the well-known *egg-box* model for the structure of the junction zones (Rees, 1969; Gidley *et al.,* 1979).

2. X-Ray Diffraction

X-ray diffraction methods (Clark and Lee-Tuffnell, 1986) provide indirect information about gels because the measurements are made on fibers, and on dried gels prepared from the hydrated materials. X-ray diffraction is the only technique for characterizing polysaccharide-ordered structures at a level approaching atomic resolution. The diffraction intensity varies with angle and the distance from the X-ray beam, as shown in Fig. 9. The diffraction pattern provides information about the symmetry and interatomic spacing in the array of molecules that makes up the fiber or film. The level of structural information that can be deduced is dependent on the level of organization within the fiber. When the ordered chains are aligned, but without any regular packing arrangement, the analysis is limited to calculations of the helix symmetry and repeat distance. If there is ordered packing of adjacent chains, extra structural information is available from which unit cell dimensions and sometimes the space grouping can be calculated. Un-

Figure 9 Fiber X-ray diffraction pattern from neutral sodium pectate. From Walkinshaw and Arnott (1981a).

fortunately, it is not uncommon for more than one sterically valid model to be consistent with the data. It is then that computer modeling becomes a useful adjunct to structural analysis from X-ray diffraction. Although not necessarily having the same structures in solution, dried fibers and films analyzed by X-rays can provide valuable insight into the types of structures that might reasonably be expected.

3. Shear Modulus and the Theory of Rubber Elasticity

Rheological properties, being those most directly related to the gel state, are basic to the study of pectin. An entry point for deriving molecular information from bulk rheological properties is the theory of rubber elasticity (Treloar, 1975) that culminates in a simple expression for the shear modulus, G, of a cross-linked polymer network:

$$G = RTc/M_c \qquad (6)$$

where c is the weight concentration of the polymer, M_c is the number average molecular weight of the polymer chains joining cross-links (active chains), R is the gas constant, and T is the absolute temperature. Implicit in this expression is the assumption that the active chains are Gaussian, i.e., they are perfectly flexible and free to adopt all possible conformations. Polysaccharide chains, including pectin, approach this condition only if they are very long (Bailey *et al.*, 1977). This means that the theory approaches validity for incipient polysaccharide gels where the junction zones are relatively weak and few, and the lengths of chains between them are at or near their maximum. Given this proviso, an extension of Eq. (6) makes it possible to calculate the size of the junction zones (the number of segments from different chains involved in their formation) and the free energy of their formation (Oakenfull, 1984). It is possible to show that

$$G = -RTc/M \cdot \{(M[J] - c)/(M_j[J] - c)\} \qquad (7)$$

where M is the number average molecular weight of the polymer, M_j is the number average molecular weight of the junction zones, and [J] is the molar concentration of junction zones. The free energy of formation of junction zones is proportional to $\ln c_m$, where c_m is the minimal concentration of sugar (or other polyol) required to induce gelation (Oakenfull and Scott, 1984). For very weak gels with well-dispersed junction zones, it is reasonable to assume that formation is an equilibrium process subject to the law of mass action. It then follows that

$$K_j = [J] M_j^n \{n(c - M_j[J]\}^{-n} \qquad (8)$$

where K_j is the association constant, and n is the number of segments from the different polymer molecules involved. By using numerical methods, [J] can be eliminated, giving a relationship between shear modulus and concentration in terms of M, M_j, K_j and n. The experimental data required are shear moduli measured at concentrations close to the gel threshold. The calculations can be carried out in the following sequence:

1. Select trial values for M, M_j, and n. It should usually be possible to estimate values within an order of magnitude by using reasonable expectations for these parameters.
2. Estimate c by extrapolation, when the shear modulus becomes zero. From Eq. 7, it can be shown that $[J] = c_0/M$. This procedure reduces the number of adjustable parameters from four to three, thereby vastly simplifying the computations.
3. Using Eq. (7) and Eq. (8), calculate G for each concentration at which experimental values of G are available.
4. Calculate the sum of squares of the differences between the experimental and calculated values of G.
5. Adjust M, M_j, and n so as to minimize the sum of squares of differences.

There are many methods for finding the minimum sum of squares rapidly with high precision (Kowalik and Osborne, 1968). For this work, with smoothly varying continuous functions, it is not necessary to go past direct search methods. All that is required is to calculate the sums of squares with different values for the parameters, in order to determine the changes needed to reduce the sum of squares sequentially until the minimum is reached.

4. The Kinetics of Gelation

A kinetic method has been developed with which it is possible to estimate the average number of polymer molecules associated at the junction zones in the gel network (Oakenfull and Morris, 1987). The method is based on measuring the rate of gelation in its earliest stages. Under these conditions, as was the case with the very weak gels, it is reasonable to assume that the cross-linking loci (L) on the polymer chains act as independent species in solution. The rate of formation of junction zones (v) is then given by

$$v = d[J]/dt = k[L]^n \tag{9}$$

where [L] is the molar concentration of cross-linking loci. Initially, few of these loci are consumed, and, to a good approximation,

$$[L] \propto c. \tag{10}$$

It follows that

$$v = d[J]/dt = k' \cdot c^n \tag{11}$$

where n is effectively the reaction order, easily calculated from the concentration dependence of the reaction rate.

V. HIGH-METHOXYL PECTIN GELATION

A. The Molecular Organization of the Junction Zones

Many of the gel-forming polysaccharides, particularly the carrageenans and agarose, form multiple helices in which the chains are intertwined, and it is these structures that cross-link the molecules to form the network (Rees, 1969; Oakenfull, 1987). Model-building computations (Rees and Wight, 1971) have shown that multiple helix formation is impossible for an α-1,4-galacturonan (pectin). The X-ray diffraction studies suggest that the HM-pectin molecules, in a threefold helical conformation, participate in an isomeric array in which the individual helices remain separate, and the structure is sustained by interchain and intermolecular hydrogen bonds (Fig. 2). Walkinshaw and Arnott (1981b) pointed out that the structure would additionally be stabilized by hydrophobic interactions between ester methyl groups (as shown in Fig. 2). However, structures deduced from solid-state measurements, while strongly indicative, do not necessarily represent the hydrated and diffuse state of the gel network.

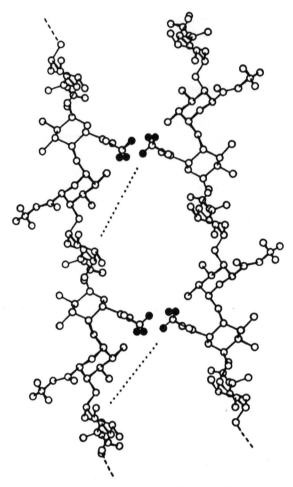

Figure 10 Structure of the junction zones in high-methoxyl pectin gels as inferred from X-ray diffraction studies. The hydrogen atoms of the hydrophobic methyl groups are represented by filled circles (●) and hydrogen bonds are indicated by dotted lines.

Oakenfull (1984) and Oakenfull and Scott (1984), arguing from an analysis of the concentration dependence of the shear modulus for very weak gels (Section IV, C, 3), have proposed that the junction zones are formed from segments of just two polysaccharide chains, as shown in Fig. 10 (see Table I). The kinetics of the initial stages of gelation showed the same result (Oakenfull and Scott, 1986). The reaction appears to be second order, suggesting that a junction zone is indeed formed from just two interacting polymer chains existing in the geometry of a threefold helix. The length of the segment needed to give sufficient stability to these ordered structures appears to increase with DE.

Oakenfull and Scott (1984) estimated the number of monosaccharide units

Table I   Number of average molecular weights and characteristics of the junction zones[a]

DE	M[b]	M_j[c]	N_j[d]	K_j[e]	n[f]
93.0	92,000	92,000	497	—	2.01
72.3	249,000	25,000	135	138	2.08
69.7	113,000	10,600	57	420	2.03
64.9	38,000	6,230	34	70	1.97

[a]Calculated from measurements of shear modulus for pectin gels containing 550 g/kg sucrose at 25°C (Oakenfull and Scott, 1984).
[b]Number average molecular weight of the polysaccharide.
[c]Number average molecular weight of the junction zones.
[d]Number of monomer units per junction zone.
[e]Association constant for the formation of junction zones (in L^{-1} mol units).
[f]Number of cross-linking loci per junction zone.

involved in a pectin junction zone (Table I). At high DE, almost the entire chain appears to be required; at the lowest DE studied (64.9%), the number of monomer units involved was 34 (17 from each chain). This analysis does not account for any possible differences in the number and distribution of rhamnose residues that would be expected to disrupt the regular helical structure. It would be interesting to see an analysis similar to that of Oakenfull and Scott (1984) for a series of pectins with systematically varied rhamnose compositions.

B. The Role of Hydrophobic Interactions in Stabilizing the Junction Zones

Morris *et al.* (1980) first suggested that the methyl groups of esterified galacturonic acid might be directly involved in the interchain associations. This suggestion was based on a comparison of the pH dependence of gelation behaviour for pectins with different DE. Walkinshaw and Arnott (1981b) subsequently used their X-ray diffraction results (Walkinshaw and Arnott, 1981a) to deduce the model structure for the junction zones (Fig. 2). Hence, evidence points to methyl ester groups as contributing to the stability of the structure.

Further evidence for the hydrophobic effect was obtained by replacing sucrose with different cosolutes to modify the interactions (Oakenfull and Scott, 1984). In the case of ethanol, the free energy of hydrophobic interaction between a pair of CH_3-groups has been shown in a model system (Oakenfull and Fenwick, 1979) to pass through a minimum (Fig. 11). Both rupture strength and the gel threshold were found to be partly proportional to the free energy of hydrophobic interaction between CH_3-groups. This is seen by comparing Fig. 11 with Fig. 12; the latter shows how the rupture strength of gels prepared under standard conditions, with a fixed concentration of pectin, varies with the concentration of ethanol. Three pectins with different degrees of esterification are shown. In each

---

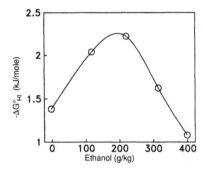

Figure 11   Effect of ethanol on the free energy of hydrophobic interaction between two CH₃-groups at 25°C. From Oakenfull and Fenwick (1979).

case there is a maximum in gel strength which coincides with maximal hydrophobic interaction.

It has been demonstrated that the hydrophobic effect on proteins can explain increased thermal stability in the presence of certain sugars or polyols (Back *et al.*, 1979). Oakenfull and Scott (1984) found that the effects of different sugars and polyols on pectin gelation were consistent with such a role. The magnitude of the effect depends on the stereochemistry of the sugar or polyol, since the spacing and orientation of the hydroxyl-groups determine how these compounds interact with water (Franks *et al.*, (1972).

When gels were prepared with the usual sucrose replaced by other sugars and polyols, the minimal concentration of sugar or polyol required to gel 3.6 g/kg pectin correlated roughly (coefficient of regression = 0.812) with the effect of the appropriate sugar or polyol on the denaturation temperature (ΔT_m) of ovalbumin (Back *et al.*, 1979; Oakenfull and Scott, 1984). ΔT_m is the difference

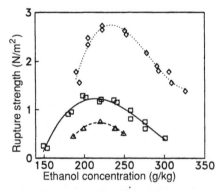

Figure 12   Rupture strength of gels prepared from HM pectins in the presence of ethanol at 25°C. Curve A, DE = 72.3%; Curve B, DE = 69.7%; Curve C, DE = 64.9%. From Oakenfull and Scott (1984).

between the denaturation temperature of the protein in aqueous buffer and that in the presence of 500 g/kg of sugar or polyol; it can be taken as an empirical measure of the effect of the sugar or polyol on hydrophobic interaction. Interestingly, sucrose is the least effective of the sugars in promoting gelation. Significantly lower concentrations were required for the other sugars, and for sorbitol.

C. The Thermodynamics of Formation of Junction Zones

The association constant for the formation of junction zones (K_j) is one of the quantities that can be estimated from shear modulus data. It provides a useful basis of thermodynamic arguments, since ΔG_j^0 can be calculated from the well-known expression

$$\Delta G_j^0 = -RT \ln K_j \qquad (12)$$

ΔG_j^0 for three HM pectins with different DE is given in Table II. The value becomes more negative with increasing DE and, at the same time, the size of the junction zones (as given by M_j) increases sharply with DE. Both quantities are linearly related to $(DE)^2$, as shown in Fig. 13. This means that they are proportional to the probability of two methyl groups coinciding when two lengths of polymer are laid alongside each other. The size and thermodynamic stability of the junction zones thus depends on the proximity of the methyl groups in parallel chains. Another interesting point is that, at a DE of about 50%, both M_j and G_j^0 extrapolate to zero. This would explain why, under the same conditions, low methoxyl pectins (DE < 50%) would fail to gel; the intermolecular forces would be too weak for stable junction zones to form. This argument depends, of course, on esterified galacturonic acid groups being randomly distributed along the galacturonan chain. This condition is not in fact strictly true, so the apparently linear relationships shown in Fig. 13 can only be approximations.

Table II   Free energy of formation of junction zones for pectin of different DE

DE(%)	n_j[a]	K_j[b]	ΔG_j^0 (kJ/mole)[c]	
			Per junction zone	Per monomer unit
72.3	135	76,600	−27.9	−0.21
69.7	57	23,310	−24.9	−0.44
64.9	34	3,880	−20.5	−0.60

[a]n_j, number of monomer units per junction zone.
[b]K_j, association constant for junction zone formation.
[c]ΔG_j^0, standard free energy of formation of junction zones.

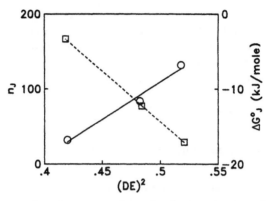

Figure 13   Average number of monomer units per junction zone (n_j, squares) and the standard free energy of formation of junction zones (ΔG_j^0; circles) plotted against the square of the degree of esterification [$(DE)^2$].

The thermodynamic data also provide a quantitative indication of the relative importance of hydrogen bonds and hydrophobic interactions in the formation of these HM pectin gels. ΔG_j^0 is considered to be made up of three terms, $viz.$, hydrophobic interaction, hydrogen bonding, and the loss of configurational entropy of the polysaccharide chains. In the form of an equation,

$$\Delta G_j^0 = (p/2)e^2 f \Delta G_{hi}^0 + (p/s)\,\Delta G_{hb}^0 - \Delta S_{conf}^0 \qquad (13)$$

where p is the number of monomer units per junction zone; e is the fraction of esterified galacturonic acid groups; f is the fraction of aligned esterified groups that contribute hydrophobic stabilization, and ΔG_{hi}^0 is the standard free energy of hydrophic interaction of a pair of CH_3-groups [-2.34 kJ/mole (Back $et\ al.$, 1979)]. ΔG_{hb}^0 is the net hydrogen bonded interaction per pair of monomer units, and ΔS_{conf}^0, the change of configurational entropy, is proportional to ln p (Flory, 1953). Consequently the free energy change per monomer unit,

$$\Delta G_j^0/p = 1.17e^2 f + 0.5\Delta G_{hb}^0 - A \ln p \qquad (14)$$

where A is a constant. The three sets of values of e, p and ΔG_j^0 from Table I then give f, ΔG_{hb}^0 and A, from which, for a typical HM pectin of DE 70%, the contributions to ΔG_j^0 (in kJ/mole) are,

$$
\begin{array}{llll}
 & \text{hydrophobic} & \text{hydrogen} & \text{loss of} \\
\Delta G_j^0 = & \text{contribution} + & \text{bonds} & + \text{ configurational entropy} \\
 & -18.6 & -37.5 & +41.1
\end{array}
$$

The hydrophobic contribution to ΔG_j^0 is only half that from hydrogen bonds, but it is essential for gelation. As previously stated, hydrogen bonding alone is insufficient to overcome the entropic barrier to gelation from the loss of freedom of motion of the polymer chains when junction zones are formed.

VI. MIXED GELS OF PECTIN AND ALGINATE

HM pectins interact with alginate to form gels that may have useful properties. Gelation occurs only at low pH, but without the added sugar required for gelation of the pectin, and without the calcium required for gelation of alginate (Toft, 1982). The gel strength and the melting temperature both depend on the ratio of alginate to pectin (Fig. 14). This system would appear to offer opportunities to adjust the textural characteristics of the gel to suit particular product applications (Morris and Chilvers, 1984). Optimal gelation occurs with pectins of high DE and alginates with high ratios of guluronic to mannuronic acid. No gelation occurs above pH 4.0. That gelation occurs through a direct and specific interaction between the two polysaccharides is indicated by the circular dichroism spectral data, showing the change in the spectrum in going from pH 7 (where no interaction occurs) to pH 3 (Fig. 15). This is quite different from that calculated from the spectra of the isolated polysaccharides (Thom *et al.*, 1982).

The nature of the junction zones in the mixed gel has been ascertained through molecular model building combined with X-ray diffraction evidence from polysaccharides. It appears that polyguluronic acid and methyl polygalacturonate can pack together in parallel, two-fold crystalline arrays. The synergistic interaction occurs, because the mixed structure permits optimal packing, devoid of cavities between the chains, compared with the less-efficient packing of the homotypical structures formed from the individual polysaccharides under the same conditions.

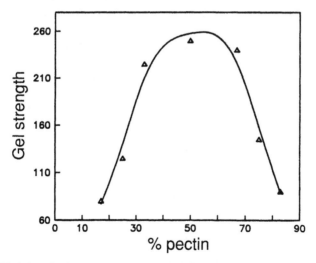

Figure 14   Variation of gel strength (Boucher) with the proportion of alginate in alginate–pectin mixed gels with a total polysaccharide concentration of 0.8% and at pH 3.5. From Toft (1982).

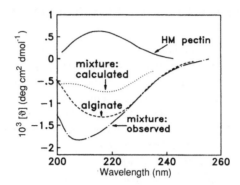

Figure 15 Spectral changes observed by subtracting the observed spectra at pH 7 from those obtained on direct acidification to pH 3 for solutions of HM pectin and alginate. The calculated spectral change for an equal mixture of the two, assuming no additional interaction occurs, is also shown and compared with the spectral change actually observed on direct acidification of the mixture. From Thom *et al.* (1982).

VII. CONCLUSIONS AND RESEARCH NEEDS

Current knowledge of the molecular basis of gelation has helped to elucidate the relationships between molecular structure, composition, and textural characteristics of a pectin gel. More information is needed, however, about the small and subtle changes in the composition of the pectin molecule, especially as they pertain to rhamnose.

The junction zones holding the polysaccharide molecules in a continuous three-dimensional network each involve association of segments from two polysaccharide chains, stabilized by a combination of hydrogen and hydrophobic bonds. The forces are individually weak, but acting in concert, they confer stability to the gel. Sucrose and other sugars stabilize the hydrophobic bonds. Rhamnose (approximately one in fifty residues) causes a kink in the polysaccharide chain, and this kinking is capable of disrupting junction-zone formation. Currently very little is known about how this affects the gelation properties of pectin. Comparative studies with pectins with different rhamnose contents would be useful.

References

Ahmed, G. E. (1981). High-methoxy pectins and their uses in jam manufacture—a literature survey. BFMIRA Scientific and Technical Surveys No. 127. Leatherhead, England.

Anger, H., and Dongowski, G. (1984). Uber die Bestimmung der Estergruppenverteilung im Pektin durch Fractionierung an DEAE-Cellulose. *Nahrung* **28**, 199–206.

Aspinall, G. O., Craig, J. W. T., and White, J. L. (1968). Lemon-peel pectin. Part 1. Fractionation and partial hydrolysis of water-soluble pectin. *Carbohydr. Res.* **7**, 442–452.

Back, J. F., Oakenfull, D. G., and Smith, M. B. (1979). Increased thermal stability of proteins in the presence of sugars and polyols. *Biochemistry* **18**, 5191–5196.

Bailey, E., Mitchell, J. R., and Blanshard, J. M. V. (1977). Free energy calculations on stiff chain constituents of polysaccharide gels. *Colloid Polym. Sci.* **255**, 856–859.

Ben-Naim, A. (1980). "Hydrophobic Interactions." Plenum, New York.

Berth, G. (1988). Studies on the heterogeneity of citrus pectin by gel-permeation chromatography on Sepharose 2B/Sepharose 4B. *Carbohydr. Polym.* **8**, 105–117.

Berth, G., Anger, H., and Linow, F. (1977). Streulichtphotometrische und viskometrische Untersuchungen an Pektinen in wassrigen Losungen zur Molmasenbestimmung. *Nahrung* **21**, 939–950.

Chen, T.-S., and Joslyn, M. A. (1967). The effect of sugars on viscosity of pectin solutions II. Comparison of dextrose, maltose, and dextrins. *J. Colloid Interface Sci.* **25**, 346–352.

Clark, A. H., and Lee-Tuffnell, C. D. (1986). Gelation of globular proteins. *In* "Functional Properties of Food Macromolecules." (J. R. Mitchell and D. A. Ledward, eds.), pp 203–272. Elsevier/Applied Science, Barking, England.

Clark, A. H., and Ross-Murphy, S. B. (1987). Structural and mechanical properties of biopolymer gels. *Adv. Polym. Sci.* **83**, 57–192.

Davis, M. A. F., Gidley, M. J., Morris, E. R., Powell, D. A., and Rees, D. A. (1980). Intermolecular association in pectin solutions. *Int. J. Biol. Macromol.* **2**, 330–332.

Deuel, H., Solms, J., and Altermat, H. (1953). Die Pektinstoffe und ihre Eigenschaften. *Vierteljahrsschr. Naturforsch. Ges. Zurich* **2**, 49–86.

Flory, P. J. (1953). "Principles of Polymer Chemistry." Cornell University Press, Ithaca, New York.

Flory, P. J., and Weaver, E. S. (1960). Helix-coil transitions in dilute aqueous collagen solutions. *J. Am. Chem. Soc.* **82**, 4518–4525.

Franks, F., Ravenhill, J. R., and Reid, D. S. (1972). Thermodynamic studies of dilute aqueous solutions of cyclic ethers and simple carbohydrates. *J. Solution Chem.* **1**, 3–9.

Gidley, M. J., Morris, E. R., Murray, E. J., Powell, D. A., and Rees, D. A. (1979). Spectroscopic and stoichiometric characterisation of the calcium-mediated association of pectate chains in gels and in the solid state. *J. Chem. Soc., Chem. Commun.*, 990–992.

Jencks, W. P. (1969). "Catalysis in Chemistry and Enzymology." McGraw-Hill, New York.

Joesten, M. D., and Schaad, L. J. (1974). "Hydrogen Bonding." Marcel Dekker, New York.

Jordan, R. C., and Brant, D. A. (1978). An investigation of pectin and pectic acid in dilute aqueous solution. *Biopolymers* **17**, 2885–2895.

Kertesz, Z. I. (1951). "The Pectic Substances." Interscience, New York.

Kowalik, J., and Osborne, M. R. (1968). "Methods for Unconstrained Optimization Problems." Elsevier, New York.

Michel, F., Doublier, J. L., and Thiboult, J. F. (1985). Etude viscométrique de la première phase de gélification des pectines hautement méthylées. *Sci. Aliments* **5**, 305–319.

Morris, E. R., Gidley, M. J., Murray, E. J., Powell, D. A., and Rees, D. A. (1980). Characterisation of pectin gelation under conditions of low water activity, by circular dichroism, competitive inhibition, and mechanical properties. *Int. J. Biol. Macromol.* **2**, 327–330.

Morris, V. J., and Chilvers, G. R. (1984). Cold-setting alginate-pectin mixed gels. *J. Sci. Food Agric.* **35**, 1370–1376.

Nelson, D. B. (1977). Pectin—A review of selected advances made in the last 25 years. *Proceedings of the International Society of Citriculture*, **3**, 739–742.

Némethy, G., and Scheraga, H. A. (1962). The structure of water and hydrophobic bonding in proteins. III. The thermodynamic properties of hydrophobic bonds in proteins. *J. Phys. Chem.* **66**, 1773–1789.

Oakenfull, D. (1984). A method for using measurements of shear modulus to estimate the size and thermodynamic stability of junction zones in noncovalently cross-linked gels. *J. Food Sci.* **49**, 1103–1104 & 1110.

Oakenfull, D. (1987). Gelling agents. *CRC Crit. Rev. Food Sci. Nutr.* **26**, 1–25.

Oakenfull, D., and Fenwick, D. E. (1979). Hydrophobic interactions in aqueous organic mixed solvents. *J. Chem. Soc., Faraday Trans.* **1**, 75, 636–645.

Oakenfull, D., and Morris, V. J. (1987). A kinetic investigation of the extent of polymer aggregation in carrageenan and furcellaran gels. *Chem. Ind.* 201–202.

Oakenfull, D., and Scott, A. (1984). Hydrophobic interaction in the gelation of high-methyoxyl pectins. *J. Food Sci.* **49,** 1093–1098.

Oakenfull, D., and Scott, A. (1986). New approaches to the investigation of food gels. In "Gums and Stabilisers for the Food Industry 3." (G. O. Phillips, D. J. Wedlock, and P. A. Williams, eds.), pp. 465–475. Elsevier, London.

Owens, H. S., Miers, J. C., and Maclay, W. D. (1948). Distribution of molecular weights of pectin proprionates. *J. Colloid Sci.* **3,** 277–291.

Rees, D. A. (1969). Structure, conformation, and mechanism in the formation of polysaccharide gels and networks. *Adv. Carbohydr. Chem.* **24,** 267–332.

Rees, D. A., and Wight, A. W. (1971). Polysaccharide conformation. Part VII. Model-building computations for α-1,4 galacturonan and the kinking function of L-rhamnose residues in pectic substances. *J. Chem. Soc. B* 1366–1372.

Richards, E. G. (1980). "An Introduction to the Physical Properties of Large Molecules in Solution." Cambridge University Press, Cambridge, England.

Smith, J. E., and Stainsby, G. (1977). Studies on pectins I. Light-scattering and M_w. *Brit. Polym. J.* **9,** 284–288.

Thom, D., Dea, I. C. M., Morris, E. R., and Powell, D. A. (1982). Interchain associations of alginate and pectins. *Prog. Food Nutr. Sci.* **6,** 97–108.

Toft, K. (1982). Interactions between pectins and alginates. *Prog. Food Nutr. Sci.* **6,** 89–96.

Treloar, L. R. G. (1975). "The Physics of Rubber Elasticity." Clarendon Press, Oxford, England.

Walkinshaw, M. D., and Arnott, S. (1981a). Conformations and interactions of pectins I. X-ray diffraction analysis of sodium pectate in neutral and acidified forms. *J. Mol. Biol.* **153,** 1055–1073.

Walkinshaw, M. D., and Arnott, S. (1981b). Conformations and interactions of pectins II. Models for junction zones in pectinic acid and calcium pectate gels. *J. Mol. Biol.* **153,** 1075–1085.

CHAPTER 6

The Chemistry of Low-Methoxyl Pectin Gelation

M.A.V. Axelos

Institut National de la Recherche Agronomique
Laboratoire de Physico-Chimie de Macromolécules
Nantes Cedex 03, France

J.-F. Thibault

Institut National de la Recherche Agronomique
Laboratoire de Biochimie et Technologie des Glucides
Nantes Cedex 03, France

I. INTRODUCTION

Gelation of the class of pectins with degree of methoxylation (DM) < 45%, known as low-methoxyl (LM) pectins, is the subject of this chapter. The mechanism of LM-pectin gelation depends on the number and sequence of consecutive carboxyl groups and on the concentration of calcium. The first requirements are intrinsic to the pectin and include not only the distribution pattern of free carboxyl groups in the homogalacturonan domains, but also the occurrence and distribution of rhamnose, acetyl (on C_2 and/or C_3 of the pyranose ring in some galacturonic acids), and amide groups, and the molecular weight of the pectin. The second requirement is extrinsic to the pectin and involves the calcium concentration, the pH, the ionic strength, and the temperature.

II. MANUFACTURE OF LOW-METHOXYL PECTINS

The high-methoxyl (HM) pectins isolated from apple pomace and citrus peel are the precursors of LM pectins, currently obtained from them by controlled deesterification in alcoholic heterogeneous medium with acid or ammonia. Acid

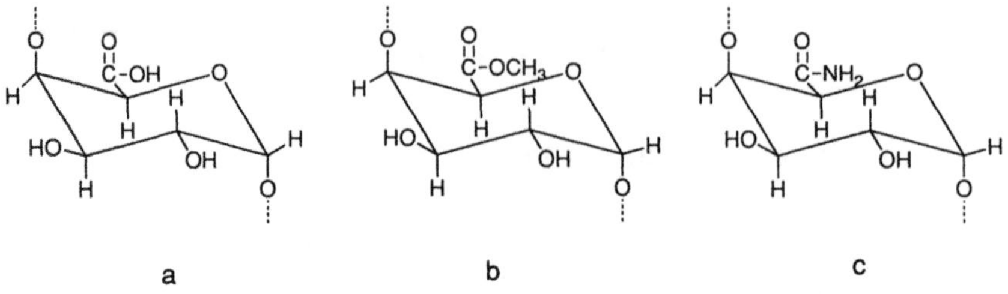

Figure 1   Functional groups: carboxyl (a); ester (b); amide (c) in low-methoxyl pectins.

deesterification terminates in a random distribution of the remaining methoxyl groups along the galacturonan backbone. The action of ammonia on HM pectins is complex, leading to amidated LM pectins, characterized by a blockwise distribution of the amide groups and a random distribution of the free carboxyl groups (Racapé *et al.*, 1989).

Another way to deesterify HM pectins is by the use of enzymes, *viz.*, pectin methylesterases. Depending on their origin (higher plants, microorganisms), different distribution (blockwise or random) of the resulting free carboxyl groups can be obtained (Kohn *et al.*, 1983). However, enzymes are used only on a bench-scale.

The functional groups encountered in LM pectins are the carboxyl (Fig. 1a), ester (Fig. 1b) and in some instances amide groups (Fig. 1c). The yield and quality of LM pectins depend on the selection and control of the extraction conditions (Taylor, 1982).

III. STRUCTURAL CONSIDERATIONS

HM-pectin gelation theory is not valid for LM pectins, although the gel characteristics are governed by a number of identical macromolecular properties, e.g., composition, size, and conformation, inherent in the polymer. Instead, LM-pectin gelation is considered as the formation of a continuous reticulum, similar to HM pectin, but of ionic cross-linkages via calcium bridges between two carboxyl groups belonging to two different chains in close proximity.

A. Polymer Conformation

Individual sugars rings in the chain are essentially rigid, and the overall conformation of the chain is determined by the rotation angles ϕ and ψ

of the glycosidic bonds. These angles, determined from the x-ray fiber-diffraction pattern of pectinic acid as well as calcium pectate are about 79° for ϕ and 90% for ψ (Walkinshaw and Arnott, 1981). These angles generate a simple 3_1 helix in contrast to the 2_1 form characterized for polyguluronic acid (Atkins *et al.*, 1973), which is stereochemically analogous to the polygalacturonic acid.

In solution, the ordered tertiary structure does not persist. The amount of interactions between the polymer and the solvent leads to a distribution of the value of ϕ and ψ, and the polysaccharide backbone adopts a worm-like conformation. Recent results from small-angle neutron and X-ray scattering on pectins of various DM gave persistence length (parameter of chain flexibility) of about 20 monomers (Durand *et al.*, 1990b). These values indicate that pectins are considerably more flexible than are xanthan molecules, for example, for which a persistence length of about 250 monomers is reported (Liu and Norisuye, 1988). They are, however, more rigid than amylose, for which the persistence length varies between 3 to 8 monomers depending on the method of determination (Ring *et al.*, 1985).

B. Mechanism and Structure of Junction Zones

Elucidation of the mechanism of LM pectin gelation relies mainly on the well-known *egg-box* model (Fig. 2) of the Rees group (Grant *et al.*, 1973). The mechanism involves junction zones created by the ordered, side-by-side associations of galacturonans, whereby specific sequences of galacturonic acid monomers in parallel or adjacent chains are linked intermolecularly through electrostatic and ionic bonding of carboxyl groups (Fig. 2). It is generally accepted that the junctions consist of dimers in 2_1 helical symmetry, similar to the 2_1 model proposed for alignates. The oxygen atoms of the hydroxyl groups, the ring oxygen atoms, and the bridging oxygen atoms of the component sugar units participate in the bonding process through their free-electron pairs (Kohn, 1987). The life of the junction depends on the strength of the electrostatic bonds. The bonds are stable when there are at least seven consecutive carboxyl groups on the interior of each participating chain (Powell *et al.*, 1982). The occurrence of methyl ester groups in the primary backbone limits the extent of such junction zones leading to formation of the gel.

Other models for LM-pectin gelation have been proposed, but they are currently unconfirmed by experimentation (Walkinshaw and Arnott, 1981). Nevertheless, all LM-pectin gels seem to develop similar if not identical junction zones (Filippov *et al.*, 1988).

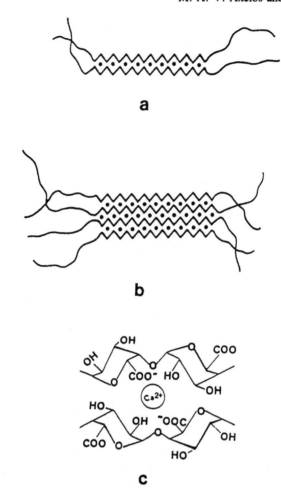

Figure 2   Schematic representation of calcium binding to polygalacturonate sequences: (a) "egg-box" dimer; (b) aggregation of dimers; (c) an "egg-box" cavity.

IV. PARAMETERS OF LOW-METHOXYL PECTIN GELATION

A. Intrinsic Parameters

1. The Effect of Molecular Weights

In addition to other prerequisites, gelation of a LM pectin is a property of its molecular weight. The rigidity of the gel is determined by the number of effective junctions formed per chain. The lower the molecular weight, i.e., the shorter the galacturonan chain, the weaker the gel (Van Deventer-Schriemer and Pilnik, 1987).

2. The Effect of Charge Distribution

Calcium cooperative binding is negligible when more than 40% of the carboxyl groups are randomly esterified (Thibault and Rinaudo, 1985). Gel-forming ability increases with decreasing DM. LM pectins with a blockwise pattern of deesterification are extremely sensitive to low calcium levels (Thibault and Rinaudo, 1985). These authors have shown that enzyme-deesterified pectins are characterized by a low activity coefficient, close to that of calcium pectate, owing to the existence in these pectin molecules of segments with a blockwise arrangement of carboxyl groups in the range of 15 units (this corresponds to a minimum of seven nonesterified residues occurring consecutively along one face of the polymer chain).

Calcium binding has been studied by a variety of methods (Morris *et al.*, 1982; Powell *et al.*, 1982; Thom *et al.*, 1982; Thibault and Rinaudo, 1986). According to Rees (1982), of the total stoichiometric requirement of bound calcium, only 50% + 5% is exchangeable with univalent counterions.

3. The Effect of Rhamnose

The occurrence in the pectin primary structure of a sugar monomer, e.g., rhamnose, whose dimensions are not compatible with the geometry of the junction zones formed by galacturonic acid monomers, precludes junction-zone formation. In citrus and apple pectins, this sugar occurs mainly in *hairy regions* (Fig. 3), defined as those regions in which neutral sugar side-chains are attached to

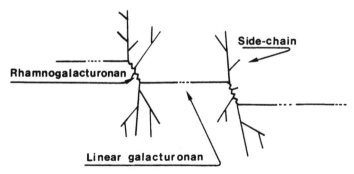

Figure 3 Schematic representation of pectin backbone showing the "hairy" regions (rhamnogalacturonan and side-chains) and the "smooth" regions (linear galacturonan).

the main chain (De Vries *et al.*, 1981). These hairy regions obstruct the molecular orientation necessary for the junction zones to develop.

4. The Effect of Acetyl Groups

The affinity of pectin for calcium ions is diminished by acetylation, because the size of the acetylated galacturonic residues is not compatible with the severe topological constraints of the chain–chain association within the junction zones (Kohn and Furda, 1968). Thus, acetyl groups impart poor gelation characteristics to LM pectins, as in the example of beet pectins that can have a degree of acetylation up to 35 mol acetyl/100 mol pectin. This effect of acetyl groups also has been reported on alginates (Skjak-Braek *et al.*, 1989).

5. The Effect of Amide Groups

Amidation increases the gelling ability of LM pectins. Amidated pectins need less calcium to gel and are less prone to precipitation by high Ca^{2+} concentration than are other forms of LM pectin (May, 1990). Racapé *et al.* (1989) suggested that the gelation of amidated pectins cannot be completely explained by the egg-box model, and that amide groups allow other types of chain association through hydrogen bonding.

B. Extrinsic Parameters

Whereas intrinsic factors dictate the nature of the LM-pectin gel, the gelation process is influenced by external factors, e.g., pectin and calcium concentrations, pH, ionic strength, content of soluble solids, and temperature.

1. The Combined Calcium–Pectin Concentration Effect

The combined influence of calcium and pectin concentration, at pH = 7, on the gelation of LM pectin (DM, 28%) sols is best illustrated in a phase diagram (Fig. 4), in which $R = 2(Ca^{2+})/(COO^-)$. R versus C, the pectin concentration, is divided into three domains (Axelos, 1990). The domain of the sol corresponds with the particular condition below R_2, in which gelation is not induced at any pectin concentration in the range 0–10 g/liter. Neither is gelation induced at any calcium concentration below a critical pectin concentration $C_0 = 1$ g/liter. This gelation threshold concentration C_0 is very close to the overlap threshold con-

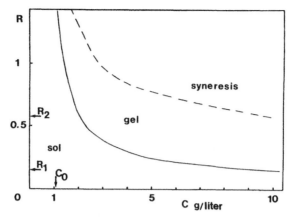

Figure 4   Phase diagram of a 28% DM pectin solution (no sugar added) in 0.05 M NaCl, at pH 7 and 20°C. R = $2(Ca^{2+})/(COO^-)$; C, pectin concentration (g/liter).

centration, the so-called C*, that denotes the transition from the dilute to the semidilute regime. This result indicates that the minimal pectin concentration required for gelation must be determined from the studies of pectin solution properties (Axelos *et al.*, 1989). Above R_1 and between C_0 and 10 g/liter pectin of dispersion, the mixture may exist in a sol, a gel, or a sol–gel equilibrium state. Above R_2 and for pectin concentrations larger than C_0, the gel is susceptible to syneresis. At high R values and pectin concentrations close to C_0, the two equilibrium curves fall in one; in this region of the phase diagram, no gelation takes place, but aggregation of the pectin occurs followed by precipitation. Hence, a phase diagram is demonstrated to have a practical utility, in arriving at the concentrations of ingredients and reagents necessary to obtain a well-defined LM-pectin gel state.

2. The Effect of pH and Ionic Strength

a. pH

Gel formation in LM-pectin solutions, with no sugar added, is slightly sensitive to pH. Indeed at low pH (about 3.5), more calcium is needed than at neutral pH to induce gelation (Fig. 5). In a phase diagram as defined in Section IV,B,1, R increases as pH decreases, but the sol–gel transition curve is much higher than that expected just by taking into account the decrease of the number of dissociated carboxyl groups due to greater acidity. At very low pH, the charge in the junction-zone cavities is neutralized by hydrogen ions, and aggregation and precipitation of the pectin occur.

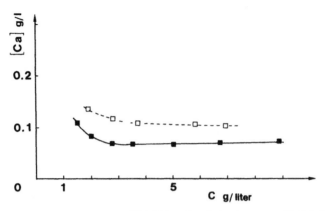

Figure 5   Sol–gel transition curves of a 28% DM pectin solution (no sugar added) in 0.1*M* NaCl, 20°C at pH 7 (—■—) and at pH 3.5 (---□---);[Ca], calcium concentration (g/liter); C, pectin concentration (g/liter).

b. Ionic Strength

It appears that an increase of the ionic strength leads to a decrease in the total amount of calcium required for gelation. For a given set of parameters, LM-pectin gels containing NaCl are more rigid than are gels without NaCl (Axelos, 1990). This effect is explained in terms of electrolyte behavior. In the presence of increasing concentrations of NaCl, the charge of the polymer is increasingly screened, allowing pectin chains to approach each other, which in turn leads to a decrease in the rate of the ionic bridging function (Axelos, 1990). It is possible that a slower rate of gelation will initiate a greater number of smaller junction zones and consequently will enhance stability and mechanical strength.

3. The Effect of Temperature

The flow of LM-pectin homogeneous gels is thermally reversible. LM pectins are therefore thermoplastic bodies. Temperature increases effect a decrease in the strength of the bonding forces within the junction. This response to heat is commonly used to prepare gels by dispersing LM pectin in aqueous calcium solutions at about 70°C and cooling them slowly. Durand *et al.* (1990a) have studied the temperature dependence of gel times and thus determined an enthalpy of the cross-linking process DH around -70 kJ/mol^{-1}. This value, four times higher than those obtained for HM pectins, is attributed to the cooperative character of the junction mechanism. The sol–gel transition temperature (T) is raised by increasing the pectin and the calcium content and by decreasing the DM. An increase in the solvent's ionic strength leads to a lowering of T.

4. Soluble-Solids Content

Implicit in Fig. 4, and stated previously, is the ability of LM pectin to form a gel with no sugar. Generally when soluble solids are added, an increase in the setting temperature is noticed, the gel strength also increases, and further, the syneresis is reduced (Christensen, 1986). The combined effect of pH and sugars promotes gelation at lower calcium level. Indeed, despite the decrease of the number of sequences of carboxyl groups available for calcium binding (Section IV,B,2,a), gelation is enhanced because of the specific effect of sugars on the water activity and hydrophobic effects. These effects are very complex; a dependence of the gel strength according to the type of sugar used has also been noticed (May and Stainsby, 1986). LM pectins are used, for example, in the manufacture of low-calorie jellies that utilize no more than 30–43% sugar. When $R < R_1$, high-viscosity dispersions suitable for dietetic beverages, juice drinks, etc., result. In these instances, firm gels and high-viscosity liquids depend on a low ratio of relatively high concentrations of pectin and CA^{2+}.

References

Axelos, M. A. V. (1990). Ion complexation of biopolymers: Macromolecular structure and viscoelastic properties of gels. *Die Makromolekulare Chemie, Macromolecular Symposia* **39,** 323–328.

Axelos, M. A. V., Thibault, J. F., and Lefebvre, J. (1989). Structure of citrus pectins and viscometric study of their solution properties. *Int. J. Biol. Macromol.* **11,** 186–191.

Atkins, E. D. T., Nieduszynski, I. A., Mackie, W., Parker, K. D., and Smolko, E. E. (1973). Structural components of alginic acid. II. The crystalline structure of poly-α-L-guluronic acid. Results of X-ray diffraction and polarized infrared studies. *Biopolymers* **12,** 1879–1887.

Christensen, S. H. (1986). Pectins. *In* "Food Hydrocolloids III" (M. Glicksman, ed.), pp. 205–203. CRC Press, Boca Raton, Florida.

De Vries, J. A., Voragen, A. G. J., Rombouts, F. M., and Pilnik, W. (1981). Extraction and purification of pectins from alcohol-insoluble solids from ripe and unripe apples. *Carbohydr. Polym.,* **1,** 117–127.

Durand, D., Bertrand, C., Clark, A. H., and Lips, A. (1990a). Calcium-induced gelation of low-methoxy pectin solutions—thermodynamic and rheological considerations. *Int. J. Biol. Macromol.* **12,** 14–18.

Durand, D., Bertrand, C., Busnel, J. P., Emery, J., Axelos, M. A. V., Thibault, J. F., Lefebvre, J., Doublier, J. L., Clark, A. H., and Lips, A. (1990b). Physical gelation induced by ion complexation: Pectin–calcium systems. *In* "Physical Networks" (W. Burchard and S. B. Ross-Murphy, eds.), pp. 283–300. Elsevier Applied Science Publishers, New York.

Filippov, M. P., Komissarenko, M. S., and Kohn, R. (1988). Investigation of ion exchange on films of pectic substances by infrared spectroscopy. *Carbohydr. Polym.* **8,** 131–135.

Grant, G. T., Morris, E. R., Rees, D. A., Smith, P. J. C., and Thom, D. (1973). Biological interactions between polysaccharides and divalent cation: The egg-box model. *FEBS Lett.* **32,** 195–198.

Kohn, R. (1987). Binding of divalent cations to oligomeric fragments of pectin. *Carbohydr. Res.* **160,** 343–353.

Kohn, R., and Furda, I. (1968). Binding of calcium ions to acetyl derivatives of pectin. *Coll. Czech. Chem. Commun.* **33,** 2217–2225.

Kohn, R., Markovic, O., and Machova, E. (1983). Deesterification mode of pectin by pectin-esterases of *Aspergillus foetidus,* tomato, and alfalfa. *Coll. Czech. Chem. Commun.* **48,** 790–797.

Liu, W., and Norisuye, T. (1988). Order–disorder conformation change of xanthan in $0.01M$ aqueous sodium chloride: Dimensional behavior. *Biopolymers* **27,** 1641–1654.

May, C. D. (1990). Industrial pectins: Sources, production, and applications. *Carbohydr. Polym.* **12,** 79–99.

May, C. D., and Stainsby, G. (1986). Factors affecting pectin gelation. *In* "Gums and Stabilisers for the Food Industry 3" (G. O. Phillips, D. J. Wedlock, and P. A. Williams, eds.), pp. 515–523. Elsevier Applied Science Publishers, New York.

Morris, E. R., Powell, D. A., Gidley, M. J., and Rees, D. A. (1982). Conformations and interactions of pectins I. Polymorphism between gel and solid states of calcium polygalacturonate. *J. Mol. Biol.* **155,** 507–516.

Powell, D. A., Morris, E. R., Gidley, M. J., and Rees, D. A. (1982). Conformations and interactions of pectins. II. Influence of residue sequence on chain association in calcium pectate gels. *J. Mol. Biol.* **155,** 517–531.

Racapé, E., Thibault, J. F., Reitsma, J. C. E., and Pilnik, W. (1989). Properties of amidated pectins II. Polyelectrolyte behaviour and calcium binding of amidated pectins and amidated pectic acids. *Biopolymers* **28,** 1435–1448.

Rees, D. A., (1982). Polysaccharide conformation in solutions and gels—recent results on pectins. *Carbohydr. Polym.* **2,** 254–263.

Ring, S. G., l'Anson, K. J., and Morris, V. J. (1985). Static and dynamic light-scattering studies of amylose solutions. *Macromolecules* **18,** 182–188.

Skjak-Braek, G., Zanetti, F., and Paoletti, S. (1989). Effect of acetylation on some solution and gelling properties of alginates. *Carbohydr. Res.* **185,** 131–138.

Taylor, A. J. (1982). Intramolecular distribution of carboxyl groups in low-methoxyl pectins—a review. *Carbohydr. Polym.* **2,** 9–17.

Thibault, J. F., and Rinaudo, M. (1985). Interactions of mono- and divalent counterions with alkali- and enzyme-deesterified pectins. *Biopolymers* **24,** 2131–2144.

Thibault, J. F., and Rinaudo, M. (1986). Chain association of pectic molecules during calcium-induced gelation. *Biopolymers* **25,** 455–468.

Thom, D., Grant, G. T., Morris, E. R., and Rees, D. (1982). Characterisation of cation binding and gelatin of polyuronates by circular dichroism. *Carbohydr. Res.* **100,** 29–42.

Van Deventer-Schriemer, W. H., and Pilnik, W. (1987). Studies of pectin degradation. *Acta Alimentaria* **16,** 143–153.

Walkinshaw, M. D., and Arnott, S. (1981). Conformation and interactions of pectins. I. X-ray diffraction analyses of sodium pectate in neutral and acidified forms. II. Models for junction zones in pectinic acid and calcium pectate gels. *J. Mol. Biol.* **153,** 1055–1073 and 1075–1085.

CHAPTER 7

Gelation of Sugar Beet Pectin by Oxidative Coupling

J.-F. Thibault

Institut National de la Recherche Agronomique
Laboratoire de Biochimie et Technologie des Glucides
Nantes Cedex 03, France

F. Guillon

Institut National de la Recherche Agronomique
Laboratoire de Technologie Appliquée à la Nutrition
Nantes Cedex 03, France

F.M. Rombouts

Agricultural University of Wageningen
Department of Food Science
Bomenweg 2, 6703 HD Wageningen, The Netherlands

I. INTRODUCTION

Pectins extracted mainly from apple pomace and citrus peels are traditionally used as gelling agents and thickeners in the food industry. Sugar beet slices have been investigated as an alternative source (Kertesz, 1951), but attempts to commercialize sugar beet pectins have failed, because of their poor gelling properties, when compared with citrus and apple pectin. The poor quality of the gel is attributed to the characteristic presence on the pectin molecules of acetylester groups (Pippen *et al.*, 1950).

Beet pulp is the solid fraction remaining after sugar extraction from beets. Dehydrated beet pulp, containing approximatively 25% of its dry weight as polygalacturonic acids, is available throughout the year, making it a potential source of commercial pectin, notwithstanding its poor gelation characteristics.

Recent studies on sugar beet pectin have revealed some new structural features, especially the presence of feruloyl groups (Rombouts *et al.*, 1983). It has been known for some time that ferulic acid, cinnamic acid, and tyrosine substituents bound to biopolymers are involved in the formation of cross-links in the presence of certain oxidants. For example, wheat flour pentosans contain feruloylester groups, and the addition of hydrogen peroxide–peroxidase to effect formation of diferulic acid results in gel formation (Geissmann and Neukom, 1973). Similar reactions were apparently used to photocross-link polymers, such as cellulose and polyvinyl alcohol, on which cinnamic acid residues were bound (Delzenne, 1969). The authors of this chapter succeeded in coupling sugar beet pectin molecules through oxidation (Rombouts *et al.*, 1983), whereby its commercial prospects may be enhanced.

In this chapter, the extraction, chemistry, and properties of sugar beet pectin, and the mechanism of oxidative coupling are discussed.

II. EXTRACTION

Very small quantities of beet pectins are solubilized by water (2.2%) and by calcium-chelating agents (1.6%) from an alcohol-insoluble residue (AIR) of sugar beet pulp (Rombouts and Thibault, 1986a), probably because much pectin has been eliminated during the leaching process (Le Quéré *et al.*, 1981; Thibault, 1988). The bulk of the pectin is solubilized by hot (85°C), dilute (0.05N) acid (HCl) (~19% of the AIR). A significant amount (11% of the AIR) is further extracted with cold (4°C), dilute (0.05N) alkali (NaOH). These two latter steps extract 95% of the galacturonic acids initially present in the AIR.

A different procedure of beet-pectin extraction was used by Selvendran (1985). Pectins were sequentially extracted from beet pulp with hot water (80°C at pH 5 for 2 hr), ammonium oxalate (80°C at pH 5 for 2 hr) and chlorite–acetic acid (70°C at pH 4 for 2 hr). The recoveries were 8.4, 7.4, and 23%, respectively.

III. THE CHEMISTRY OF BEET PECTIN

A. Composition

Beet pectins have fairly low molecular weights in the order of 15,000 to 48,000 (Michel *et al.*, 1985; Phatak *et al.*, 1988; Dea and Madden, 1986), which could partly explain their low gelling power. They are relatively high in neutral sugars (arabinose, galactose, and rhamnose) amounting to 6 to 24%, relative to apple and citrus pectins.

The composition of Selvendran's extract (Selvendran, 1985) differed by a lower content of galactose and a higher content of mannose. According to the author, the pectins extracted with water and oxalate originated from the middle lamellae, whereas those solubilized by chlorite–acetic acid originated from the primary cell wall. The latter pectins are very poor in galacturonic acids but are rich in neutral sugars, especially in arabinose; they contain ester-linked ferulic and *p*-coumaric acids, and it was thought that degradation of phenolic compounds by chlorite–acetic acid, a delignifying reagent, was responsible for their presence.

The degree of acetylation of the beet pectins is high (up to 35%), especially in the acid-soluble fraction; it is such high concentrations that might preclude the molecular organization necessary for gelation (Rombouts and Thibault, 1986a).

Sugar beet pectins contain proteins rich in hydroxyproline (1–2%), even after their purification by ion-exchange chromatography. In addition, feruloyl groups esterified to the pectins are present in small quantities ($< 1\%$) (Rombouts and Thibault, 1986a), unlike apple, citrus, cherry, and potato pectins (Rombouts *et al.*, 1983). Feruloyl groups are known to be present in spinach (Fry, 1983), a member of the same botanical family as sugar beet.

The presence of feruloyl groups in the sugar-beet pectins was shown by spectrophotometry (Rombouts and Thibault, 1986a) to give a bathochromic shift of the double absorption peak at 300 nm and 325 nm at pH 4.8 toward a single peak at 375 nm at pH 10. This shift is typical of esters of phenolic compounds containing cinnamic acid. The feruloyl group was also shown by high-performance thin-layer chromatography (Rombouts and Thibault, 1986a) and high-pressure liquid chromatography (Guillon and Thibault, 1988) to occur, after hydrolysis of the phenol esters. Ferulic acid is the only known phenolic acid ester-linked to the pectin.

B. Fine Structure

Chemical analysis, described in the previous section, showed that beet pectins are rich in galacturonic acids, arabinose, and galactose, with a high content of acetic acid and a low but significant content of ferulic acid. In order to have a better insight into their chemical structure, highly purified enzymes were specifically used to degrade them (Rombouts and Thibault, 1986b). The results obtained with depolymerizing enzymes (endo-polygalacturonase and pectate-lyase) after chemical deesterification confirmed what is known already for other pectins from apples (De Vries *et al.*, 1982) and cherries (Thibault, 1983), *i.e.*, the neutral sugar side-chains attached to the rhamnogalacturonic acid backbone occur in blocks, the so-called hairy fragments, leaving large parts of the main chain unsubstituted (*smooth* regions). The location of the acetylester groups has been studied by analyzing the products of pectin hydrolysis after sequential treatment with pectin methylesterase and endo-polygalacturonase, and with exo-arabinanase and endo-galactanase. The results show that 80–90% of the acetyl groups are located in the smooth regions through the number two carbon (C-2) and/or C-3 of the galacturonic acid units, and are fairly regularly distributed along the chains (Rombouts and Thibault, 1986c). In contrast, the feruloyl ester groups were mostly recovered from the hairy fragments, while only part of the hydroxyproline-rich material was found there (Rombouts and Thibault, 1986b; Guillon and Thibault, 1988).

The fine structure of the neutral sugar side-chains and the exact location of the feruloyl groups in the hairy fragments were examined by chemical and enzymic approaches. By methylation, it was shown that acid- and alkali-soluble pectins as well as their hairy fragments, have closely related chemical structures. The backbone of the hairy fragments consists of α 1,4-linked galacturonic acid residues with 1,2-linked L-rhammnosyl residues. The C-4 of the rhamnosyl residues is the main point of attachment of the side-chains. Arabinose residues, mainly in the furanose form, are terminal, 1,5- and 1,3,5-linkages. Galactose residues are mainly 1,4-linked with few branch points on C-3, but 1,3-, 1,6-and 1,3,6-linkages are also found. Thus, the presence of types I and II (arabino)-galactans is indicated (Guillon and Thibault, 1988, 1989; Guillon *et al.*, 1989). These structural features are consistent with the results obtained from ^{13}C-nuclear magnetic resonance (NMR) that have shown sugar beet pectin to be a branched, 1,4-linked α-D-galacturonan containing a small proportion or rhamnose with side-chains composed of 1,5-linked α-L-arabinofuranosyl residues and β-galactosyl residues, probably 1,4-linked (Keenan *et al.*, 1985).

More information on the arrangement of the arabinose and galactose residues in the hairy fragments was obtained from mild acid hydrolysis and enzymic degradation (Guillon and Thibault, 1989; Guillon *et al.*, 1989). Hydrolysis by an arabinofuranosidase was a two-step process where first α-1,3 bonds were cleaved, leaving a linear α-1,5 arabinan, which was then hydrolyzed. A similar pattern was observed with highly branched arabinans (Rombouts *et al.*, 1988). An endo-arabinanase, hydrolyzing preferentially linear arabinan, shows low activity in the initial hairy fragments, but enhanced activity when the hairy fragments were incubated first or at the same time with arabinofuranosidase. The fact that β-galactosidase and endo-galactanase degrade only slightly the hairy fragments, and that the galactanase activity was not increased by removing arabinose units, suggests that the galactanase is not hindered by arabinose side residues. It was concluded that galactose units occur rather as short chains. This is confirmed by the high value of the ratio of terminal to 1,4-linked galactose, and by the absence of oligomers of galactose in the products resulting from a mild acid hydrolysis. More drastic treatment such as 0.1M trifluoroacetic acid (TFA) at 100°C for 1 hr did not remove all the galactosyl residues, suggesting the attachment of some galactose residues to galacturonic acid residues. Side-chains are therefore composed mainly of highly branched α-1,5-linked arabinans, the C-3 of the arabinosyl residues being the branching points, of β-1,4-linked galactans with few branching points on the C-3 and of β-1,3-, and of β-1,6-linked galactans.

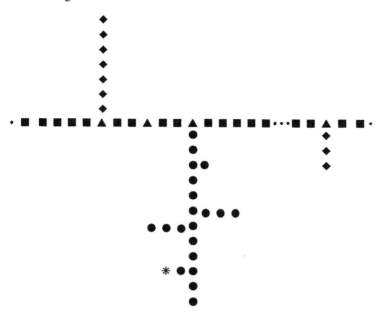

Figure 1 Tentative structure of the hairy fragments of sugar beet pectin (■ galacturonic acid, ▲ rhamnose, ● arabinose, ◆ galactose, * feruloyl group).

The location of the feruloyl groups was also investigated. Part (30%) are ester-linked to arabinose residues, as shown by the fact that they were removed together with arabinose after mild acid hydrolysis with 0.05 *M* TFA, or after degradation with endo-arabinanase and arabinofuranosidase. The arabinose–ferulic acid compounds were not characterized, because of their low concentration. Possibly feruloyl groups are located at the nonreducing arabinopyranosyl termini, as described for spinach pectins (Fry, 1983). A subsequent part (~10%) of the feruloyl groups were released with galactose by the endogalactanase, but a substantial amount (~ 30%) was obtained by a more-severe acidic treatment such as 0.1*M* TFA. The remainder was probably linked to some residual galactosyl units. The results suggest that all the feruloyl groups are not in equally exposed domains of the pectins, and therefore, are not equally accessible (Thibault, 1988). A general scheme of the possible structure of hairy fragments from beet pectins is shown in Fig. 1.

C. Effect of Oxidizing Agents

Oxidation experiments show that oxidants do not have the same effects on beet pectins. The addition of oxidants, e.g., potassium periodate, potassium permanganate, sodium chlorite, potassium ferricyanide, and hydrogen peroxide, to a solution of sugar beet pectin causes a continuous decrease in the reduced viscosity. In contrast, hydrogen peroxide–peroxidase and ammonium persulfate increase the viscosity of the medium. The former increases the reduced viscosity instantaneously to a value that decreases very slightly with time, whereas the latter causes a continuous increase in viscosity to a maximum (Thibault and Rombouts, 1986).

Gels or solutions with markedly increased viscosity may be produced by addition of hydrogen peroxide–peroxidase or by persulfate, depending on the experimental conditions (mainly pH and concentration of pectin and reagents), and also on the chemical structure of the initial pectin. All these aspects, including the mechanism of the reaction with persulfate ions, have been studied (Thibault *et al.*, 1987; Thibault, 1986, 1988).

D. Evidence of Participation of Feruloyl Groups

Experimentation indicated that feruloyl groups are fundamentally involved in the cross-linking reaction (Thibault and Rombouts, 1986). The results showed that

1. the contents of galacturonic acids and total neutral sugars were not changed during the reaction. In contrast, the concentration of feruloyl groups decreased continuously, whereas the reduced viscosity increased simultaneously after

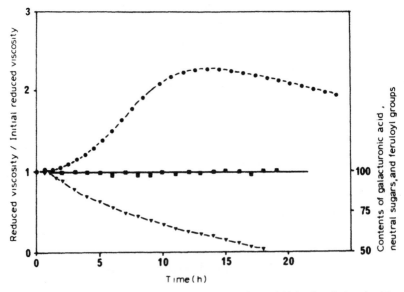

Figure 2 Changes with time in the ratio of reduced viscosity to initial reduced viscosity (●), of the contents of galacturonic acid and total neutral sugars (■), and of feruloyl groups (▼) expressed as percentage of the initial values of a 0.44% solution of pectin in 0.01*M* ammonium persulfate at 25°C.

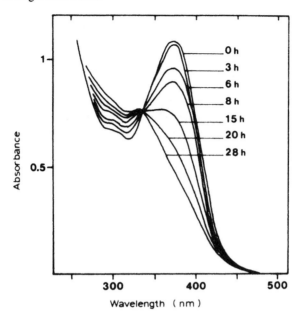

Figure 3 Changes with time of the UV spectra of a 0.44% solution of pectin in 0.01M ammonium persulfate at 25°C; the solution (1 ml) was mixed with 0.2M glycine–sodium hydroxide buffer (pH 10, 3 ml). The spectra were recorded immediately.

an induction period to a maximal value and then decreased slowly (Fig. 2). Hence, the increase in viscosity was clearly related to a polymerization process. The decrease is difficult to study, since there is a complex dependence on polymer concentration, molecular weight and shape, and on pH and ionic strength, all being variables of the reaction time;

2. the absorption spectra of the pectin under alkaline conditions as a function of reaction time showed a decrease of the peak at 375 nm, and all the curves passed through an isosbestic point at 330 nm (Fig. 3). This observation confirmed the specific participation of feruloyl groups in the reaction; and

3. gel-permeation chromatography of the pectins on Sepharose CL-2B showed an increasing amount of material rich in galacturonic acid and in neutral sugars eluting in the void volume, where the feruloyl groups also accumulated. At $K_{av} = 0.85$, a peak was left from which feruloylated material progressively disappeared. This result proved that the polymerization is specific for feruloylated pectins, and that some of the pectin molecules were not feruloylated.

E. Mechanism of Oxidative Coupling

Hydrogen peroxide–peroxidase and persulfate are known to create free radicals. Hydrogen peroxide, by itself, does not initiate free-radical formation, but does so in conjunction with peroxidase. Persulfate decomposes in aqueous solution to give radicals, and is therefore used in the presence of reducing agents or singly for aqueous polymerization of acrylamide, methacrylamide, acrylonitrile, and methylmethacrylamide (Thomson, 1983). That the reaction of sugar beet pectin with persulfate ions involves the formation of free radicals is confirmed by the inhibition observed when 1-propanol, acetate, citrate, or phosphate ions are added to scavenge free radicals. The intervention of free radicals was also shown by kinetic data obtained with feruloylated pectins and pure ferulic acid as a reference (Thibault *et al.*, 1987).

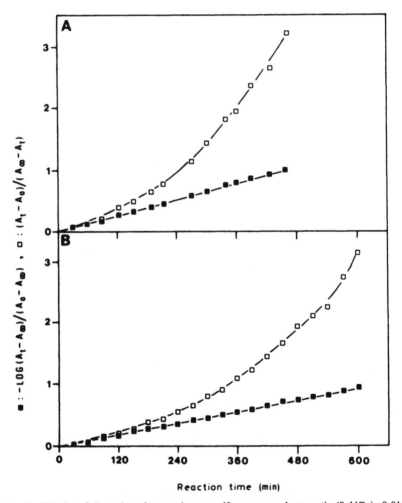

Figure 4 Kinetics of the action of ammonium persulfate on sugar beet pectin (0.44% in 0.01M ammonium persulfate at 40°C). (A) ferulate (0.25 mM in 0.05 M ammonium persulfate at 25°C). (B), analyzed in terms of pseudo-first order (■) and pseudo-second order (□) reactions with respect to feruloyl residues and ferulate.

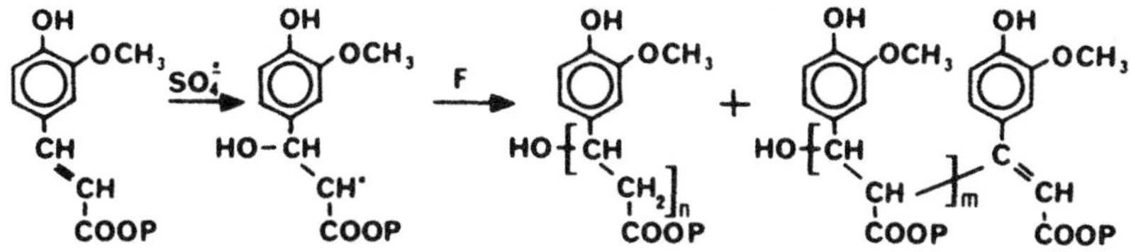

Figure 5 Hypothetical reaction scheme for the cross-linking of sugar beet pectins with ammonium persulfate (F, ferulate or feruloyl group; P, H or pectin chain).

The reaction of persulfate ions with sugar beet pectin (or ferulic acid) follows a pseudo-first order law with respect to pectin (or ferulate) and not a pseudo-second order law (Fig. 4); consequently, the reaction with persulfate is not an intermolecular condensation of feruloyl residues as suggested for feruloyalted arabinoxylan treated with hydrogen peroxide–peroxidase (Geissmann and Neukom, 1973). Furthermore, the fact that ferulic acid alone reacted with persulfate, giving the same kinetic parameters (reaction order, rate constants, and energy of activation) as sugar beet pectins, was a strong indication that the cross-linking of sugar beet pectin involves only the feruloyl residues.

^{1}H-NMR studies on ferulic acid demonstrated that the aromatic nuclei were not modified, but that the double bonds were involved in the reaction (Thibault et al., 1987). A broadening of the signals from aromatic protons was observed and ascribed to a polymerization process. This inference was in accord with the gel-permeation chromatography data that indicated the production of oligomers of ferulates, with degree of polymerization up to 10. This polymerization was also confirmed by the rate of disappearance of ferulic acid, being proportional to the ferulic acid concentration and to the square root of the persulate concentration, as in a classical free-radical polymerization. The mechanism in Fig. 5 is proposed for the reaction of persulfate ions with pectins and ferulic acid. The polymerization of feruloyl groups in pectic molecules is probably sterically hindered, leading mainly to dimerization. pH plays an important role in the reaction. Gelation occurs only in the range of pH $= 3.8$–5.7 (Fig. 6). Apparently, a low pH enhances the rate of disappearance of ferulic acid as well as feruloyl groups, but not of the cross-linking reaction. This may be explained by the decomposition of persulfate ions to sulfate ion radicals that may react with water to produce the hydroxyl radical and oxygen according to the following equations:

$$S_2O_8^{2-} \rightarrow 2SO_4^{-\cdot} \tag{1}$$

$$2SO_4^{-\cdot} + 2H_2O \rightarrow 2HO^{\cdot} + 2HSO_4^{-} \tag{2}$$

$$2HO^{\cdot} \rightarrow H_2O + 1/2\ O_2 \tag{3}$$

The reaction can be initiated either by the sulfate ion radical or by the hydroxyl radical. Low pH favors nonradical decomposition of persulfate ions. The increase in the rate of disappearance of feruloyl groups under acidic conditions can be due to oxidation of these residues. Analysis of the pH-dependence of the reaction

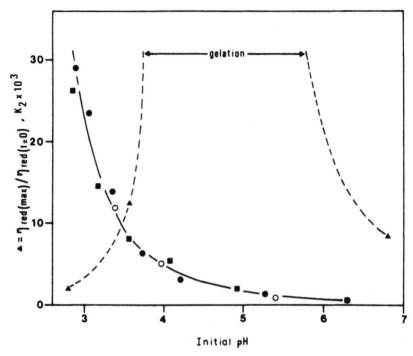

Figure 6 Effect of the initial pH on the second-order rate constant for pectins [●,0.44% of pectin (Na$^+$); ■, 1% pectin (Na$^+$); ○, 0.44% pectin (Ca^{2+})]; and on the ratio (▲) of maximum reduced viscosity to initial reduced viscosity of a 1% solution of pectin (Na$^+$) in the presence of 0.01 M ammonium persulfate at 25°C.

is complicated by the polyelectrolyte nature of pectins, inasmuch as the molecules can adopt an extended conformation in a fully ionized state, and a coil conformation in the acid form. However, since the rate constant for fully ionized pectins is independent of the counterion (sodium, calcium) and of the ionic strength, and is not changed by enzymic depolymerization, polymer conformation may not play an important role in the reaction (Thibault *et al.*, 1987).

F. Influence of the Structure of the Side-Chains

Neutral sugars in the side-chains of a pectin that was incapable of gelling were removed by cold (25°C) or hot (100°C) dilute (0.05M) TFA hydrolysis, and with highly purified arabinofuranosidase, endo-arabinanase, α-galactosidase and endo-galactanase. These studies were for the purpose of correlating structural features of the hairy regions with gelling power. The results showed that pectin capable of gelling was obtained only after cold acid hydrolysis, or after the elimination of arabinose residues by the arabinofuranosidase (Guillon and Thibault, 1987, 1990). Structural analysis of the resulting pectins showed that the treatment with cold acid, similar to the treatment with arabinofuranosidase, leads mainly to a slight decrease of the terminal arabinose units, while the other structural characteristics are not changed. In contrast, hydrolysis with β-galactosidase or extensive degradation of the arabinans by the combined action of arabinofuranosidase and endo-arabinanase or by hot dilute acid did not significantly improve the gelling power.

IV. PROPERTIES OF CROSS-LINKED BEET PECTIN

Pectin gelation is a function of molecular weight. Sugar beet pectin has a lower average molecular weight than do apple and citrus pectins. Depending on the concentration of the reactants and on the reaction time, the cross-linking of beet pectin may be used to obtain higher molecular weights and gels with properties different from those of apple and citrus gels. In Table I is shown the effect of cross-linking with hydrogen peroxide and peroxidase on viscosity–average molecular weight of a beet pectin. It is seen that the molecular weight may be increased by 210%, and that gels are obtained when the pectins are above a critical concentration (Rombouts *et al.*, 1983). Striking increases in reduced viscosity can also be obtained with persulfate. As with the hydrogen peroxide–peroxidase system, gels are obtained when the pectin concentration is increased above 0.6% (Fig. 7) (Thibault and Rombouts, 1986).

Solutions of cross-linked beet pectin with increased molecular weight show flow properties varying from roughly Newtonian to shear-thinning, accompanied by a tremendous increase in viscosity (Fig. 8) (Doublier and Thibault, 1985).

Table I   Influence of the concentration of pectin, peroxidase, and hydrogen peroxide on the cross-linking reaction with sugar beet pectin

Pectin concentration (g/liter)	Peroxidase concentration (mg/liter)	H_2O_2 concentration (mmol/liter)	Gelation	% increase molecular weight
6.26	0.25	2	+	
5.01	0.25	2	+	
3.76	0.25	2	−	
2.51	0.25	2	−	
5.45	0.67	2	−	
5.61	0.1	2	−	140
3.48	1.7	8.3	+	
3.5	2.0	2	+	
3.01	8.3	4	−	150
4.88	0.07	0.7	−	210
3.19	0.04	1.3	−	180

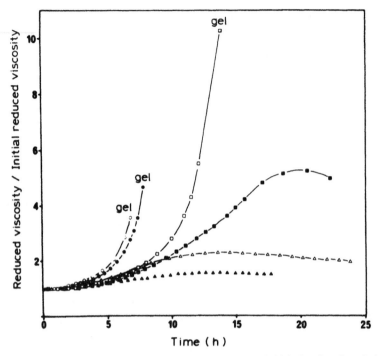

Figure 7   Changes with time in the ratio of reduced viscosity to initial viscosity of a solution of pectin at various concentrations (▲, 0.25; △, 0.44; ■, 0.6; □, 0.8; ●, 1; ○, 1.2%) in 0.01 M ammonium persulfate at 25°C.

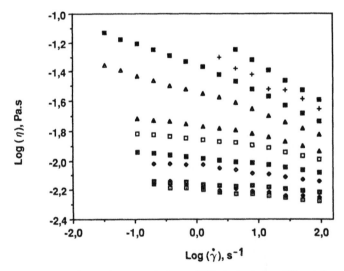

Figure 8   Flow curves of (1.2%) pectin solution at 25°C obtained after different times of cross-linking with M/100 ammonium persulafate (from bottom to top: 100, 155, 205, 280, 320, 370, 400, 437, 465, 485, and 500 min.).

Table II Effects of the nature of the counterion and of the ionic strength of the external solution on the swelling (ml/g) of cross-linked pectins

Ionic form	Feruloyl-modified residues (%)					
	74.6	74.1	70	69.4	67.1	66.7
H^+	45	45	60	50	55	55
Ca^{2+}	80	75	100	75	100	125
Na^+	95	110	120	110	140	180
Na^+ in 0.001 M NaCl	100	105	120	115	125	160
Na^+ in 0.01 M NaCl	70	75	90	80	90	125
Na^+ in 0.1 M NaCl	45	40	50	50	60	75

Gels prepared with beet pectins are less brittle and more elastic than are gels prepared with a commercial amidated pectin (Rombouts *et al.*, 1983).

Beet-pectin gels are distinguished from other gels by their chemical irreversibility. Thus, the cross-linked pectins from such gels can be isolated after drying. The xerogels obtained do not dissolve in water but, instead, swell to many times their initial volume, and thereby exhibit remarkable water-absorption capacities in the order of 50 to 180 ml water per gram of xerogel (Thibault, 1986). The swelling also depends on the degree of neutralization of the carboxyl groups, on the nature of the counterions, and on the ionic strength of the medium. The acid forms of the cross-linked pectins swell to a rather limited extent (45–60 ml/g). When they are fully neutralized by sodium hydroxide, their bed volumes increase by factors of 2 and 3. The maximum value was ~180 ml/g (Table II). K^+, Li^+ and Na^+ did not have a profound effect. A more limited swelling was observed when the samples were neutralized with calcium hydroxide, because calcium ions are more tightly bound than are sodium ions. Shrinkage was not observed, presumably because of the low charge density of the beet pectins and the acetylation of the galacturonic acid residues. When the ionic strength was increased, the swelling of the modified pectins decreased, and reached values observed in the acid form when the solution was 0.1M in NaCl (Table II). This fact reflects the role of the screening of the fixed charges, resulting in a reduction of the electrostatic repulsion and therefore a reduction of the volume.

V. CONCLUSION

The study of the gelation of beet pectin led to the main conclusions that not all the pectin is able to form a gel, and that no simple relationship exists between the gelling capacity and chemical composition of the pectin (Thibault, 1988; Guillon and Thibault, 1990).

The feruloyl groups in the periphery of the arabinose side-chains are of special importance in the gelling process of beet pectins. They must be in exposed domains of the side-chains, making them accessible for cross-linking by per-sulfate ions. The extraction of beet pectins from sugar-beet must liberate these areas without extensive degradation of the arabinans.

The oxidative-coupling reaction with beet and other feruloylated pectins leads to an increasing apparent molecular weight of the pectin, and ultimately to gel formation. This then is yet another way to construct a food gel, in addition to the acid–sugar–pectin and the calcium-pectate mechanisms. The cross-linked pectins have an extremely high water-absorption capacity, a property that should lead to applications as cloud stabilizers in beverages, for example, and as humectants in food and nonfood products.

References

Dea, I.C.M., and Madden, J.K. (1986). Acetylated pectic polysaccharides of sugar beet. *Food Hydrocoll.* **1**, 71–88.

Delzenne, G.A. (1969). Synthesis and photocrosslinking of light-sensitive polymers. *Eur. Polym. J. (Suppl)*, 55–91.

De Vries, J.A. Rombouts, FM., Voragen, A.G.J., and Pilnik, W. (1982). Enzymic degradation of apple pectins. *Carbohydr. Polym.* **2**, 25–33.

Doublier, J.L., and Thibault, J.F. (1985). Unpublished data.

Fry, S.C. (1983). Feruloylated pectic substances from the primary cell wall: Their structure and possible functions. *Planta* **157**, 111–123.

Geissmann, T., and Neukom, H. (1973). On the composition of the water-soluble wheat flour pentosans and their oxidative gelation. *Lebensm. Wiss. u. Technol.* **6**, 59–62.

Guillon, F., and Thibault, J.F. (1987). Characterization and oxidative cross-linking of sugar beet pectins after mild acid hydrolysis and arabanases and galactanases degradation. *Food Hydrocoll.* **5/6**, 547–549.

Guillon, F., and Thibault, J.F. (1988). Further characterization of acid- and alkali-soluble sugar-beet pulp pectins. *Lebensm. Wiss. u. Technol.* **21**, 198–205.

Guillon, F., and Thibault, J.F. (1989). Methylation analysis and mild acid hydrolysis of the hairy fragments from sugar-beet pectins. Structural investigations of the neutral sugar-side chains of sugar beet pectins, part 1. *Carbohydr. Res.* **190**, 85–96.

Guillon, F., and Thibault, J.F. (1990). Oxidative crosslinking of chemically and enzymatically modified sugar-beet pectin. *Carbohydr. Polym.* **12**, 353–374.

Guillon, F., Thibault, J.F., Rombouts, F.M., Voragen, A.G.J., and Pilnik, W. (1989). Enzymic hydrolysis of the hairy fragments from sugar-beet pectins, Structural investigations of the neutral sugar-side chains of sugar-beet pectins, Part 2, *Carbohydr. Res. 190*, 97–108.

Keenan, M.H.J., Belton, P.S., Mattew, J.A., and Howson, S. J. (1985). A ^{13}C-n.m.r. study of sugar-beet pectin. *Carbohydr. Res.* **138**, 168–170.

Kertesz, Z.I. (1951). Beet pectin. *In* "The Pectic Substances" (Z.I. Kertesz, ed.), pp. 463–466. Interscience, New York.

Le Quéré, J.M., Baron, A., Segard, E., and Drilleau, J.F. (1981). Modification of sugar-beet pectin during processing. *Sci. Aliments* **1**, 501–511.

Michel, F., Thibault, J.F., Mercier, C., Heitz, F., and Pouillaude, F. (1985). Extraction and characterization of pectins from sugar-beet pulp. *J. Food Sci.* **50**, 1499–1500.

Phatak, L., Chang, K.C., and Brown, G. (1988). Isolation and characterization of pectin in sugar-beet pulp. *J. Food Sci.* **53**, 830–833.

Pippen, G.L., Mac Cready, R.M., and Owens, H.S. (1950). Gelation properties of partially ace-tylated pectins. *J. Amer. Chem. Soc.* **72**, 813–816.

Rombouts, F.M., and Thibault, J.F. (1986a). Feruloylated pectic substances from sugar-beet pulp. *Carbohydr. Res.* **154**, 177–188.

Rombouts, F.M., and Thibault, J.F. (1986b). Enzymatic and chemical degradation and the fine structure of pectins from sugar-beet and pulp. *Carbohydr. Res.* **154**, 189–204.

Rombouts, F.M., and Thibault, J.F. (1986c). Sugar-beet pectins: Chemical structure and gelation through oxidative coupling. *In* ACS Symposium Series, No. 310, Chemistry and Functions of Pectins (M.L. Fishman and J.J. Jen, eds.), pp. 49–60. American Chemical Society, Washington, DC.

Rombouts, F.M., Thibault, J.F., and Mercier, C. (1983). Procédé de modification des pectines de betterave, produits obtenus et leur applications. French Patent No. 83 07208; European Patent No. 603 318; U.S. Patent No. 4 6 72 034.

Rombouts, F.M., Voragen, A.G.J., Searle-Van Leeuwen, M.F., Geraeds, C.C.J.M., Schols, H.A., and Pilnik, W. (1988). The arabinanases of *Asperigillus niger*—purification and characterisation of two α-L-arabinosidases and an endo-1,5-α-arabinanase. *Carbohydr. Polym.* **9**, 25–48.

Selvendran, R.R. (1985). Developments in the chemistry and biochemistry of pectic and hemicellulosic polymers. *J. Cell Sci. (Suppl.)* **2**, 51–88.

Thibault, J.F. (1983). Enzymatic degradation and β-elimination of the pectic substances in cherry fruits. *Phytochemistry* **22**, 1567–1571.

Thibault, J.F. (1986). Some physicochemical properties of sugar-beet pectins modified by oxidative cross-linking. *Carbohydr. Res.* **155**, 183–192.

Thibault, J.F. (1988). Characterization and oxidative cross-linking of sugar-beet pectins extracted from cossettes and pulps under different conditions. *Carbohydr. Polym.* 8, 209–223.

Thibault, J.F., Garreau, C., and Durand, D. (1987). Kinetics and mechanism of the reaction of ammonium persulfate with ferulic acid and sugar-beet pectins. *Carbohydr. Res.* **163**, 15–27.

Thibault, J.F., and Rombouts, F.M. (1986). Effects of some oxidizing agents and especially ammonium persulfate on sugar-beet pectins. *Carbohydr. Res.* **154**, 205–216.

Thomson, R.A.M. (1983). Methods of polymerisation for preparation of water-soluble polymers. *In* "Chemistry and Technology of Water-Soluble Polymers" (C.A. Finch, ed.), pp. 31–70. Plenum, New York.

CHAPTER 8

Pectinesterase

T. Sajjaanantakul
Department of Food Science and Technology
Faculty of Agro-Industry
Kasetsart University
Bangkok, Thailand

Leigh Ann Pitifer
Department of Food Science and Technology
Cornell University
Geneva, New York

I. INTRODUCTION

From the perspective of the food processor and consumer, the ability to dictate biochemical outcomes in food from the incorporation of pectinesterase (PE), also called pectylhydrolase and pectinmethylesterase, was an important development in the fruit and vegetable industry. In this chapter, the occurrence, extraction, assay, properties, and applications of PE are discussed.

II. OCCURRENCE

PE is synthesized by plants, molds, and bacteria. Only a few yeasts have been reported to produce it. It is not known to be synthesized by animals.

A. Plant Pectinesterase

PE is present in many plants and microorganisms (Versteeg, 1979). It was first identified in carrot by Fremy (1840), but this discovery was overlooked until 1925 when Kopaczewski (1925) reported its unique requirement of alkali salts. The isolates reported to date deesterify consecutive galacturonate monomers along the chain. More recent investigations were undertaken by Lineweaver and colleagues (Lineweaver and Ballou, 1943, 1945; Lineweaver and Jansen, 1951). Plant PE is inactivated below pH 7.

B. Microbial Pectinesterase

Microbial PE is an extracellular, inducible, or constitutive enzyme (depending on the origin). The substrate may be pectin, pectic acid, D-galacturonate or other carbohydrate material (Phaff, 1947; Bateman and Beer, 1965; Ogundero, 1988). The literature (for example, Perley and Page, 1971; Waggoner and Dimond, 1955; Starr and Nasuno, 1967; Ogundero, 1988) emphasizes specific enhancers, substrates, conditions, etc., for optimal production of PE by individual microorganisms.

C. Plant versus Microbial Pectinesterase

There are marked differences between microbial and plant PE. Fungal PE is generally more resistant to chemical agents and has an isoelectric point (pI) and pH optimum in the acidic range, in contrast to the alkaline pH range for plants. Microbial PE is more sensitive to heat and is less affected by salt (Lineweaver and Jansen, 1951; Kertesz, 1955). Some fungal PE catalyze reactions in a random manner. The properties of some microbial PE have been summarized by Fogarty and Kelly (1983).

III. INDUSTRIAL PRODUCTION

The industrial production of pectic enzymes was reviewed by Fogarty and Ward (1974). In most instances, PE is produced by submerged or semisolid fermentation; however, it can be produced by surface-culture techniques, using *Scler-*

multifermentans was reported (Macmillan and Vaughn, 1964; Lee *et al.*, 1970). The PE in crude enzyme extracts can be purified, and the activity increased by chromatographic techniques (Brady, 1976; Delincee, 1978; Rombouts *et al.*, 1979; Versteeg *et al.*, 1980).

PE extraction is complicated by its association with the cell wall, whence it is dissolved or desorbed by an alkaline (pH $\geq$ 7), 5% (0.86M) solution of NaCl (MacDonnell *et al.*, 1945; Jansen *et al.*, 1960a; Nakagawa *et al.*, 1971; Pressey and Avants, 1972). Plant PE should be extracted above pH 7.0, in order to avoid its inactivation. Hultin and Levine (1963) reported that PE of banana pulp was completely extracted with water, while the salt solution showed activity after 10 successive extractions. They suggested the possibility that limited binding sites were available to PE on the particulate matter. Suspensions of orange cell wall at pH 4 were capable of binding added PE up to 15 times the amount of that naturally present (Jansen *et al.*, 1960a). Goldberg (1984) extracted *ionically bound* PE from mung bean cell wall with 1M NaCl, then *covalently bound* PE with the assistance of cellulase and pectinase, stating that the binding strength of the PE to the cell walls increased with the maturity of the hypocotyl cell. The data confirmed that PE is an extracellular enzyme, inasmuch as more than 95% of the activity was located in the cell-wall fraction, and only 1–2% was located in the cytoplasmic fraction. Wicker *et al.* (1988), having extracted a thermostable PE from orange only after cell-wall degradation by pectinase and cellulase, surmised that the enzyme was entrapped in the cell-wall matrix by divalent cation bridges. The results supported the inference that PE is in loose association with cell-wall materials, and that minor amounts are tightly bound and extracted only after cell-wall degradation.

Purified PE may be obtained by first salting out the enzyme with ammonium sulfate, or by precipitating it with acetone and redissolving and dialyzing the precipitate before ion-exchange and size-exclusion chromatography (Markovic and Slezarik, 1969; Markovic, 1974; Delincee and Radola, 1970). Different columns in series, e.g., DEAE-Sephadex A-50, DEAE-Biogel A, SP-Sephadex C50 and Sephacryl-S 200 (Baron *et al.*, 1980), usually help to isolate PE in a very pure state.

IV. COMPOSITION

Analysis of tomato and orange PE showed the presence of comparatively high concentrations of alanine, glycine, aspartic acid, and glutamic acid, and the absence of cysteine, cysteic acid, and hydroxyproline (Markovic and Slezarik, 1969; Nakagawa *et al.*, 1970; Manabe, 1973a). Versteeg (1979) found a small quantity of cysteic acid in orange PE. The extraction of a lipid fraction from a purified PE suggested to Lee and Macmillan (1968) that PE was a lipoprotein. An electrophoresis band was stained only faintly by aniline black (for protein), but it gave a prominent color with oil red O (for lipid) (Lee and Macmillan, purified PE suggested to Lee and Macmillan (1968) that PE was a lipoprotein. An electrophoresis band was stained only faintly by aniline black (for protein), but it gave a prominent color with oil red O (for lipid) (Lee and Macmillan, 1968). Delincee (1976) suspected a glycoprotein. Theron *et al.* (1977a, b) found a glycoprotein in plum and tomato PE, but a similar isolate of Markovic (1974) from tomato was neither a lipoprotein nor a glycoprotein, nor was a banana PE a lipoprotein (Markovic *et al.*, 1975). Moustacas *et al.* (1986) argued that a purified soybean PE was not a glycoprotein, because its polyacrylamide gel pattern did not show color after staining for glycoprotein with the Zacharius stain (Zacharius *et al.*, 1969). Seymour *et al.* (1989) detected a carbohydrate moiety in purified PE from grapefruit pulp by dansyl hydrazine staining. The conflicting data are evidence that the composition of PE has not yet been completely elucidated. The enzyme's close association with lipo- and glyco-proteins in the plasmalemma membrane or the primary and secondary cell walls is a complicating factor in any compositional analysis.

V. PROPERTIES AND CHARACTERISTICS

A. Methanol Generation

Methanol is a major product of the action of PE on pectin (Fig. 1). The amount generated during fruit and vegetable fermentation depends on the variety, pectin content, and level of active PE in the substrate. The accumulation is not normally

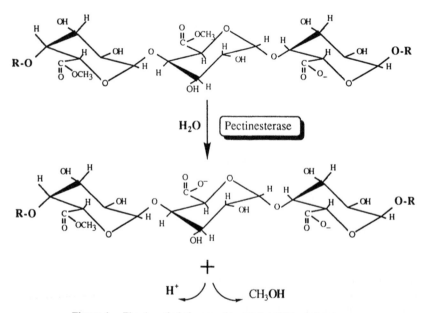

Figure 1   The demethylation reaction catalyzed by pectinesterase.

harmful to humans. Methanol content may be subject to legal limits in some countries (Al-Delaimy *et al.*, 1966; Lee *et al.*, 1975b, 1979b; Drzazga *et al.*, 1987). Heat treatment before fermentation inactivates PE and prevents methanol accumulation. Low-temperature blanching might not suffice to prevent methanol fermentation (Bartolome and Hoff, 1972b). According to Ishii and Yokotsuka (1972), the increase observed in apple juice clarified by commercial pectinase might have been owing to PE in the enzyme preparation.

B. Specificity

PE's hydrolytic action is specific for the methylester groups of partially methoxylated polygalacturonic acid (MacDonnell *et al.*, 1950; Deuel and Stutz, 1958), but not toward other polyuronides, as, for example, fully methoxylated pectin or methyl alginate. Neither does it hydrolyze the glycyl and glyceryl esters, ethyl acetate, and ethyl oxalate (Deuel, 1947; Mills, 1949; MacDonnell *et al.*, 1950; Deuel and Stutz, 1958; Manabe, 1973b). Ethyl polygalacturonates are hydrolyzed at a rate that is 6–16 % slower than that of the methyl galacturonates. The allyl, propyl, and propargyl esters are hydrolyzed at very slow rates. Pectin containing a 65–75 % degree of esterification gives the highest activity.

Kertesz (1937) first suggested that macromolecular size may be important in the enzymic deesterification of pectin. Dimers and trimers of D-galacturonic acid methyl esters are not hydrolyzed, but methylesters of polygalacturonic acid with a degree of polymerization (DP) $\geq$ 10 are (McCready and Seegmiller, 1954). Markovic *et al.* (1983) demonstrated that the shortest polymer chain of partially esterified pectin that can be hydrolyzed by tomato PE contained DP = 5; for *Aspergillus foetidus* PE, the DP = 2. The reaction rate for both enzymes increased with increasing molecular weight of the esterified oligomers.

C. Mode of Action

Most commercial pectinases are a mixture of PE, polygalacturonase, and polygalacturonate lyase, and pectin degradation therefore entails depolymerization as well as demethylation.

Purified PE appears to act on polyuronides containing an unesterified carboxyl group adjacent to a methylated carboxyl group. It does not act on the ester bond of a monomer that is between two adjacent, esterified monomers. Also it does not completely deesterify pectin; the action stops at a certain degree of esterification (Mills, 1949; Evans and McHale, 1968; Speirs *et al.*, 1980; Kohn *et al.*, 1983; King *et al.*, 1986), possibly limited by the effect of small substrate polymer size.

1. Plant Pectinesterase

Plant PE deesterifies pectin linearly, creating blocks of free carboxyl groups, ultimately resulting in a calcium sensitive form of pectin. The sequential action is believed to start on the methylester groups next to free carboxyl groups (Speiser *et al.*, 1945; Lineweaver and Jansen, 1951; McCready and Seegmiller, 1954; Solms and Deuel, 1955; Heri *et al.*, 1961; Kohn *et al.*, 1968, 1983). This belief is based partly on the fact that an alkali-deesterified (randomly deesterified) pectin is a better substrate for orange PE than is nonrandomly esterified pectin of the same DP (Deuel and Stutz, 1958).

Lee and Macmillan (1970), studied the cooperative action of *C. multifermentans* exo-polygalacturonate lyase and tomato PE on pectin. The former enzyme cleaves glycosidic bonds sequentially from the reducing end. About 50% of the PE activity was believed to occur at the reducing end and 50%, within the galacturonan chain. Lyase activity increased immediately after addition of PE (Fig. 2). A complex and coordinate action by PE and polygalacturonase lyase on pectin was reported for *C. multifermentans* (Miller and Macmillan, 1970; Sheiman *et al.*, 1976). The PE from this microorganism was also believed to deesterify pectin from the reducing end (Lee *et al.*, 1970). Miller and Macmillan

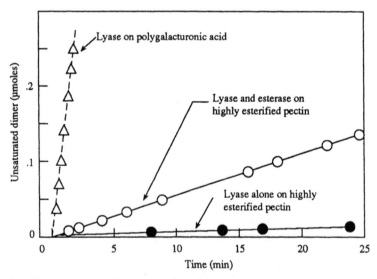

Figure 2   Effect of tomato pectinesterase on the activity of polygalacturonate lyase acting on highly esterified pectin. Reaction mixtures contained 0.5% polygalacturonic acid methyl glycoside (solid lines), 0.005 *M* CaCl$_2$, 0.033 *M* phosphate buffer (pH 7.0), and 0.25 unit of lyase in a final volume of 3.0 ml. In addition, the reaction mixture represented by the middle curve contained 0.031 unit of tomato pectinesterase in a final volume of 3.3 mol. With permission from Lee and Macmillan (1970).

(1971) stated that more than half the activity of PE from *Fusarium oxysporum* took place at the reducing end, while the remaining activity took place at other loci along the pectin molecule. Versteeg (1979) questioned these conclusions, suggesting that if the PE had a high turnover rate and acted toward the reducing end, the observed increase in the lyase activity would occur. No studies with other plant PE have been designed to proportionate the activity between the reducing end and the interior of the pectin chain, as in the case of tomato PE.

Although plant PE, acting linearly along the pectin chain, results in heterogeneity with respect to methoxyl content (Heri *et al.*, 1961), the presence of a block distribution of methoxyl groups in pectins of some plants (large radish, carrot, cauliflower, and orange) was, however, recently reported (Anger and Dongowski, 1985; Tuerena *et al.*, 1984). This homogeneity may be owing to the PE action in plant tissues (Kohn *et al.*, 1985).

2. Microbial Pectinesterase

The action of microbial PE may be blockwise or random. Markovic and Kohn (1984) indicated that *Trichoderma reesei* PE deesterified pectin in a blockwise manner. Ishii *et al.* (1979, 1980), Baron *et al.* (1980) and Kohn *et al.* (1983) reported a random deesterification by PE from *A. foetidus, Aspergillus japonicus, Aspergillus niger, F. oxysporum, Penicillium chrysogenum, Sclerotinia archidis,* and *Sclerotinia libertiana*.

D. Kinetics of Pectinesterase Activity

PE generally follows Michaelis–Menten kinetics, but it has isozymes with different pH optima and kinetic properties. The Michaelis constant (K_m) and inhibition constant (K_i) decrease with increasing pH (Versteeg *et al.*, 1978). Orange PE has K_m = 1.5–4.0 mM galacturonate (Solms and Deuel, 1955), similar at the upper limit to those of tomato PE with K_m = 4.0 mM (Lee and Macmillan, 1968) and apple PE with K_m = 4.9 mM (Castaldo *et al.*, 1989). *C. multifermentans* PE was reported to have K_m = 2.5 mM (Lee *et al.*, 1970). Goldberg (1984) found that PE generated from mature mung bean hypocotyl cells exhibited an acidic pH shift, an increase in binding strength to the cell wall, and a higher apparent K_m, which implied that the PE associated with new growing cells exhibited a higher affinity to the substrate. The activation energy, i.e., the energy needed for binding the substrate to the reaction site, approximates 5800–6000 cal/mol (Owens *et al.*, 1944; Speiser *et al.*, 1945; Lee and Wiley, 1970). It should be remembered that the activity of PE is affected by a number of factors, e.g., pH, cations, temperature, and purity of the enzyme.

E. Effect of Cations

Kopaczewski (1925) first reported the effects of salts on PE activation (Lineweaver and Jansen, 1951). Enhancement of activity occurred in the presence of either monovalent or divalent cations, but not anions (Lineweaver and Ballou, 1945). There is an optimal cation concentration for optimal activity under specific conditions. The activation is more prominent at lower pH, and is independent of enzyme and substrate concentrations (Lineweaver and Ballou, 1945; Pithawala et al., 1948). Lee and Macmillan (1968) reported that cations increased PE activity but were not a requirement for that activity. High salt concentrations depress the activity (Lineweaver and Ballou, 1945; Pithawala et al., 1948; Van Buren et al., 1962). Interestingly, Lineweaver and Ballou (1945) showed that alfalfa PE activity was inhibited by $0.05\ M\ Ca^{+2}$, but not by Mg^{+2}, even at a concentration of $0.1\ M$. Nakagawa et al. (1970) reported that tomato PE was inhibited by $MgCl_2$ above $0.10\ M$, and by NaCl above $1\ M$, at pH 7.0. Suppression of tomato PE activity by $0.01\ M\ MgCl_2$ and other cations ($0.5\ M$ NaCl, $0.10\ M$ KCl, $0.01\ M\ CaCl_2$) was demonstrated by Roeb and Stegemann (1975), who also observed that Fusarium sulphureum PE was completely inhibited by $0.001\ M\ MgCl_2$. These last authors suggested that the differences in sensitivity to $MgCl_2$ could be used for differential diagnosis of plant and fungal PE. Coniothyrium diplodiella PE was inhibited by tri- and tetravalent cations ($FeCl_3$, $SnCl_4$) (Endo, 1964). According to Roeb and Stegemann (1975), changes in the physical properties of pectin that limit the availability of the substrate to the enzyme, e.g., changes wrought by cross-linking, may be responsible for the inhibition. Differences in the effects initiated by different cations suggest that they may play different roles in PE activity. The effect seems also to depend on the source and hence the nature of the PE itself, and on pH.

Pectic acid has an inhibitory effect on PE that is lessened by cations, a mode of action assumed by Lineweaver and Ballou (1945) to be a cation–carboxyl complexation of the acid. They also suggested that cations at higher pH were less effective because at higher pH, PE is less protonated and thus has less tendency to form an inactive complex with an anion inhibitor. A diagram of the effect of cations and pH on PE activity, proposed by Lineweaver and Ballou (1945), is drawn in Fig. 3. This mechanism was later supported by information on the pI of PE and its behavior with pectic acid under different pH and salt concentrations (Termote et al., 1977; Versteeg et al., 1978). Generally, microbial PE is less affected by cations than is plant PE. The pI of tomato PE was reported to range between 7 and 9.3, with a major isozyme having a pI 8.6 (Delincee, 1976).

PE exhibits a cationic function in the absence of salt at pH 4.75, since it is readily adsorbed on carboxymethylcellulose and negatively charged diatomaceous earths (Celite) (Lineweaver and Ballou, 1945; MacDonnell et al., 1945; Waggoner and Dimond, 1955). In the presence of salt, the adsorption is counteracted by anions (Lee and Macmillan, 1968; Theron et al., 1977b).

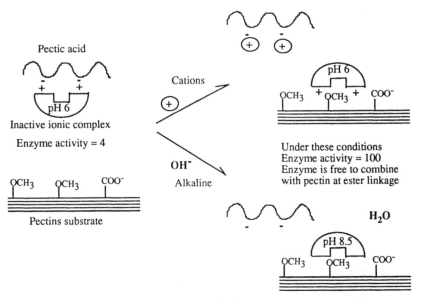

Figure 3 A diagram of the effect of cations and pH on pectinesterase activity. With permission from Lineweaver and Ballou (1945).

In the activation of PE, divalent cations were 5–20 times more effective than monovalent cations. Goldberg (1984) demonstrated that the K_m in the presence of Mg^{+2} was approximately 15 times lower than that in the presence of Na^+ and K^+, and that Mg^{+2} enhanced the enzyme affinity for pectin. Inasmuch as Vijayalakshmi et al. (1979) indicated an involvement of the terminal amino end-group in the protein of PE at the active site, cations may well remove an anionic inhibitor from this amino group, and thereby permit the enzyme to be activated. Other cations that have an activation effect are ammonium on alfalfa PE (Lineweaver and Ballou, 1945) and lithium and ammonium on fungal PE (Endo, 1964; Yoshihara et al., 1977). A onefold to twofold increase in activity due to cation activation was observed for several microbial PE (Baker and Walker, 1962; Endo, 1964; Oi and Satomura, 1965; Miller and Macmillan, 1970; Kimura et al., 1973; Baron et al., 1980).

F. Inhibitors and the Active Site

Besides the inhibitory effect of pectic acid and a high concentration of cations previously discussed, PE activity is also inhibited by a block of free carboxyl groups in the pectin chain (Lineweaver and Ballou, 1945). In many cases, the inhibition is a competitive type of reaction (Lee and Macmillan, 1968; Termote *et al.*, 1977; Versteeg *et al.*, 1978; Moustacas *et al.*, 1986). Inhibition was less pronounced in the presence of cations (Lineweaver and Ballou, 1945). Lee and Macmillan (1968) reported K_i = 7 mM D-galacturonic acid for purified tomato PE. Termote *et al.* (1977) found that a minimal DP of 8 to 15 was required for pectate to inhibit orange PE. Inhibition of *T. reesei* PE by pectic acid with K_i = 2.54 mM D-galactopyranosiduronate was reported by Markovic *et al.* (1985). Chang *et al.* (1965), Hultin *et al.* (1966), and Lee and Wiley (1970) reported a noncompetitive inhibition of PE by sugars in papaya, banana, and apple. All PE in fruits are inhibited by high sugar concentrations. Inhibition amounted to as much as a 40% loss of activity at 15% sucrose (Lee and Wiley, 1970). Low-molecular-weight polyols, e.g., glycerol, glucose, and maltose, inhibited banana PE (Brady, 1976).

Partial inhibition of purified tea leaf PE by catechins was found by Ramaswamy and Lamb (1958). According to Hall (1966), tannic acid was the most potent inhibitor of the phenolic compounds tested. These included gallic, shikimic, chlorogenic, and caffeic acids. The inhibition of an apple PE by phenolic substances and a pear leuco-anthocyanin was noted by Pollard *et al.* (1958). They suggested that the effect might be a consequence of protein precipitation, inasmuch as addition of gelatin to the phenolic substrate decreased the inhibition. The concentration of the phenolic compounds and the substrate were important factors in the inhibition. The inhibition by phenolic compounds is reversible (Fuchs, 1965; Hall, 1966).

Anionic detergents inhibit PE activity (McColloch and Kertesz, 1947; Hultin and Levine, 1963; Pressey and Avants, 1972). This type of inhibition depends on the chain length of the detergent fatty acids. Maximal inhibition occurs with a chain length of 14 carbons.

Tomato PE was inhibited by iodine but not by organophosphates and carbamates leading to the conclusion that the enzyme was not a serine esterase (Markovic and Patocka, 1977). Iodine inhibition is non-competitive and irreversible. Iodine had very little effect on crude PE, but completely inhibited the purified form. Thus, the degree of inhibition would seem to depend on the purity of the enzyme.

One molecule of iodine was claimed to interact with one catalytic site on PE (Markovic and Patocka, 1977). Since a concentration of 10^{-3} M iodine oxidized sulfhydryl groups but did not inhibit activity, and no cysteine or methionine, and only traces of histidine were detected in tomato PE (Markovic and Slezarik, 1969; Nakagawa *et al.*, 1970), it is likely that the active site on PE contains a tyrosine residue. Markovic and Patocka (1977) are of the opinion that inhibition might be due to iodination of the tyrosine residue, however, possible involvement of histidine at the active site cannot be ruled out.

The role of tyrosine and possibly histidine at the active site was supported by Markovic and Machova (1985), when immobilization of PE through preferential binding *via* the tyrosine hydroxyl group resulted in complete loss of activity. Vijayalakshmi *et al.* (1979) suggested that the terminal NH_2 of the PE is also involved in the active site, because immobilization through amino end coupling resulted in an inactive enzyme complex.

Some fungal PE are inhibited by ammonium (Kimura *et al.*, 1973) and mercury ions (Yoshihara *et al.*, 1977). In contrast to plant PE, they are more resistant to inhibition by detergents and chemical compounds such as iodoacetic acid, phenylthiourea (McColloch and Kertesz, 1947), and tannins (Mills, 1949).

G. Molecular Forms and Optimal pH

Purified PE was found to have molecular weights in the range of 22,000 to 37,000. These numbers vary with the source and the analytical method of determining them (Markovic and Slezarik, 1969; Miller and Macmillan, 1971; Miyairi *et al.*, 1975; Brady, 1976; Delincee, 1976; Theron *et al.*, 1977b; Markovic, 1978; Versteeg *et al.*, 1978). A PE isolated from Navel orange was reported at 54,000 (Versteeg, 1979), and another from apple, at 55,000 (Castaldo *et al.*, 1989).

Hultin and Levine (1963) reported three forms of PE from banana pulp. Subsequently, multiple forms and isozymes have been reported from bananas (Hultin *et al.*, 1966; Brady, 1976; Markovic *et al.*, 1975), carrots (Markovic, 1978), oranges (Evans and McHale, 1968; Versteeg *et al.*, 1978, 1980), plums (Theron *et al.*, 1977a), and tomatoes (Lee and Macmillan, 1968, 1970; Pressey and Avants, 1972; Delincee, 1976).

Cross-linked pectate affinity chromatography, developed by Rexova-Benkova (1972) and Rexova-Benkova and Tibensky (1972), and modified by Rombouts *et al.* (1979), is capable of separating (two) PE isozymes in orange. The PE I form of Versteeg *et al.* (1978) had a lower optimal pH and was more active in that range than was the PE II form. K_m, K_i and temperature stability were also different. The two forms were constituted by a single polypeptide with molecular weight of approximately 36,200. The pI of orange PE I and PE II were 10.05 and 11, respectively. It is noteworthy that soybean PE displayed two interconvertible ionization states, based on the classical Dixon model (Moustacas *et al.*, 1986).

The optimal pH for PE activity depends on the PE origin. Plant PE shows a broad optimum between pH 6.5 and 9.5. The optimum can shift to the acid side of the pH scale in the presence of cations (Lineweaver and Ballou, 1945) and at proper salt concentrations (Hultin and Levine, 1963; Lee and Macmillan, 1968). Above pH 8.5, PE activity cannot be determined accurately without correction for alkali deesterification.

Optimal fungal PE activity is in the pH range 4.0–5.2, with the exception of *Acrocylindrium sp.* (Kimura *et al.*, 1973) and *Rhizoctonia solani* (Bateman, 1963) PE that have their optimum between 6.0 and 7.5. Bacterial PE show an optimal pH in the range 7.0–9.0 (Miller and Macmillan, 1970). It is conceivable that multiple forms of the same PE may be responsible for differences in optimal pH, as well as in pI, cationic activation, heat stability, substrate affinity, cell-wall binding affinity, etc. Rexova-Benkova and Markovic (1976) listed optimal pH for some plant and microbial PE.

H. Heat Activation and Inactivation

Purified PE is more sensitive to heat than is crude PE (Pollard and Kieser, 1951). Acidic pH and low fruit-pulp content increase the thermal response (Kertesz, 1939; Joslyn and Sedky, 1940; Rouse and Atkins, 1952; Van Buren et al., 1962; Nath and Ranganna, 1977). PE from F. oxysporum, a mold, is much more heat stable than is PE from C. multifermentans, a bacterium (Miller and Macmillan, 1971).

Owusu-Yaw et al. (1988) were able to inactivate PE in orange juice completely by lowering the pH to 2.0 with hydrochloric acid. The process avoided the disadvantages of heat treatment, but exposed the juice to a loss of ascorbic acid and sensory attributes.

Activation of PE in tissue, caused by relatively low temperatures (50–76°C), was observed in carrot (Lee et al., 1979a), cauliflower (Hoogzand and Doesburg, 1961), potato (Bartolome and Hoff, 1972b), and snap bean (Van Buren et al., 1962). There was a doubling of the rate in several fruits and vegetables at 50°C relative to 30°C (Vas et al., 1967). Potato PE was inactivated below 50°C; above 70°C, it was destroyed (Bartolome and Hoff, 1972b). The optimal temperature is near 55°C (Hultin and Levine, 1963; Lee and Wiley, 1970).

A purified PE was completely inactivated by heating between 65 and 90°C for 1 to 5 min (Nakagawa et al., 1970; Lee and Wiley, 1970; Markovic, 1978; Versteeg et al., 1980). A heat-inactivated PE did not regenerate upon frozen storage (Van Buren et al., 1962). In the 20–50°C interval, the Q_{10} (rate change with 10°C increase) for snap bean PE was 1.4 (Van Buren et al., 1962), similar to that from apple (Lee and Wiley, 1970), tomato (McColloch and Kertesz, 1947), and commercial, fungal PE (Calesnick et al., 1950). Thermostable forms of orange PE, inactivated by heating for about 1 min at greater than 90°C, were isolated by Versteeg et al. (1980) and Wicker et al. (1988).

Nakagawa et al. (1970) interpreted an inflection at 27°C in an Arrehnius plot as evidence of either two different isozymes or thermal inactivation by two different mechanisms. Nonlinearity of the heat inactivation line from PE extracted from orange juice pulp was demonstrated by Wicker and Temelli (1988). They reported D values (time to inactivate 90% of the enzyme) at 90°C equaling 0.225 sec, and 32 sec, and Z values (increase in temperature needed to have a 10 times faster rate of heat inactivation) equaling 10.8°C and 6.5°C, for heat-sensitive and heat-stable PE, respectively. The Z values were in agreement with those of the purified orange isozymes (Versteeg et al., 1980). Similar Z values have been recommended for citrus juice processing (Eagerman and Rouse, 1976). The

Table I Thermal activation of pectinesterase

Source	Temp. (°C)	Time (min)	pH	Reference
Plants				
Carrot tissue	76	10	6.2	Lee *et al.* (1979a)
Cauliflower	70	15	6.0	Hoogzand and Doesburg (1961)
Green bean	71	9.5	—	Kaczmarzyck *et al.* (1963)
Snap bean	76	2–4	6.5[a]	Van Buren *et al.* (1960)
Tomato, crude PE	50–60	10	6.5	Speiser *et al.* (1945)
Tomato	100	30 sec	—	Hsu *et al.* (1965)
Mold				
Coniothyrium diplodiella	45	15	4.4	Endo (1964)

[a] average natural pH.

temperature–time relationships governing PE activity are listed in Tables I and II, in which pH effects and differences in heat stability are given for crude, purified, and native PE. Observe that microbial PE has a lower optimal temperature and is more heat labile than is plant PE.

Activation in intact tissue by heating was suggested by Taylor (1982) for commercial production of low-methoxyl pectin. In cases where citrus pomace is dried for feed, addition of calcium hydroxide has been recommended to activate PE, in order to coagulate calcium pectate, thus facilitating water removal by a screw press (Pilnik and Rombouts, 1978).

I. Radiation Effect

Gamma-irradiation (γ) with ^{60}Co, followed by heating, was found to be an effective way to inactivate commercially prepared PE (Delincee, 1970). However, tomato juice and whole fruits provided protection from a 1000 krad dose (Vas *et al.*, 1968). The same dosage applied to cherries effected a decrease in PE activity (Somogyi and Romani 1964b). Higher dosages cause undesirable fruit and fruit-juice flavors. Paradoxically, 100–300 krads resulted in higher PE activity in cherries and mangoes (Somogyi and Romani, 1964b; Dennison and Ahmed, 1967). At such low levels of irradiation, activity might increase or decrease, depending on the location of PE in the fruit.

VI. ASSAY

PE activity was once measured by analysis of calcium–pectin gels (Fremy, 1840), and by viscosity changes in pectin solutions (Weurman, 1954). These methods

Table II   Thermal inactivation of pectinesterase

Source	Temp. (°C)	Time (min)	pH	Reference
Plants				
Apple, crude	80	10	6.5	Lee and Wiley (1970)
crude	90	5	6.5	Lee and Wiley (1970)
purified	65	5	3.5	Pollard and Kieser (1951)
purified	90	1	6.5	Castaldo *et al.* (1989)
Apple juice	80	15	3.8	Pollard and Kieser (1951)
Grapefruit juice	90.5	3 sec	3.5	Rouse and Atkins (1952)
10% pulp	90	3 sec	3.5	
38° Brix	90.6	12 sec	3.3	Atkins and Rouse (1954)
Orange juice				
5% pulp	90.5	15 sec	4.1	Rouse and Atkins (1952)
10% pulp	90.5	1	4.1	Rouse and Atkins (1952)
42° Brix	90.6	12 sec	3.6	Atkins and Rouse (1954)
Orange, purified PE I	70	50 sec	4.0	Versteeg *et al.* (1980)
purified PE II	60	50 sec	4.0	Versteeg *et al.* (1980)
purified high MW	90	50 sec	4.0	Versteeg *et al.* (1980)
Snap bean, crude	75	15	5.5	Van Buren *et al.* (1962)
crude	75	5	5.0	Van Buren *et al.* (1962)
Tea leaf, partially purified	80	5	7.0	Ramaswamy and Lamb (1958)
Tobacco leaves, crude	80	5	8.0	Holden (1946)
Tomato juice	80	45 sec	4.4	Kertesz (1939)
	55	2	2.5	Kertesz (1939)
	50	2	1.1	Kertesz (1939)
Tomato	100	90 sec		Hsu *et al.* (1965)
Tomato, purified	38	30	7.0	Miller and Macmillan (1970)
purified	65	5	7.0	Nakagawa *et al.* (1970)
purified PE	80	45.5 sec	8.0	Pithawala *et al.* (1948)
Bacterium Clostridum multifermentans	38	30	7.0	Miller and Macmillan (1970)
Molds				
Coniothyrium diplodiella	55	10	5.0	Endo (1964)
Sclerotinia libertiana	80	5	4.0	Oi and Satomura (1965)
Commercial fungal PE	62	30	3.5	Calesnick *et al.* (1950)

are sensitive to a number of factors, e.g., macromolecular size and ionic strength, and consequently they came into disrepute. PE activity is now measured more accurately by methanol content of the hydrolyzate, or by titration of the free carboxyl groups.

A. Titration and Manometry

Free carboxyl groups appear as a result of PE deesterification, and carboxyl groups are easily decarboxylated. Either of these two processes is applicable, through titrimetry and manometry, to the determination of PE activity (Kertesz, 1937; Kiermeier, 1949; Mills, 1949). Results from the two methods compared favorably (Glasziou and Inglis, 1958). Titrimetry and its modifications (Lineweaver and Ballou, 1945; Rouse and Atkins, 1955; Vas *et al.*, 1967; Lee and Macmillan, 1968; Delincee and Radola, 1970) are the preferred methods, because of their simplicity.

In a typical PE assay, a pectin dispersion (0.4–1.0%) containing a definite concentration of NaCl (0.05–1.0 M) is held at constant temperature (30°C) and pH, which is kept constant by titrating the medium with 0.005–0.5 M NaOH. Alkali consumption with time is measured, no longer with pH indicators (Kertesz, 1937; Lineweaver and Ballou, 1945), but with a glass-electrode. The PE unit is expressed in milliequivalents of ester hydrolyzed per unit time per mg of enzyme at a specified temperature and pH (pH 7). Above pH 8.0, a correction for alkaline deesterification must be made. The computed activity should be evaluated with caution, because the deesterification reaction is nonstoichiometric.

Cations, we have understood, play a significant role in PE activity and hence in the PE-activity assay. Tannins and other phenolic compounds should be absent. Polyvinylpyrrolidone, 2-mercaptobenzothiazole or another phenol inhibitor is added to the crude extract to inhibit polyphenoloxidase (Nagel and Patterson, 1967; DeSwardt and Maxie, 1967; Palmer and Roberts, 1967; Brady, 1976; Buescher and Furmanski, 1978; Ben-Arie *et al.*, 1979).

B. pH and Color Indicators

In some instances, excepting those in which the substrate pH is close to the pK_a, the initial rate of PE activity can be monitored through a decrease in pH with time of an unbuffered pectin dispersion (Somogyi and Romani, 1964a). Qualitatively, color changes in a pH indicator like methyl red, can be used as a rapid screening and identificaiton method for PE (Hultin *et al.*, 1966; Markovic and Slezarik, 1969; Versteeg *et al.*, 1978). Quantitatively, color changes are measured by a spectrophotometer (Brady, 1976) using *p*-nitrophenyl acetate (Bradford *et al.*, 1976; Zimmerman, 1978) or bromothymol blue (Hagerman and Austin, 1986).

Agar-plate methods of detecting PE are based on chemical reactions between pectin methylester groups, alkaline hydroxylamine, and ferric chloride to form a red color complex (McComb and McCready, 1958; Gee *et al.*, 1959). The presence of PE is shown by colorless zones instead of the otherwise red coloration. Similar changes in bromothymol blue were used by Zimmerman (1978) to prescreen PE inhibitors. This bromothymol blue technique was adapted to gel electrophoresis and isoelectric focusing; the modification has been referred to as the printing technique (Delincee and Radola, 1970; Markovic, 1974). Delincee (1976) combined the alkaline hydroxylamine reaction and the printing technique to increase the sensitivity of the PE assay. Roeb and Stegeman (1975) included pectin in polyacrylamide electrophoresis gels and stained with methylene blue to detect PE bands.

C. Methanol Analysis

Methanol, a product of the PE deesterification of pectin, may be oxidized to formaldehyde and quantitated spectrophotometrically (Whright, 1927; Wood and Siddiqui, 1971). This method was automated by Vijayakshmi et al. (1976) for routine measurement of PE in commercial enzyme preparations. Kalvons and Bennett (1986) increased the sensitivity of the Wood and Siddiqui (1971) modification by employing an alcohol oxidase to convert the released methanol to formaldehyde. Pifferi et al. (1985) measured formaldehyde, generated from methanol by oxidation, with Besthorn's hydrazone, and claimed a sixfold increase in sensitivity over the use of acetylacetone (Wood and Siddiqui, 1971). Citrate, glucose, and tartaric acid interfere with the oxidation reactions. Gouch and Simpson (1970) and Lee and Wiley (1970) described methanol assays by gas–solid chromatography in the determination of PE activity. Bartolome and Hoff (1972a) first converted methanol to methyl nitrile before gas chromatography, and obtained a sensitivity of 3 ppm of sample. Other modifications since have further lowered the sensitivity threshold (Krop et al., 1974; Lee et al., 1975a; Baron et al., 1978; McFeeters and Armstrong, 1984).

D. Radioisotope Assay

Radioisotope techniques incorporating biosynthesized [14C] methyl-labeled pectin (Kauss et al., 1969) and pectin esterified with labeled [14C] diazomethane (Milner and Avigad, 1973), as PE substrates, are useful for elucidating mechanisms and biosynthetic pathways in PE metabolism, but they are not applicable to food research in which a natural pectin substrate is necessary.

VII. THE ROLE OF PECTINESTERASE IN PLANT GROWTH

The role of PE in plant growth is not clear. Kauss and Hassid (1967) discovered that PE trans-methylated pectin from S-adenosyl-L-methionine and performed a demethylating function. Cell-wall PE, activated by preincubating cell-wall fractions at pH 8 and adjusting to pH 5, facilitated growth (Nari et al., 1986). Investigations by Glasziou and Inglis (1958) and Glasziou (1959) implicated auxins such as 2,4-dichlorophenoxyacetic acid and β-indolylacetic acid in PE binding to the plant cell wall. It was pointed out that PE could be activated or inhibited by 3-indolylacetic acid, dependent on auxin concentration. The binding of PE to plant cell walls was not substantiated (Bryan and Newcomb, 1954; Jansen *et al.*, 1960b; Nakagawa *et al.*, 1971).

Goldberg (1984) observed a correlation between decreasing cell-wall growth potential and decreasing PE activity in mung bean hypocotyl. No significant correlation was observed in the presence of high cation concentrations.

Recently, Ricard and Noat (1986) proposed the theory of ionic control of the complex, multienzyme system and the dynamics of plant cell-wall synthesis and extension. This theory emphasizes the significance of an electric potential difference ($\Delta\psi$) between the inside and outside of the cell wall. With a large $\Delta\psi$, reflecting a high, local proton concentration, activation of enzymes involved in cell-wall extension, e.g., glycosyltransferases (optimal pH = 4–5), could occur. This kind of activation can be described as being acid induced. Extension of the cell wall and incorporation of neutral cell-wall material would result in a decrease of the wall charge density (local proton concentration) and therefore a decline in $\Delta\psi$. If PE with an alkaline pH optimum is now activated, its action in restoring a high $\Delta\psi$ would permit continuing growth of the cell wall. By their mathematical model, a slight change in local pH produces a drastic change in the wall's charge density, triggering wall-enzyme activity. Calcium bound to the cell wall diminishes the steep charge transition and suppresses cell growth. This suppression is discussed by Demarty *et al.* (1984). Figure 4, adapted from Ricard and Noat

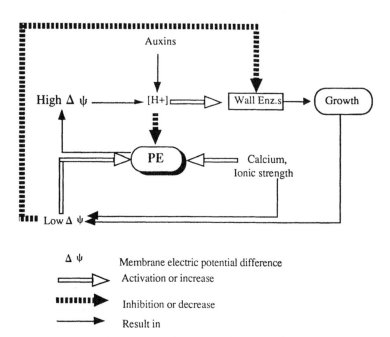

Figure 4 A tentative model of pectinesterase and electrostatic and ionic control of cell growth. With permission from Ricard and Noat (1986).

(1986), illustrates the electrostatic and ionic control of the cell-wall enzyme system. It also illustrates the significance of PE in altering $\Delta\psi$. A more detailed understanding of changes in PE characteristics, e.g., K_m, pH profiles, covalent binding to the wall, cationic sensitivity, and localization, would aid in a better understanding of the role of PE in wall morphology during cell growth and senescence.

What is known with certainty about PE in plant tissue is that its level of activity varies with species, variety, maturity, and source. Different concentrations were found for different parts of citrus (MacDonnell et al., 1945; Rothschild et al., 1974; Tahir et al., 1975). The reason for and causes of changes in activity during fruit ripening are less certain, because the changes do not appear to have a common pattern. An increase in activity was found with maturity of apple (Jacquin, 1955; Lee, 1969), banana (Hultin and Levine, 1963), cacao (Gamble, 1973), cherry (Al-Delaimy et al., 1966), grape (Lee et al., 1979b), guava (Shastri and Shastri, 1975), mango (Mattoo and Modi, 1969), orange (Rouse and Atkins, 1953; Tahir et al., 1975), pear (Jacquin, 1955), and tomatoes (Kertesz, 1938; Pithawala et al., 1948; Bell et al., 1951; Hobson, 1963). At the very least, there was an insignificant change with maturing avocado (Zauberman and Schiffmann-Nadel, 1972; Rouse and Barmore, 1974), banana (DeSwardt and Maxie, 1967; Palmer, 1971; Brady, 1976), cucumber (Bell et al., 1951), peach (Shewfelt, 1965), pear (Nagel and Patterson, 1967; Ben-Arie et al., 1979), and pea (Collins, 1970).

The increase in banana PE activity during ripening, observed by Hultin and Levine (1963), was challenged by DeSwardt and Maxie (1967), in that no precautions were taken against the high concentration of phenolic compounds in the young fruit. Although Nagel and Patterson (1967) found a decrease in specific activity of PE per gram of pear tissue, they indicated that it might actually increase, if the inhibition by phenolic compounds was prevented.

No significant correlation was found between activity and firmness retention of fresh and canned peaches (Shewfelt, 1965), canned snap beans (Summers, 1989), and fresh tomatoes (Hall and Dennison, 1960; Hobson, 1962; Sawamura et al., 1978). An exception was the avocado, for which Barmore and Rouse (1976) suggested that PE activity could be used to predict changes in firmness during controlled atmosphere storage. Summers (1989) reported that firm-podded snap bean cultivars exhibited higher PE activity than soft-podded types.

VIII. IMMOBILIZED PECTINESTERASE

Immobilization methods for pectic enzymes are discussed by Kminkova and Kucera (1983). Weibel et al. (1975) covalently coupled PE to porous glass beads and found a significant decrease in activity and a lower pH activity profile, when compared with uncoupled PE. They suggested that this characteristic was due to the residual, negative charge on glass beads and/or conformational changes of the enzyme.

Partially purified fungal PE was immobilized on a nylon polyisonitrile derivative (Vijayalakshmi et al., 1979) using the covalent coupling technique described by Goldstein et al. (1974). Compared with its uncoupled counterpart, the coupled enzyme was more stable and was more sensitive to salt, had a lower activity, a slightly lower optimal pH and optimal temperature.

Markovic and Machova (1985) compared tomato PE bonded covalently to activated Sepharose 4B to tomato PE adsorbed on polyethylene terephthalate. The activity of the former was 7.5% of the latter, a value almost twice as high as that obtained by Vijayalakshmi et al. (1979). A PE from *A. foetidus* yielded 11.5% activity, relative to the covalent bonding technique, and 23.1%, relative to the adsorption technique. The low activity of the immobilized enzymes is believed to be a consequence of stearic hindrance and limited diffusion of pectin to the enzyme.

IX. FOOD APPLICATIONS

The major application of PE is in the clarification and stabilization of clouds in fruit and vegetable juices. Other practical uses are controlling texture in fruit and vegetable products, increasing juice yield, facilitating maceration, and exploiting the properties of low-methoxyl pectin. A mixture of microbial PE, PG, and pectin lyase rapidly degrades pectin. PE may be applied either by addition of a commercial preparation or by manipulation of the native PE *in situ*.

A. Cloud Stabilization

Native PE will deesterify pectin in fruit juices. In the presence of calcium ions, the hydrocolloidal matter will coagulate, with ensuing sedimentation, loss of cloudiness and flavor intensity, and increased susceptibility to oxidation in the juice (Joslyn and Pilnik, 1961; Krop, 1974). Inactivation by heating or freezing is therefore necessary for some juices. The degree of heat inactivation required to stabilize the juice depends on the juice pH and soluble-solids content. PE is more easily inactivated at low pH and low solids content (Rouse and Atkins, 1952; Nath and Ranganna, 1977; Owusu-Yaw et al., 1988). As the pH increases (from 4.0 to 7.0), the rate of inactivation increases (Marshall et al., 1985).

A commercial pectinase preparation with a high ratio of PG to PE was most effective in stabilizing a cloud (Baker and Bruemmer, 1972); 0.001 unit/ml PG was required for every unit/ml of PE (Krop and Pilnik, 1974). Pectic acid hydrolysates (DP, 8–15) stabilized a cloud, i.e., retarded precipitation (Termote et al., 1977). Precipitaiton is facilitated by hydrolysis that achieves a DP lower than 21% (Baker, 1979). The highest cloud stabilization of citrus juice occurred at DP 12. A stable juice was observed when pectin present was degraded to 8 to 10 degrees of polymerization (Termote *et al.*, 1977). PE isozymes are not equally inhibited by pectic acid oligomers (Versteeg *et al.*, 1978).

Versteeg *et al.* (1980) found that one Navel orange PE isozyme, the *high-molecular-weight PE*, accounted for only 5% of the total PE activity, but was largely responsible for gelation in concentrated juice made by the cut-back process. This PE was comparatively very heat stable. Conditions for heat inactivation were 50 sec at 90°C, which is similar to the recommended pasteurization process for orange juice (Eagerman and Rouse, 1976).

A freshly prepared juice was stabilized by adding calcium, grinding the pulp, and allowing *in situ* PE to act (Pilnik and Rombouts, 1978).

Unlike PE stabilization, other methods of maintaining a cloud in juices, e.g., by the use of calcium chelating agents and polyphenolic inhibition of PE (Kew and Veldhuis, 1961), must consider their impact on taste and product identity, and the legal ramifications pertaining to their inclusion in food (Pilnik and Rombouts, 1978).

B. Clarification

The common practice of adding commercial pectic enzymes to extracted, lime, lemon, and apple juice, for example, begets a series of reactions that ultimately terminate in coagulation and sedimentation of the cloud-forming particles. The filtered, supernatant liquid is the clarified juice (Pilnik and Rombouts, 1978; Pollard and Kieser, 1959; Endo, 1965a; Yamasaki *et al.*, 1967). Apple-cloud particles are believed to consist of positively charged protein–carbohydrate complexes surrounded by negatively charged pectin molecules. The particles eventually coagulate under the influence of electrostatic interaction (Endo, 1965b; Yamasaki *et al.*, 1964, 1967). The concentration of apple juice necessitates the prior hydrolysis of pectin and the addition of calcium carbonate or sodium chloride (Pollard and Kieser, 1951).

C. Texture

Processing conditions have an important effect on the texture of processed fruits and vegetables. An increase in firmness (texture) of apple slices (Wiley and Lee, 1970), cauliflower (Hoogzand and Doesburg, 1961), carrot (Lee *et al.*, 1979a), potato (Bartolome and Hoff, 1972b), canned tomato (Hsu *et al.*, 1965), and green beans (Kaczmarzyk *et al.*, 1963; Van Buren *et al.*, 1960) was observed, after PE was activated by blanching. One possible mechanism for this effect is gelatin of the resulting low-methoxyl pectin by traces of calcium. Demethylated pectin, being less prone than its methylated precursor to degradation by β-elimination during thermal processing, might have simply contributed to firmer final texture (BeMiller and Kumari, 1972; Sajjaanantakul *et al.*, 1989).

Kaczmarzyk *et al.* (1963) did not find an increase in firmness of blanched beans (82°C, 3 min) when they were soaked in a solution of PE and/or calcium lactate at 54°C for up to 1 hr. It is possible, they suggested, that there already was extensive deesterification by bean PE before incubation, and also that there was limited diffusion of added PE into the bean pods. Similar observations were made on apple slices (Wiley and Lee, 1970).

D. Maceration and Liquefaction

Maceration and liquefaction of fruits and vegetables are useful operations in the production of nectars, vegetable juices, and baby foods, and native PE in combination with PG can be manipulated to accomplish this (Sulc and Vujicic, 1973). Pressey and Avants (1982), Burns and Pressey (1988), and Koch and Nevins (1989) reported a PE enhancement of the solubilization of tomato pectin and release of protoplasts. Similar findings were reported on the preparation of a cloud-stable carrot juice (Anastasakis *et al.*, 1987).

In cold-break tomato juice processing, disintegrated tomato pulp is held for a time to allow PE to act on the native pectin, before the depolymerizing action of PG. The serum obtained is concentrated more easily than is the untreated, more viscous pulp. In the hot-break process, the juice is heated rapidly at a high temperature to inactivate PE and PG; a high-viscosity juice having properties similar to single-strength juice is obtained (McColloch *et al.*, 1950; Wagner *et al.*, 1969).

E. Gelation

The influence of PE on gelation and on the initial rate of gelation of apple pectin was demonstrated by Yamaoka *et al.* (1983) and Baron *et al.* (1981), respectively. Recalling that PE functions in alternative gelatin mechanisms, the action of PE, whereby the galacturonan chain does not decrease in length, is more apt to promote gelation than is alkaline or acid hydrolysis (Tuerena *et al.*, 1984).

PE has been implicated in gelation of papaya puree (Chang *et al.*, 1965), tea leaf (Lamb and Ramaswamy, 1958), and apple, strawberry and orange juice (Oi and Satomura, 1965). Fungal PE (pH optimum, 4.0–5.5) has the potential for use in gelling low-sugar fruit products. The generation of low-methoxyl pectin *in situ* endows PE with the ability to thicken canned foods (Speirs *et al.*, 1980).

X. SUMMARY

PE is widespread in higher plants, molds, and bacteria. It plays an important albeit uncertain role in the dynamics of cell growth and in the texture of fruits and vegetables. It exhibits classical Michaelis–Menten kinetics. Its substrate specificity, pH activity profile, and conditions for activation and inhibition are well documented. Isozymes are largely responsible for its broad pH activity range and variable response to external stimuli, notably heat. Not so well documented is its biochemical purpose relative to pectin. Some questions need to be answered: Does PE initiate action at the reducing or nonreducing end-group? Is one macromolecular conformation favored in preference to another? Is synthesis or transmutation the dominant reaction pathway of pectin hydrolysis by plant PE in plant tissue? The answers provided will one day engender superior processing methodologies for plant tissues so vital to humans as food.

References

Al-Delaimy, K. A., Borgstrom, G., and Bedford, C. L. (1966). Pectic substances and pectic enzymes of fresh and processed Montmorency cherries. *Q. Bull. Mich. State Univ.* **49,** 164–171.

Anastasakis, M., Lindamood, J. B., Chism, G. W., and Hansen, P. M. T. (1987). Enzymatic hydrolysis of carrot for extraction of a cloud-stable juice. *Food Hydrocoll.* **1,** 247–261.

Anger, H., and Dongowski, G. (1985). Verteilung der freien carboxylgruppen in nativen pektinen aus obst und gemuse. *Die Nahrung* **29,** 397–404.

Atkins, C. D., and Rouse, A. H. (1954). Time–temperature–concentration relationships for heat inactivation of pectinesterase in citrus juices. *Food Technol.* **8,** 498–500.

Baker, R. A. (1979). Clarifying properties of pectin fractions separated by ester content. *J. Agric. Food Chem.* **27,** 1387–1389.

Baker, R. A., and Bruemmer, J. H. (1972). Pectinase stabilization of orange juice cloud. *J. Agric. Food Chem.* **20,** 1169–1173.

Barker, K. R., and Walker, J. C. (1962). Relationship of pectolytic and cellulolytic enzyme production by strains of *Pellicularia filamentosa* to their pathogenicity. *Phytopath.* **52,** 1119–1125.

Baron, A., Calvez, J., and Drilleau, J.-F. (1978). Dosage direct et indirect du methanol par chromatographie en phase gazeuse—Application pratique. *Ann. Fals. Exp. Chim.* **71,** 29–34.

Baron, A., Prioult, C., and Drilleau, J.-F. (1981). Gelatin of apple pectin. 1. Experimental method and study of the influence of pectin esterase concentration on pectin gelation. *Sc. aliments* **1,** 81–89.

Baron, A., Rombouts, F., Drilleau, J.-F., and Pilnik, W. (1980). Purification et proprietes de la pectinesterase produite par *Aspergillus niger*. *Lebensm.-Wiss. u. Technol.* **13,** 330–333.

Barmore, C. R., and Rouse, A. H. (1976). Pectinesterase activity in controlled-atmosphere stored avocado. *J. Am. Soc. Hort. Sci.* **101,** 294–296.

Bartolome, L. G., and Hoff, J. E. (1972a). Gas chromatographic methods for the assay of pectin methylesterase, free methanol, and methoxy groups in plant tissues. *J. Agric. Food Chem.* **20,** 262–266.

Bartolome, L. G., and Hoff, J. E. (1972b). Firming of potatoes: Biochemical effects of preheating. *J. Agric. Food Chem.* **20,** 266–270.

Bateman, D. F. (1963). Pectolytic activities of culture filtrates of *Rhizoctonia solani* and extracts of *Rhizoctonia*-infected tissues of bean. *Phytopath.* **53,** 197–204.

Bateman, D. F., and Beer, S. V. (1965). Simultaneous production and synergistic action of oxalic acid and polygalacturonase during pathogenesis by *Sclerotium rolfsii*. *Phytopath.* **55**, 204–211.

Bell, T. A., Etchells, J. L., and Jones, I. D. (1951). Pectinesterase in the cucumber. *Arch. Biochem.* **31**, 431–441.

BeMiller, J. N., and Kumari, G. V. (1972). β-Elimination in uronic acids: Evidence for an ElcB mechanism. *Carbohydr. Res.* **25**, 419–428.

Ben-Arie, R., Sonego, L., and Frenkel, C. (1979). Changes in pectic substances in ripening pears. *J. Am. Soc. Hort. Sci.* **104**, 500–505.

Bradford, M. M., McRorie, R. A., and Williams. W. L. (1976). Involvement of esterases in sperm penetration of the corona radiata of the ovum. *Biol. Reprod.* **15**, 102–106.

Brady, C. J. (1976). The pectinesterase of the pulp of banana fruit. *Aust. J. Plant Physiol.* **3**, 163–172.

Bryan, W. H., and Newcomb, E. H. (1954). Stimulation of pectin methylesterase activity of cultured tobacco pith by indoleacetic acid. *Physiol. Plant.* **7**, 290–297.

Buescher, R. W., and Furmanski, R. J. (1978). Role of pectinesterase and polygalacturonase in the formation of woolliness of peaches. *J. Food Sci.* **43**, 264–266.

Burns, J. K., and Pressey, R. (1988). Enhancement of the release of protoplasts and pectin from tomato locular gel by pectin methylesterase. *J. Am. Soc. Hort. Sci.* **113**, 624–626.

Calesnick, E. J., Hills, C. H., and Willaman, J. J. (1950). Properties of a commercial fungal pectase preparation. *Arch. Biochem.* **29**, 432–440.

Castaldo, D., Quagliuolo, L., Servillo, L., Balestrieri, C., and Giovane, A. (1989). Isolation and characterization of pectin methylesterase from apple fruit. *J. Food Sci.* **54**, 653–655, 673.

Chang, L. W., Morita, L. L., and Yamamoto, H. Y. (1965). Papaya pectinesterase inhibition by sucrose. *J. Food Sci.* **30**, 218–222.

Collins, J. L. (1970). Pectin methylesterase activity in southern peas (*Vigna sinensis*). *J. Food Sci.* **35**, 1–4.

Delincee, H. (1970). The effects of combined heat and irradiation treatment on pectin methylesterase. *Confructa* **15**, 366–375.

Delincee, H. (1976). Thin-layer isoelectric focusing of multiple forms of tomato pectinesterase. *Phytochemistry* **15**, 903–906.

Delincee, H. (1978). Irradiation of industrial enzyme preparations. II. Characterization of fungal pectinesterase by thin-layer isoelectric focusing and gel filtration. *J. Food Biochem.* **2**, 71–85.

Delincee, H., and Radola, B. J. (1970). Some size and charge properties of tomato pectin methyl-esterase. *Biochem. Biophys. Acta* **214**, 178–189.

Demarty, M., Morvan, C., and Thellier, M. (1984). Calcium and the cell wall. *Plant Cell Environ.* **7**, 441–448.

Dennison, R. A., and Ahmed, E. M. (1967). Irradiation effects on the ripening of Kent mangoes. *J. Food Sci.* **32**, 702–705.

DeSwardt, G. H., and Maxie, E. C. (1967). Pectin methylesterase in the ripening banana. *S. Afr. J. Agric. Sci.* **10**, 501–506.

Deuel, H. (1947). Uber glykolester der pektinsaure. *Helv. Chim. Acta* **30**, 1523–1534.

Deuel, H., and Stutz, E. (1958). Pectic substances and pectic enzymes. In "Advances in Enzymology." (F. F. Nord, ed.), Vol. 20, pp. 341–382. Interscience, New York.

Drzazga, B., Mitek, M., and Betlejewski, W. (1987). The effect of depectinization in apple wine production on the wines' quality with special attention paid to methanol content. *Confructa* **31**, 107–113.

Eagerman, B. A., and Rouse, A. H. (1976). Heat inactivation temperature–time relationships for pectinesterase inactivation in citrus juices. *J. Food Sci.* **41**, 1396–1397.

Echandi, E., and Walker, J. C. (1957). Pectolytic enzymes produced by *Sclerotinia sclerotiorum*. *Phytopath.* **47**, 303–306.

Endo, A. (1964). Studies on pectolytic enzymes of molds. Part XII. Purification and properties of pectinesterase. *Agric. Biol. Chem. (Tokyo)* **28**, 757–764.

Endo, A. (1965a). Studies on pectolytic enzymes of molds. Part XIII. Clarification of apple juice by the joint action of purified pectolytic enzymes. *Agric. Biol. Chem.* **29**, 129–136.

Endo, A. (1965b). Studies on pectolytic enzymes of molds. Part XVI. Mechanism of enzymatic clarification of apple juice. *Agric. Biol. Chem.* **29**, 229–233.

Evans, R., and McHale, D. (1978). Multiple forms of pectinesterase in limes and oranges. *Phytochemistry* **17**, 1073–1075.

Fogarty, W. M., and Kelly, C. T. (1983). Pectic enzymes. In "Microbial Enzymes and Biotechnology" (W. M. Fogarty, ed.), pp. 131–182. Applied Science Publishers, New York.

Fogarty, W. M., and Ward, O. P. (1974). "Progress in Industrial Microbiology," D. J. D. Hockenhull, ed. Vol. 13, pp. 59–119, Churchill-Livingston, Edinburgh.

Fremy, E. (1840). Recherches: Sur la pectine et l'acide pectique. *J. Pharm. Chim.* **26**, 368–393.

Fuchs, A. (1965). On the gelation of pectin by plant extracts and its inhibition. *Acta Bot. Neerl.* **14**, 315–322.

Gamble, W. (1973). The presence of pectin methylesterase in cacao pulp. *Experientia* **29**, 421.

Gee, M., Reeve, R. M., and McCready, R. M. (1959). Reaction of hydroxylamine with pectinic acids. Chemical studies and histochemical estimation of the degree of esterification of pectic substances in fruit. *J. Agric. Food Chem.* **7**, 34–38.

Glasziou, K. T. (1959). The localization and properties of pectin methylesterase of *Avena coleoptiles*. *Physiol. Planta.* **12**, 670–680.

Glasziou, K. T., and Inglis, S. D. (1958). The effect of auxins on the binding of pectin methylesterase to cell walls. *Aust. J. Biol. Sci.* **11**, 127–141.

Goldberg, R. (1984). Changes in the properties of cell-wall pectin methylesterase along the *Vigna radiata* hypocotyl. *Physiol. Plant.* **61**, 58–63.

Goldstein, L., Freeman, A., and Sokolovsky, M. (1974). Chemically modified nylons as supports for enzyme immobilization. *Biochem. J.* **143**, 497–509.

Gouch, T. A., and Simpson, C. F. (1970). Variation of performance of porous polymer bead columns in gas chromatography. *J. Chromatogr.* **51**, 129–137.

Hagerman, A. E., and Austin, P. J. (1986). Continuous spectrophotometric assay for plant pectin methyl esterase. *J. Agric. Food. Chem.* **34**, 440–444.

Hall, C. B. (1966). Inhibition of tomato pectinesterase by tannic acid. *Nature,* **12**, 717–718.

Hall, C. B., and Dennison, R. A. (1960). The relationship of firmness and pectinesterase activity of tomato fruits. *Proc. Am. Soc. Hort. Sci.* **75**, 629–631.

Heri, W., Neukom, H., and Deuel, H. (1961). Chromatographie von Pektinen mit verschiedener Verteilung der Methylestergruppen auf den Fadenmolekeln. *Helv. Chim. Acta.* **44**, 1945–1949.

Hobson, G. E. (1962). Pectic enzymes in the ripening of normal and abnormal tomato fruit. *Food Sci. Technol., Proc. Int. Congr.* **1**, 461–466.

Hobson, G. E. (1963). Pectinesterase in normal and abnormal tomato plants. *Biochem. J.* **86**, 358–365.

Holden, M. (1946). Studies on pectase. *Biochem. J.* **40**, 103–108.

Hoogzand, C., and Doesburg, J. J. (1961). Effect of blanching on texture and pectin of canned cauliflower. *Food Technol.* **15**, 160–163.

Hsu, C. P., Deshpande, S. N., and Desrosier, N. W. (1965). Role of pectin methylesterase in firmness of canned tomatoes. *J. Food Sci.* **30**, 583–588.

Hultin, H. O., and Levine, A. S. (1963). On the occurrence of multiple molecular forms of pectinesterase. *Arch. Biochem. Biophys.* **101**, 396–402.

Hultin, H. O., Sun, B., and Bulger, J. (1966). Pectin methyl esterase of banana. Purification and properties. *J. Food Sci.* **31**, 320–327.

Ishii, S., Kiho, K., Sugiyama, S., and Sigimoto, H. (1979). Low-methoxyl pectin prepared by pectinesterase from *Aspergillus japonicus*. *J. Food Sci.* **44**, 611–614.

Ishii, S., Kiho, K., Sugiyama, S., and Sigimoto, H. (1980). Novel pectin esterase, process for its production, and process for producing demethoxylated pectin by the use of said pectin esterase. U.S. Patent 4,200,694.

Ishii, S., and Yokotsuka, T. (1972). Clarification of fruit juice by pectin trans-eliminase. *J. Agr. Food Chem.* **20**, 787–791.

Jacquin, P. (1955). Pectin methylesterase and pectinic substances in the apple and pear and their importance in the fabrication of ciders and perries. *Ann. Tech. Agric.* **4**, 67–99.

Jansen, E. F., Jang, R., and Bonner, J. (1960a). Binding of enzyme to *Avena coleoptile* cell walls. *Plant Physiol.* **35**, 567–574.

Jansen, E. F., Jang, R., and Bonner, J. (1960b). Orange pectinesterase binding and activity. *Food Res.* **25**, 64–72.

Joslyn, M. A., and Pilnik, W. (1961). Enzymes and Enzyme Activity. In "The Orange, Its Biochemistry and Physiology." (W. B. Sinclair, ed.), pp. 373–435. University of California Press, Berkeley, California.

Joslyn, M. A., and Sedky, A. (1940). Effect of heating on the clearing of citrus juices. *Food Res.* **5**, 223–232.

Kaczmarzyk, L. M., Fennema, O., and Powrie, W. D. (1963). Changes produced in Wisconsin green snap beans by blanching. *Food Technol.* **17**, 123–126.

Kauss, H., and Hassid, W. Z. (1967). Enzymic introduction of the methyl ester groups of pectin. *J. Biol. Chem.* **242**, 3449–3453.

Kauss, H., Swanson, A. L., Arnold, R., and Odzuck, W. (1969). Biosynthesis of pectic substances. Localization of enzymes and products in a lipid–membrane complex. *Biochim. Biophys. Acta* **192**, 55–61.

Kertesz, Z. I. (1937). Pectic enzymes. I. The determination of pectin methoxylase activity. *J. Biol. Chem.* **121**, 589–598.

Kertesz, Z. I. (1938). Pectic enzymes. II. Pectic enzymes of tomatoes. *Food Res.* **3**, 481–487.

Kertesz, Z. I. (1939). Pectic enzymes. III. Heat inactivation of tomato pectin-methoxylase (pectase). *Food Res.* **4**, 113–116.

Kertesz, Z. I. (1955). Pectic enzymes. In "Methods in Enzymology," (S. P. Colowick and N. O. Kaplan, ed.), Vol. I. pp. 158–162. Academic Press, New York.

Kew, T. J., and Veldhuis, M. K. (1961). Cloud stabilization in citrus juice. U.S. Patent 2,995,448.

Kiermeier, F. (1949). Zur manometrischen bestimmungsmethode der pektase. *Annalen der Chemie.* (*Justus Liebigs*) **561**, 232–238.

Kimura, H., Uchino, F., and Mizushima, S. (1973). Purification and properties of pectinesterase from *Acrocylindrium*. *Agric. Biol. Chem.* **37**, 1209–1210.

King, K., Mitchell, J. R., Norton, G., and Caygill, J. (1986). *In situ* deesterification of lime pectin. *J. Sci. Food Agric.* **37**, 391–398.

Klavons, J. A., and Bennett, R. D. (1986). Determination of methanol using alcohol oxidase EC.1.1.3.13 and its application to methyl ester content of pectin. *J. Agric. Food Chem.* **34**, 597–599.

Kminkova, M., and Kucera, J. (1983). Comparison of pectolytic enzymes covalently bound to synthetic ion exchangers using different methods of binding. *Enzyme Microb. Technol.* **5**, 204–208.

Koch, J. L.and Nevins, D. J. (1989). Tomato fruit cell wall. I. Use of purified tomato polygalacturonase and pectin methylesterase to identify developmental changes in pectins. *Plant Physiol.* **91**, 816–822.

Kohn, R., Dongowski, G., and Bock, W. (1985). The distribution of free and esterified carboxyl groups within the pectin molecule after the influence of pectin esterase from *Aspergillus niger* and oranges. *Die Nahrung* **29**, 75–85.

Kohn, R., Furda, I., and Kopec, Z. (1968). Distribution of free carboxyl groups in the pectin molecule after treatment with pectin esterase. *Coll. Czech. Chem. Commun.* **33**, 264–269.

Kohn, R., Markovic, O., and Machova, E. (1983). Deesterification mode of pectin by pectin esterase of *Aspergillus foetidus*, tomatoes, and alfalfa. *Coll. Czech. Chem. Commun.* **48**, 790–797.

Kopaczewski, W. (1925). Sur la coagulation de la pectine. *Bull. Soc. Chim. Biol.* **7**, 419–428.

Krop. J. J. P. (1974). "The mechanism of cloud-loss phenomena in orange juice." Ph.D. thesis, Agricultural Univ., Wageningen, The Netherlands.

Krop, J. J. P., and Pilnik, W. (1974). Cloud loss studies on citrus juices: Cloud stabilization by a yeast-polygalacturonase. *Lebensm.- Wiss. u. Technol.* **7**, 121–124.

Krop, J. J. P., Pilnik, W., and Faddegon, J. M. (1974). The assay of pectinesterase by a direct gas chromatographic methanol determination—Application to cloud-loss studies in citrus juices. *Lebensm.-Wiss. u. Technol.* **7**, 50–52.

Lamb, J., and Ramaswamy, M. S. (1958). Studies on the fermentation of Ceylon tea. XI. Relations between the polyphenol oxidase activity and pectin methylesterase activity. *J. Sci. Food Agric.* **9**, 51–56.

Lee, C. Y., Acree, T. E., and Butts, R. M. (1975a). Determination of methyl alcohol in wine by gas chromatography. *Anal. Chem.* **47**, 747–748.

Lee, C. Y., Robinson, W. B., Van Buren, J. P., Acree, T. E., and Stoewsand, G. S. (1975b). Methanol in wines in relation to processing and variety. *Am. J. Enol. Viticult.* **26**, 184–187.

Lee, C. Y., Bourne, M. C., and Van Buren, J. P. (1979a). Effect of blanching treatments on the firmness of carrots. *J. Food Sci.* **49**, 615–616.

Lee, C. Y., Smith, N. L., and Nelson, R. R. (1979b). Relationship between pectin methylesterase activity and the formation of methanol in Concord grape juice and wine. *Food Chem.* **4**, 143–148.

Lee, M., and Macmillan, J. D. (1968). Mode of action of pectic enzymes. I. Purification and certain properties of tomato pectinesterase. *Biochemistry* **7**, 4005–4010.

Lee, M., and Macmillan, J. D. (1970). Mode of action of pectic enzymes. III. Site of initial action of pectinesterase on highly esterified pectin. *Biochemistry,* **9**, 1930–1934.

Lee, M., Miller, L., and Macmillan, J. D. (1970). Similarities in the action patterns of exopolygalacturonate lyase and pectinesterase from *Clostridium multifermentans. J. Bacteriol.* **103**, 595–600.

Lee, Y. S. (1969b). "Measurement, characterization, and evaluation of pectinesterase in apple fruits." Ph.D. thesis, Univ. of Maryland, College Park, Maryland.

Lee, Y. S., and Wiley, R. C. (1970). Measurement and characterization of pectinesterase in apple fruits. *J. Am. Soc. Hort. Sci.* **95**, 465–468.

Lineweaver, H., and Ballou, G. A. (1943). Properties of alfalfa pectinesterase (pectase). *Fed. Proc.* **2**, 66.

Lineweaver, H., and Ballou, G. A. (1945). The effect of cations on the activity of alfalfa pectinesterase (pectase). *Arch. Biochem.* **6**, 373–387.

Lineweaver, H., and Jansen, E. F. (1951). Pectic enzymes. In "Advances in Enzymology" (F. F. Nord, ed.), Vol. 11. pp. 267–295. Interscience, New York.

MacDonnell, L. R., Jansen, E. F., and Lineweaver, H. (1945). The properties of orange pectinesterase. *Arch. Biochem.* **6**, 389–401.

MacDonnell, L. R., Jay, R., Jansen, E. F., and Lineweaver, H. (1950). The specificity of pectinesterases from several sources with some notes on purification of orange pectinesterase. *Arch. Biochem.* **28**, 260–273.

Macmillan, J. D., and Vaughn, R. H. (1964). Purification and properties of polygalacturonic acid trans-eliminases produced by *Clostridium multifermentans. Biochemistry* **3**, 564–572.

Manabe, M. (1973a). Purification and properties of *Citrus natsudaidai* pectinesterase. *Agric. Biol. Chem.* **37**, 1487–1491.

Manabe, M. (1973b). Saponification of ester derivatives of pectic acid by *Natsudaidai pectinesterase.* (Studies on the derivatives of pectic substances, Part V). *J. Agric. Chem. Soc. Japan* **47**, 385–390.

Markovic, O. (1974). Tomato pectin esterase—characterization of one of its multiple forms. *Coll. Czech. Chem. Commun.* **39**, 908–913.

Markovic, O. (1978). Pectinesterase from carrot (*Daucus carrota* L.). *Experientia* **34**, 561–562.

Markovic, O., Heinrichova, K., and Lenkey, B. (1975). Pectolytic enzyme from banana. *Coll. Czech. Chem. Commun.* **40**, 769–774.

Markovic, O., and Kohn, R. (1984). Mode of pectin deesterification by *Trichoderma reesei* pectinesterase. *Experientia* **40**, 842–843.

Markovic, O., and Machova, E. (1985). Immobilization of pectin esterase from tomatoes and *Aspergillus foetidus* on various supports. *Coll. Czech. Chem. Commun.* **50**, 2021–2027.

Markovic, O., Machova, E., and Slezarik, A. (1983). The action of tomato and *Aspergillus foetidus* pectinesterases on oligomeric substrates esterified with diazomethane. *Carbohydr. Res.* **116**, 105–111.

Markovic, O., and Patocka, J. (1977). Action of iodine on the tomato pectinesterase. *Experientia* **33**, 711–713.

Markovic, O., and Slezarik, A. (1969). Isolation and partial characterization of pectinesterase from tomatoes. *Coll. Czech. Chem. Commun.* **34**, 3820–3825.

Markovic, O., Slezarik, A., and Labudova, I. (1985). Purification and characterization of pectinesterase and polygalacturonase from *Trichoderma reesei*. *FEMS Lett.* **27**, 267–271.

Marshall, M. R., Marcy, J. E., and Braddock, R. H. (1985). Effect of total solids level on heat inactivation of pectinesterase in orange juice. *J. Food Sci.* **50**, 220–222.

Mattoo, A. K., and Modi, V. V. (1970). Biochemical aspects of ripening and chilling injury in mango fruit. *Proc. Conf. Trop. Subtrop. Fruits, 1969. Trop. Prod. Inst. (London).* 111–115.

McColloch, R. J., and Kertesz, Z. I. (1947). Pectic enzymes. VIII. A comparison of fungal pectin methylesterase with that of higher plants, especially tomatoes. *Arch. Biochem.* **13**, 217–229.

McColloch, R. J., Nielsen, B. W., and Beavans, E. A. (1950). Factors influencing the quality of tomato paste. II. Pectic changes during processing. *Food Technol.* **4**, 339–343.

McComb, E. A., and McCready, R. M. (1958). Use of hydroxamic acid reaction for determining pectinesterase activity. *Stain Technol.* **33**, 129–131.

McCready, R. M., and Seegmiller, C. G. (1954). Action of pectic enzymes on oligogalacturonic acids and some of their derivatives. *Arch. Biochem. Biophys.* **50**, 440–450.

McFeeters, R. F., and Armstrong, S. A. (1984). Measurement of pectin methylation in plant cell wall. I. *Anal. Biochem.* **139**, 212–217.

Miller, L., and Macmillan, J. D. (1970). Mode of action of pectic enzymes. II. Further purification of exopolygalacturonate lyase and pectinesterase from *Clostridium multifermentans*. *J. Bacteriol.* **102**, 72–78.

Miller, L., and Macmillan, J. D. (1971). Purification and pattern of action of pectinesterase from *Fusarium oxysporum* f. sp. vasinfectum. *Biochemistry* **10**, 570–576.

Mills, G. B. (1949). A biochemical study of *Pseudomonas prunicola* Wormald. 1. Pectin esterase. *Biochem. J.* **44**, 302–305.

Milner, Y., and Avigad, G. (1973). A sensitive radioisotope assay for pectin methylesterase activity. *Anal. Biochem.* **51**, 116–120.

Miyairi, K., Okuno, T., and Sawai, K. (1975). Purification and physicochemical properties of pectinesterase of apple fruits (Ralls). *Bull. Fac. Agric. Hirosaki Univ.* **24**, 22–30.

Moustacas, A. M., Nari, J., Diamantidis, G., Noat, G., Crasnier, M., Borel, M., and Ricard, J. (1986). Electrostatic effects and the dynamics of enzyme reactions at the surface of plant cells. 2. The role of pectin methyl esterase in the modulation of electrostatic effects in soybean cell walls. *Eur. J. Biochem.* **155**, 191–197.

Nagel, C. W., and Patterson, M. E. (1967). Pectic enzymes and development of the pear (*Pyrus communis*). *J. Food Sci.* **32**, 294–297.

Nakagawa, H., Sekiguchi, K., Ogura, N., and Takehana, H. (1971). Binding of tomato pectinesterase and β-fructofuranosidase of tomato cell wall. *Agric. Biol. Chem.* **35**, 301–307.

Nakagawa, H., Yanagawa, Y., and Takehana, H. (1970). Studies on the pectolytic enzyme. Part V. Some properties of the purified tomato pectinesterase. *Agric. Biol. Chem.* **34**, 998–1003.

Nari, J., Noat, G., Diamantidis, G., Woudstra, M., and Ricard, J. (1986). Electrostatic effects and the dynamics of enzyme reactions at the surface of plant cells. 3. Interplay between limited cell-wall autolysis, pectin methyl-esterase activity and electrostatic effects in soybean cell wall. *Eur. J. Biochem.* **155**, 199–202.

Nath, N., and Ranganna, S. (1977). Time–temperature relationship for thermal inactivation of pectinesterase in mandarin orange (*Citrus reticulata* Blanco) juice. *J. Food Technol.* **12**, 411–419.

Ogundero, V. W. (1988). Pectinesterase production by *Rhizopus stolonifer* from postharvest soft rots of potato tubers in Nigeria and its activity. *Die Nahrung* **32**, 59–65.

Oi, S., and Satomura, Y. (1965). Increase in viscosity and gel formation of fruit juice by purified pectinesterase. *Agric. Biol. Chem.* **29**, 936–942.

Owens, H. S., McCready, R. M., and Maclay, W. D. (1944). Enzymic preparation and extraction of pectinic acids. *Ind. Eng. Chem.* **36**, 936–938.

Owusu-Yaw, J., Marshall, M. R., Koburger, J. A., and Wei, C. I. (1988). Low-pH inactivation of pectinesterase in single-strength orange juice. *J. Food Sci.* **53**, 504–507.

Palmer, J. K. (1971). The banana. In "Biochemistry of Fruits and Their Products" (A. C. Hulme, ed.), Vol. 2, pp. 65–105. Academic Press, New York.

Palmer, J. K., and Roberts, J. B. (1967). Inhibition of banana polyphenoloxidase by 2-mercapto-benzothiazole. *Science* **157**, 200–201.

Perley, A. F., and Page, O. T. (1971). Differential induction of pectolytic enzymes of *Fusarium roseum* (LK.). emend. Snyder and Hansen. *Can. J. Microbiol.* **17**, 415–420.

Phaff, H. J. (1947), The production of exocellular pectic enzymes by *Penicillium chrysogenum* I. On the formation and adaptive nature of polygalacturonase and pectinesterase. *Arch. Biochem.* **13**, 67–81.

Pifferi, P. G., Malacarne, A., Lanzarni, G., and Casoli, U. (1985). Spectrophotometric method for the determination of pectin methylesterase activity by Besthorn's hydrazone. *Chemie Mikrobiol. Technol. der Lebensm.* **9**, 65–69.

Pilnik, W., and Rombouts, F. M. (1978). Pectic enzymes. In "Polysaccharides in Food" (J. M. V. Blanchard and J. R. Mitchell, ed.), pp. 109–126. Butterworths, London.

Pithawala, H. R., Sovur, G. R., and Screenivas, A. (1948). Characterization of tomato pectinesterse. *Arch. Biochem.* **17**, 235–248.

Pollard, A., and Kieser, M. E. (1951). The pectase activity of apples. *J. Sci. Food Agric.* **2**, 30–36.

Pollard, A., and Kieser, M. E. (1959). Pectin changes in cider fermentations. *J. Sci. Food Agric.* **10**, 253–260.

Pollard, A., Kieser, M. E., and Sissons, D. J. (1958). Inactivation of pectic enzymes by fruit phenolics. *Chem. Ind.* No. 30, 952.

Pressey, R., and Avants, J. K. (1972). Multiple forms of pectin methylesterase in tomatoes, *Phytochemistry* **11**, 3139–3142.

Pressey, R., and Avants, J. K. (1982). Solubilization of cell walls by tomato polygalacturonases: Effects of pectinesterase. *J. Food Biochem.* **6**, 57–74.

Ramaswamy, M. S., and Lamb, J. (1958). Studies on the fermentation of Ceyon tea. X. Pectic enzymes in tea leaf. *J. Sci. Food Agric.* **9**, 46–56.

Rexova-Benkova, L. (1972). On the character of the interaction of endopolygalacturonase with cross-linked pectic acid. *Biochem. Biophys. Acta* **276**, 215–220.

Rexova-Benkova, L., and Markovic, O. (1976). Pectic enzymes. In "Advances in Carbohydrate Chemistry and Biochemistry." (R. S. Tipson and D. Horton, ed.), Vol. 33, pp. 323–385. Academic Press, New York.

Rexova-Benkova, L., and Tibensky, V. (1972). Selective purification of *Aspergillus niger* endo-polygalacturonase by affinity chromatography on cross-linked pectic acid. *Biochem. Biophys. Acta* **268**, 187–193.

Ricard, J. and Noat, G. (1986). Electrostatic effects and the dynamics of enzyme reactions at the surface of plant cells. 1. A theory of the ionic control of a complex multienzyme system. *Eur. J. Biochem.* **155**, 183–190.

Roeb, L., and Stegemann, H. (1975). Pectin methylesterases: Isozymes, properties, and detection on polyacrylamide gels. *Biochem. Physiol. Pflanzen* **168**, S.607–615.

Rombouts, F. M., Wissenburg, A. K., and Pilnik, W. (1979). Chromatographic separation of orange pectinesterase isozymes on pectates with different degree of cross-linking. *J. Chromatogr.* **168**, 151–161.

Rothschild, G., Moyal, Z., and Karsentry, A. (1974). Pectinesterase activity in the component parts of different Israeli citrus fruit varieties. *J. Food Technol.* **9**, 471–475.

Rouse, A. H., and Atkins, C. D. (1952). Heat inactivation of pectinesterase in citrus juices. *Food Technol.* **6**, 291–294.

Rouse, A. H., and Atkins, C. D. (1953). Maturity changes in pineapple, oranges and their effect on processed frozen concentrate. *Proc. Fla. State Hort. Soc.* **66**, 268–273.

Rouse, A. H., and Atkins, C. D. (1955). Pectinesterase and pectin in commercial citrus juices as determined by methods use at the citrus experiment station. *Fla. Agric. Exp. Sta. Bull.* **570**, 1–19.

Rouse, A. H., and Barmore, C. R. (1974). Changes in pectic substances during ripening of avocadoes. *Hort. Science* **9**, 36–37.

Sajjaanantakul, T., Van Buren, J. P., and Downing, D. L. (1989). Effect of methyl ester content on heat degradation of chelator-soluble carrot pectin. *J. Food Sci.* **54**, 1272–1277.

Sawamura, M., Knegt, E., and Bruinsm, J. (1978). Levels of endogenous ethylene, carbon dioxide, and soluble pectin, and activities of pectin methylesterase and polygalacturonase in ripening tomato fruits. *Plant Cell Physiol.* **19**, 1061–1069.

Seymour, T., Wicker, L., and Marshall, M. R. (1989). Purification and properties of pectinesterase enzymes in Marsh white grapefruit pulp. Abstract No. 267, presented at 50th Annual Meeting of Inst. of Food Technologists, Chicago, IL, June 25–29.

Shastri, P. N., and Shastri, N. V. (1975). Studies of pectin methylesterase activity during development and ripening of guava fruit. *J. Food Sci. Technol.* **12**, 42–43.

Sheiman, M. I., Macmillan, J. D., Miller, L., and Chase, T., Jr. (1976). Coordinated action of pectinesterase and polygalacturonate lyase complex of *Clostridium multifermentans. Eur. J. Biochem.* **64**, 565–572.

Shewfelt, A. L. (1965). Changes and variations in the pectic constitution of ripening peaches as related to product firmness. *J. Food Sci.* **30**, 573–576.

Solms, J., and Deuel, H. (1955). Uber den Mechanisms der enzymatischen Verseifung von Pektinstoffen. *Helv. Chim. Acta* **38**, 321–329.

Somogyi, L. P., and Romani, R. J. (1964a). A simplified technique for the determination of pectin methylesterase activity. *Anal. Biochem.* **7**, 498–501.

Somogyi, L. P., and Romani, R. J. (1964b). Irradiation-induced textural changes in fruits and its relation to pectin metabolism. *J. Food Sci.* **29**, 366–371.

Speirs, C. I., Blackwood, G. C., and Mitchell, J. R. (1980). Potential of fruit waste containing *in vivo* deesterified pectin as a thickener in canned products. *J. Sci. Food Agric.* **31**, 1287–1294.

Speiser, R., Eddy, C. R., and Hills, C. H. (1945). Kinetics of deesterification of pectin. *J. Phys. Chem.* **49**, 563–579.

Starr, M. P., and Nasuno, S. (1967). Pectolytic activity of phytopathogenic *Xanthomonas. J. Gen. Microbiol.* **46**, 425–433.

Sulc, D., and Vuljicic, B. (1973). Untersuchungen der wirksamkeit von enzymprapäraten auf pektinsubstraten und frucht- und gemusemaischen. *Flussiges. Obst.* **40**, 79–83.

Summers, W. L. (1989). Pectinesterase and D-galacturonase activities in eight snap bean cultivars. *Hort. Science* **24**(3), 484–486.

Tahir, M. A., Chaudhary, M. S., and Malik, A. A. (1975). Pectinesterase (PE) activity in component parts of Jaffa oranges during ripening. *Pak. J. Sci. Res.* **27**, 59–60.

Taylor, A. J. (1982). Intramolecular distribution of carboxyl groups in low methoxyl pectins—a review. *Carbohydr. Polymer.* **2**, 9–17.

Termote, F., Rombouts, F. M., and Pilnik, W. (1977). Stabilization of cloud in pectinesterase active orange juice by pectic acid hydrolysates. *J. Food Biochem.* **1**, 15–34.

Theron, T., Devilliers, O. T., and Schmidt, A. A. (1977a). Purification of pectin methylesterase from Santa Rosa plums. *Agrochemophysica* **9**, 7–12.

Theron, T., Devilliers, O. T., and Schmidt, A. A. (1977b). Isolation and purification of pectic methylesterase from Marvel tomatoes. *Agrochemophysica* **9**, 93–96.

Tuerena, C. E., Taylor, A. J., and Michell, J. R. (1984). Carboxy distribution of low-methoxy pectin deesterification *in situ. J. Sci. Food Agric.* **35**, 797–804.

Van Buren, J. P., Moyer, J. C., and Robinson, W. B. (1962). Pectin methylesterase in snap beans. *J. Food Sci.* **27**, 291–294.

Van Buren, J. P., Moyer, J. C., Wilson, D. E., Robinson, W. B., and Hand, D. B. (1960). Influence of blanching conditions on sloughing, splitting, and firmness of canned snap beans. *Food Technol.* **14**, 233–236.

Vas, K., Nedbalek, M., Scheffer, H., and Kovacs-Proszt, G. (1967). Methodological investigation on the determination of some pectic enzymes. *Fruchtsaft-Ind.* **12**, 164–184.

Vas, K., Kovacs-Proszt, G., and Scheffer, H. (1968). Effect of ionizing radiation on pectin methylesterase of some fruits and vegetables. *Int. J. Appl. Rad. Isotop.* **19**, 273–281.

Versteeg, C. (1979). "Pectinesterases from the orange fruit—their purification, general characteristics and juice-cloud destabilizing properties." Ph.D. thesis. Agricultural Univ. Wageningen, The Netherlands.

Versteeg, C., Rombouts, F. M., and Pilnik, W. (1978). Purification and some characteristics of two pectinesterase isozymes from orange. *Lebensm.-Wiss. u. -Technol.* **11**, 267–274.

Versteeg, C., Rombouts, F. M., Spaansen, C. H., and Pilnik, W. (1980). Thermostability and orange juice cloud–destabilizing properties of multiple pectinesterases from orange. *J. Food Sci.* **45**, 969–972.

Vijayalakshmi, M. A., Picque, D., Jaumouille, R., and Segard, E. (1979). Immobilized pectinesterase. In "Food Processing Engineering," 2nd International Congress on Engineering and Food, 8th European Food Symposium, (P. Linko, ed.), pp. 152–158. Applied Science, London.

Vijayalakshmi, M. A., Sarris, J., and Varoquaux, P. (1976). Determination automatique de l'acticite pectin-esterasique dans une preparation pectinolytique commerciale. *Lebensm.-Wiss. u. -Technol.* **9**, 21–23.

Wagner, J. R., Miers, J. C., Sanshuck, D. W., and Becker, R. (1969). Consistency of tomato products. 5. Differentiation of extractive and enzyme-inhibitory aspects of the acidified hot-break process. *Food Technol.* **23**, 247–250.

Waggoner, P. E., and Dimond, A. E. (1955). Production and role of extracellular pectic enzymes of *Fusarium oxysporum* f. *lycopersici. Phytopathology* **45**, 79–87.

Weibel, M. K., Barrios, A., Delotto, R., and Humphrey, A. E. (1975). Immobilized enzymes: Pectinesterase covalently coupled to porous glass particles. *Biotech. Bioeng.* **17**, 85–98.

Weurman, C. (1954). Pectase in Doyenne Boussoch pears and changes in the quality of the enzyme during development. *Acta Bot. Neerl.* **3**, 100–107.

Whright, L. O. (1927). Comparison of sensitivity of various tests for methanol. *Ind. Eng. Chem., Ind. Ed.* **19**, 750–752.

Wicker, L., and Temelli, F. (1988). Heat inactivation of pectinesterase in orange juice pulp. *J. Food Sci.* **53**, 162–164.

Wicker, L., Vassallo, M. R., and Echeverria, E. J. (1988). Solubilization of cell wall–bound, thermostable pectinesterase from Valencia orange. *J. Food Sci.* **53**, 1171–1174, 1180.

Wiley, R. C., and Lee, Y. S. (1970). Modifying texture of processed apple slices. *Food Technol.* **24**, 1168–1170.

Wood, P. J., and Siddiqui, I. R. (1971). Determination of methanol and its application to measurement of pectin ester content and pectin methylesterase activity. *Anal. Biochem.* **39**, 418–428.

Yamaoka, T., Tsukada, K., Takahashi, H., and Yamauchi, N. (1983). Purification of a cell wall–bound pectin gelatinizing factor and examination of its identity with pectin methylesterase. *Bot. Mag. Tokyo* **96**, 139–144.

Yamasaki, M., Kato, A., Chu, S.-Y., and Arima, K. (1967). Pectic enzymes in the clarification of apple juice. Part II. The mechanism of clarification. *Agric. Biol. Chem.* **31**, 552–560.

Yamasaki, M., Yasui, T., and Arima, K. (1964). Pectic enzymes in the clarification of apple juice. Part I. Study in the clarification reaction in a simplified model. *Agric. Biol. Chem.* **28**, 779–787.

Yoshihara, O., Matsuo, T., and Kaji, A. (1977). Purification and properties of acid pectinesterase from *Corticium rolfsii. Agric. Biol. Chem.* **41**, 2335–2341.

Zacharius, R. M., Zell, T. E., Morrisson, J. M., and Woodlock, J. J. (1969). Glycoprotein staining following electrophoresis on acrylamide gels. *Anal. Biochem.* **30**, 148–152.

Zauberman, G., and Schiffmann-Nadel, M. (1972). Pectin methylesterase and polygalacturonase in avocado fruit at various stages of development. *Plant Physiol.* **49**, 864–865.

Zimmerman, R. E. (1978). A rapid assay for pectinesterase activity which can be used as a prescreen for pectinesterase inhibitors. *Anal. Biochem.* **85**, 219–223.

CHAPTER 9

The Polygalacturonases and Lyases

J. K. Burns

Citrus Research and Education Center
IFAS, University of Florida
Lake Alfred, Florida

I. THE POLYGALACTURONASES

A. Introduction

The polygalacturonase enzymes catalyze the hydrolytic cleavage of the O-gly-cosyl bond of α-D-$(1 \rightarrow 4)$polygalacturonan. The pattern of degradation proceeds in either a random (endo-polygalacturonase, EC 3.2.1.15) or terminal fashion (exo-polygalacturonase, EC 3.2.1.67). Viscometric analysis in tandem with reducing-group analysis is a sensitive determinant of the two hydrolytic patterns. Within short reaction times, random cleavage of polygalacturonate results in large decreases in viscosity with only small percentages of glycosidic bonds hydrolyzed, whereas terminal cleavage is characterized by little viscosity change with substantial percentages of glycosidic bonds broken (Pressey, 1986a; Rexova-Benkova and Markovic, 1976; Tam, 1983).

B. Endo-polygalacturonase

1. Substrates and Products of Hydrolysis

The optimal substrate of endo-polygalacturonase (endo-PG) is polygalacturonan (Barmore and Brown, 1981; Bartley et al., 1982; Pressey and Avants, 1982b; Robertsen, 1987; Walton and Cervone, 1990). In general, substrates with a low degree of esterification are preferred (Rexova-Benkova and Markovic, 1976). As the degree of esterification increases (usually above 20%), V_{max} decreases and K_m increases (Archer, 1979). Relative activities of the enzyme with various pectin sources as substrate can be twofold to tenfold less than with polygalacturonic acid. Highly methylated pectin and polymethylpolygalacturonic acid are not hydrolyzed. Enzymatic deesterification of pectin contained within isolated cell walls resulted in an increase in relative activity comparable to that with polygalacturonic acid as substrate (Pressey and Avants, 1982b). In contrast, chemical deesterification of pectin resulted in decreased enzymatic activity as compared to that of native pectin (Bartley et al., 1982). Structural changes in the pectic polymer as well as pH shifts in the assay media with chemical deesterification were attributed to the decrease in activity.

High-molecular-weight polymers are hydrolyzed at faster rates than are low-molecular-weight substrates. The affinity of the enzyme for the substrate is greater with polymers of greater chain length, as indicated by decreasing K_m and increasing V_{max} (Kimura et al., 1973). Only small variations in rate of reaction have been observed with substrates of degree of polymerization (DP) of 24 and above; but below this value, rate of hydrolysis decreases (Liu and Luh, 1980; Pressey andAvants, 1973a). In Saccharomyces fragilis, three rates of polymer degradation were observed. Large polymers were split at a fast rate. The second phase of degradation was characterized by the hydrolysis of pentamers and tetramers at a slower rate, and finally, the third phase consisted of very slow degradation of trimer to dimer and monomer (Demain and Phaff, 1954).

Oligomers, decreasing from 24 galacturonosyl residues, have been measured as released products in the initial stages of large-polymer degradation (Cervone et al., 1989; De Lorenzo et al., 1990; Liu and Luh, 1980; Marcus et al., 1986; Robertsen, 1987). The result of exhaustive digestions yields several diminutive end-products, the final sizes of which depend on the enzyme source. Trigalacturonic acid is the smallest oligomer hydrolyzed. Digalacturonic acid is never attacked, and monogalacturonic acid accumulates as a result of polymer degradation in all cases. The pattern of oligomer degradation and accumulation of various end-products has suggested that, in the species examined, several types of substrate–enzyme binding sites occur, each with different active catalytic group positions (Rexova-Benkova and Markovic, 1976). In each case, the bond adjacent to the nonreducing end of the substrate is protected from hydrolysis. The presence of digalacturonic acid in cells of tomato, Erwinia, and Saccha-

romyces (Demain and Phaff, 1954; Nasuno and Starr, 1966; Patel and Phaff, 1960); trigalacturonic acid in *Acrocylindrium* (Kimura *et al.*, 1973); and tetra-galacturonic acid in *Rhizopus* (Liu and Luh, 1980) indicates that in these species no further hydrolysis of the oligomer end-product occurs.

2. Enzyme Characterization

a. pH Optimum

The pH optimum of endo-PG has been reported to be between 3.6 and 5.5 (Ali and Brady, 1982; Archer and Fielding, 1979; Barmore and Brown, 1981; Barmore *et al.*, 1984; Chan and Tam, 1982; Hart *et al.*, 1990; Knegt *et al.*, 1988; Pressey and Avants, 1973a,b, 1982b; *Prusky et al.,* 1989; Robertsen, 1987). The pH optimum becomes more acidic as the molecular weight of the substrate decreases (Barash and Eyal, 1970; Pressey and Avants, 1971). In tomato, hydrolysis of polygalacturonic acid III (molecular mass, 3.2 kDa) is optimal at pH 4.2 (Pressey and Avants, 1973a,b; Pressey and Avants, 1982b), whereas pH optima as low as 2.0 have been reported for smaller oligomers (Patel and Phaff, 1960). Changes in pH optima have been reported with different buffers and changes in ionic strength of the reaction media (Knegt *et al.*, 1988; Pressey and Avants, 1971, 1973a).

b. Metal Requirement

Endo-PG has no metal requirement (Kimura *et al.,* 1973; Pressey and Avants, 1973a), although it was once thought to have a monovalent cation requirement (Patel and Phaff, 1960; Pressey and Avants, 1971). The effect of cations such as Na^+ has now been attributed to increased availability of the substrate by dispersion of the polygalacturonans in solution (Pressey and Avants, 1973b; Pressey and Avants, 1982b). Cations such as Ca^{2+}, which interact with the polygalacturonans to form aggregates, are inhibitory at concentrations as low as 0.4 mM. The aggregates thus formed are considered to be unavailable for enzymatic attack and may actually bind to and immobilize the enzyme (Pressey and Avants, 1973a).

c. Isolation and Purification

Isolation and purification of the enzyme from fungal sources has been achieved with a variety of conventional techniques. Usually the organism is grown in culture on polygalacturonan carbon sources. The culture fluid is then concentrated by acetone or ammonium sulfate precipitation, ultrafiltration, or *in vacuo* (Marcus *et al.*, 1986; Prusky *et al.*, 1989; Robertsen, 1987; Walton and Cervone, 1990). The enzyme has also been isolated from infected plant tissue (Barash *et al.*, 1984). Anion-exchange, cation-exchange, affinity, and gel-filtration column chromatography, as well as preparative electrofocusing and native-polyacryl-amide gel electrophoresis (PAGE) are followed by molecular weight determinations with gel-filtration chromatography, thin-layer electrofocusing, or sodium dodecyl sulfate (SDS)-PAGE. Isoelectric points are determined by isoelectric focusing. The enzyme has also been identified by immunodiffusion analysis (Barash et al., 1984).

In tomato, extraction of endo-PG has received considerable attention because of the timing of its appearance in relation to fruit ripening and softening (Grierson and Tucker, 1983; Knegt et al., 1988; Pressey, 1986c; Tucker et al., 1980). There is no doubt that two major isozymes of endo-PG are found in tomato (Pressey and Avants, 1973a). The two forms, PG1 and PG2, can readily be separated by either anion-exchange or gel-filtration column chromatography (Pressey, 1986b; Pressey and Avants, 1973a).

Isozyme recoveries can be greatly affected by homogenization and extraction conditions (Pressey, 1986a), and this has led to confusion with regard to the role of each isozyme in the ripening process. If pericarp tissue is homogenized in water and adjusted to pH 3 (Pressey, 1986b) or left at the endogenous tissue pH of approximately 4 (Knegt et al., 1988), only 2% of the total polygalacturonase activity is lost, and many of the soluble sugars are removed. Subsequent extraction of the pelleted cell-wall residue at pH 6.0 with 1.0 to 1.25 M NaCl releases PG1 and PG2. Concentration of the clarified homogenate by ultrafiltration (Pressey, 1986b; Knegt et al., 1988) is superior to ammonium sulfate precipitation (Crookes and Grierson, 1983; Moshrefi and Luh, 1983; Tucker et al., 1980), since loss of total polygalacturonase activity, reflected in loss of PG1, occurs with the salt. In addition, activity can be lost during dialysis, in which a precipitate (formed when partitioned against low-ionic-strength solutions) can bind PG1 (Pressey, 1986b).

d. Molecular Weight and Isoelectric Point (pI)

The molecular weight of the recovered fungal enzyme is dependent on source. Enzymes from Geotrichum, Cladosporium, and Rhizoctonia have molecular weights between 34,500 and 40,000 (Barash et al., 1984; De Lorenzo et al., 1990; Marcus et al., 1986; Robertsen, 1987). Endo-PG from Cochliobolus has a molecular-weight range of 27,000 to 41,000, dependent on the method of determination (Walton and Cervone, 1990). Molecular weights in the range of 50,000 to 68,000 have been reported for Rhizopus, Mucor, Aureobasidium, and Colletotrichum (Archer and Fielding, 1979; Prusky et al., 1989). Isozymes occur in the larger-molecular-weight sources and, in several cases, the isozymes were shown to be glycoproteins, differing only in their carbohydrate component. Isoelectric points of the enzyme and the isozymes are reported to vary from 2.6 to 10, dependent on the carbon source of the medium in which the organism is grown.

Both tomato isozymes of endo-PG are glycoproteins (Ali and Brady, 1982; Moshrefi and Luh, 1983). The molecular weight of PG1 prepared by conventional column chromatography and/or SDS-PAGE electrophoresis has been reported to be between 84,000 and 115,000. A molecular weight for PG1 of 199,500 has also been reported. PG1 was found to be formed by binding of PG2 to a glycoprotein *converter* found in cell walls of several plants (Pressey, 1984; Tucker *et al.*, 1981). Binding of the converter confers heat stability to PG2, and this has been the basis of distinguishing in assay the identity of the isozymes (Pressey, 1984).

Less discrepancy is found with the molecular weight of PG2, reported in the range of 42,000 to 46,000 (Ali and Brady, 1982; Biggs *et al.*, 1986; Moshrefi and Luh, 1983; Pressey and Avants, 1973a; Tucker *et al.*, 1980, 1981). *In vitro* translation product analysis and immunological evidence indicates that the molecular weight of PG2 is between 45,000 and 48,000 (DellaPenna *et al.*, 1987; Grierson *et al.*, 1985). Further fractionation by SDS-PAGE of column-purified PG2 obtained from overripe fruit yields two additional isozymes, PG2A and PG2B, with molecular weights of 43,000 and 46,000, respectively. These two isozymes are thought to differ from one another only by slight alterations in the carbohydrate moiety of the glycoprotein. The isoelectric points of PG1, PG2A, and PG2B are 8.6, 9.4, and 9.4, respectively (Ali and Brady, 1982).

Endo-PG has been identified in several other fruit sources. For example, the enzyme has been isolated from mango (Roe and Bruemmer, 1981), papaya (Chan and Tam, 1982; Lazan *et al.*, 1989), peach (Pressey and Avants, 1973b; Pressey and Avants, 1978), pear (Bartley *et al.*, 1982; Pressey and Avants, 1976), pepper (Jen and Robinson, 1984), and watermelon (Elkashif and Huber, 1988). In tissues in which it was characterized, the enzyme was optically active at pH 4.5 to 5.0 (Chan and Tam, 1982; Jen and Robinson, 1984; Pressey and Avants, 1976). The molecular weight, reported in papaya, was 164,000 (Chan and Tam, 1982).

e. Inhibitors

Protein inhibitors of endo-PG activity are known. Polygalacturonase-inhibiting proteins, isolated from cell walls of a number of higher plant sources, have been shown to bind polygalacturonases from fungi, thereby greatly reducing activity (Cervone *et al.*, 1987b; Cervone *et al.*, 1990; De Lorenzo *et al.*, 1990; Walton and Cervone, 1990). No binding occurs between the inhibitor and enzyme from higher plant or bacterial sources. Dissociation of the inhibitor–enzyme complex is affected by pH and salt concentration and results in complete restoration of activity (Cervone *et al.*, 1987b). Epicatechin, a nonprotein inhibitor found in avocado, was found to decrease activity of endo-PG isolated from *Colletotrichum*, with $K_i = 0.29$ mM (Prusky *et al.*, 1989).

C. Exo-Polygalacturonase

1. Substrates and Products of Hydrolysis

Deesterified pectic polymers or polygalacturonans are preferred substrates for exo-polygalacturonase (exo-PG). For enzyme isolated from vegetative tissues of higher plants, a fourfold increase in hydrolysis rate was measured with pectate as substrate when compared to pectin (Pressey and Avants, 1977). No hydrolysis of citrus pectin was observed with the enzyme isolated from peach (Pressey and Avants, 1973b). Chemical deesterification of the pectic polymer resulted in an increase in the rate of hydrolysis when exo-PG from tomato was used (Dick and Labavitch, 1989). However, no effect of deesterification was seen on the substrate hydrolysis rate when pear exo-PG was used.

Substrate size has an effect on enzyme activity. In general, rate of hydrolysis and substrate affinity increase as DP increases. However, in most cases, highest V_{max} and lowest K_m values are observed with substrates of intermediate size. V_{max} and K_m values become less favorable when substrate sizes are above or below the intermediate-size optimum (Pressey, 1987; Pressey and Avants, 1973b, 1975a,b, 1976, 1977; Pressey and Reger, 1989).

The enzyme attacks the substrate at the nonreducing end, resulting in the release of the major end-product of hydrolysis, galacturonic acid (Chan and Tam, 1982; Konno et al., 1989; Pressey, 1987; Pressey and Avants, 1973b, 1975a,b, 1976, 1977; Pressey and Reger, 1989). For enzyme isolated from Erwinia chrysanthemi, only digalacturonic acid accumulates after digestion of the substrate and is therefore considered the major end-product for this source (Collmer et al., 1982). Exo-PG does not digest long-chain polymers to completion. Only 18 and 24% of the total glycosidic bonds of pectate were cleaved with enzyme isolated from peach and carrot, respectively, whereas 50% of the total bonds were hydrolyzed with enzyme from E. chrysanthemi (Collmer et al., 1982; Pressey and Avants, 1975b). Evidence suggests that the enzyme does not remain with the first substrate molecule attacked, but rather dissociates and reassociates with other substrate molecules. This multichain mechanism results in a gradual disappearance of the substrate and appearance of lower-molecular-weight forms as hydrolysis proceeds.

2. Enzyme Characterization

a. pH Optimum and Metal Requirement

The pH optimum of exo-PG is between 4.6 and 6.0 and, in contrast to endo-PG, is independent of substrate size (Chan and Tam, 1982; Collmer et al., 1982; Ghazali and Leong, 1987; Konno et al., 1989; Pressey, 1987; Pressey and

Avants, 1973b, 1975a,b, 1976, 1977; Pressey and Reger, 1989). An optimum of 0.4 to 0.5 mM Ca^{2+} is required for activity in enzyme from most sources examined, and additions of chelating agents such as ethylenediaminetetraacetic acid (EDTA) or citrate reduce activity to near zero. Higher Ca^{2+} concentrations result in substrate insolubilization with subsequent reduced enzyme activity. Sr^{2+}, Cd^{2+}, and Mg^{2+} can substitute for the Ca^{2+} requirement, but generally are not so effective. Ba^{2+}, Hg^{2+}, Cu^{2+}, Mn^{2+}, and Cd^{2+} are reported to be inhibitory to exo-PG isolated from some sources. High concentrations of monovalent cations, especially Na^+, are also inhibitory (Konno et al., 1989; Pressey, 1987; Pressey and Avants, 1973b, 1975a,b, 1976, 1977; Pressey and Reger, 1989). No metal requirement was reported for exo-PG isolated from carrot (Konno et al., 1989; Pressey and Avants, 1975b) or Erwinia (Collmer et al., 1982).

b. Isolation and Purification

Concentration of homogenates or filtrates by ammonium sulfate precipitation or ultrafiltration, followed by conventional ion-exchange chromatography, gel-filtration chromatography, preparative-PAGE, and electrofocusing techniques have been commonly employed for isolation and partial enzyme purification (Collmer et al., 1982; Dick and Labavitch, 1989; Konno et al., 1989; Pressey, 1987; Pressey and Reger, 1989). Separation of exo- from endo-PGs has been achieved by exploiting their differential solubilities in salt (Chan and Tam, 1982) or different chromatographic binding behaviors (Pressey, 1987; Pressey and Avants, 1973b; Robertsen, 1987).

c. Molecular Weight and pI

A wide range of molecular weights has been reported for exo-PG. The enzyme from papaya has a reported molecular weight of 33,600 (Chan and Tam, 1982), while the enzyme from all other examined sources ranges from 47,000 to 68,000 (Collmer et al., 1982; Konno et al., 1989; Pressey, 1987; Pressey and Avants, 1973b, 1975a,b, 1977; Pressey and Reger, 1989). Konno et al. (1989) have determined the pI of the carrot enzyme to be 6.2, whereas the measured pI of enzyme from Erwinia was 8.3 (Collmer et al., 1982).

D. Production and Biology of the Polygalacturonases

Polygalacturonases are produced by higher and lower plants and bacteria and are associated with the cell wall. The role of the enzyme in microbial pathogenesis (Collmer and Keen, 1986), fruit ripening (Brady, 1987), abscission (Riov, 1974), and growth (Pressey, 1986a) has been either established or proposed. Ripening tomato fruit and pollen appear to be the richest sources of endo- and exo-PGs from higher plants, respectively (Pressey, 1986a; Pressey and Reger, 1989). Endo-PG is a major extracellular enzyme secreted by bacteria or fungi in response to a suitable substrate in culture or *in planta* (Collmer and Keen, 1986).

Many plant-pathogenic bacteria and fungi produce endo-PG in culture in response to an inductive carbon source such as pectin, polygalacturonic acid, or isolated plant cell walls (Collmer and Keen, 1986; and references therein). Correlations exist between the presence of endo-PG and plant cell-wall degradation, tissue maceration, and symptom development. Partially purified enzyme can effectively macerate host tissue and release cell-wall polymers and oligomers (Prusky *et al.*, 1989; Robertsen, 1987), and similarly, symptoms appear as a result of enzyme injections (Barash *et al.*, 1984). Whether the enzyme is the cause of pathogenesis or merely a result of disease progression has not been established, however. *Aspergillus nidulans*, a saprophyte with limited pathogenic potential, produces levels of endo-PG equal to or exceeding those of established plant pathogens (Dean and Timberlake, 1989a). The role of the enzyme in pathogenesis is therefore questioned and awaits further clarification, possibly by the development of mutants altered specifically at the level of the polygalacturonase gene.

Exo-PG is also produced by plant pathogens in culture and in infected tissues. In contrast to endo-PG, it is produced in much less quantity and does not effectively macerate tissue (Bugbee, 1990; Collmer and Bateman, 1982; Collmer *et al.*, 1982; Robertsen, 1987; Willis *et al.*, 1987). A role for the enzyme has been suggested for *E. chrysanthemi*, in which dimers released from polymer hydrolysis are assimilated with subsequent induction of pectate lyase (Collmer and Bateman, 1982; Collmer *et al.*, 1982).

The action of fungal endo-PG has been shown to activate several plant-defense responses by releasing pectic oligomers as a result of the ongoing digestion of host cell walls. Necrosis, lignification, phytoalexin production, and induction of protease inhibitor proteins are elicited responses of host cells to released oligomers (Bishop *et al.*, 1984; Cervone *et al.*, 1987a; Jin and West, 1984; Lee and West, 1981; Robertsen, 1986, 1987; Walker-Simmons *et al.*, 1984). The responses are elicited by small oligomers of various lengths, usually greater than DP = 4, and are thought to act by the regulation of gene expression (Collinge and Slusarenko, 1987). Seemingly to ensure an adequate supply of proper-sized oligomers, many plants contain endo-PG–binding proteins in their cell walls which, when released by the action of fungal enzyme digestion, will bind to and subsequently inhibit the enzyme (Cervone *et al.*, 1987b, 1989). Although residual enzyme activity still exists after binding, it is sufficient to provide a supply of elicitor-active oligomers with a half-life many times greater than that of oligomers hydrolyzed in the absence of the inhibitor (Cervone *et al.*, 1989).

Products of polygalacturonase-mediated hydrolysis have also been shown to elicit ethylene production. Infiltration of pectic fragments of DP > 8 into pre-climacteric tomato fruit resulted in the stimulation of ethylene production (Brecht and Huber, 1988). Although galacturonic acid also elicits ethylene production in tomato (Baldwin and Pressey, 1990; Kim *et al.*, 1987), the response was not as striking as with fragments of DP = 2–10 (Baldwin and Pressey, 1988b). Infiltration of the polygalacturonase enzymes into tissues also elicits an ethylene response, presumably by the release of active pectic oligomers (Baldwin and Pressey, 1988a, 1989).

The role of endo-PG in ripening has been extensively studied, especially in tomato fruit where it is particularly abundant (Pressey, 1986a). Although exo-PG is present in tomato fruit, the amount and activity is quite low and unchanged during ripening (Pressey, 1987). Endo-PG was once thought to have an initiative function in tomato fruit ripening (Tigchelaar *et al.*, 1978). Later work discounted this theory, however, by confirmation of ethylene synthesis before the appearance of endo-PG (Grierson and Tucker, 1983).

Appearance of endo-PG during ripening of tomato pericarp is a result of *de novo* synthesis of the enzyme (Tucker and Grierson, 1982). No endo-PG mRNA, endo-PG protein, or enzyme activity was detected in immature or mature green fruit (Biggs and Handa, 1989). The endo-PG gene becomes transcriptionally active in the late mature green stage of ripening (DellaPenna *et al.*, 1989). Endo-PG mRNA, endo-PG protein, and endo-PG activity peak at the breaker, ripe, and red-ripe stages, respectively (Biggs and Handa, 1989). Activation of transcription at the late mature green stage suggests that this is an important regulation point. Levels of endo-PG mRNA and further processing of the endo-PG protein indicate that regulation also occurs at the translational and posttranslational levels (Biggs and Handa, 1989; Biggs *et al.*, 1986; DellaPenna *et al.*, 1989).

The *converter*, a glycoprotein of MW 102,000, binds PG2 to form PG1 (Pressey, 1984; Tucker *et al.*, 1981). The role of the converter in the ripening process has not been resolved. Endo-PG first appears during ripening as PG1, but later in the ripening process, PG2 becomes the predominant species. The PG converter may be responsible for anchoring PG2 to the cell-wall substrate (Knegt *et al.*, 1988). Alternatively, PG1 may be an artifact of extraction generated as PG2 and converter are solubilized simultaneously. Therefore, PG2 may be the only polygalacturonase *in vivo* (Pressey, 1988).

Previous research has implied a relationship between endo-PG activity, cell-wall uronide solubilization, and loss of firmness during ripening in tomato (Ahrens and Huber, 1990; Brady *et al.*, 1985; Crookes and Grierson, 1983; Huber, 1983; Pressey and Avants, 1982a; Themmen *et al.*, 1982). Similar relationships exist in mango (Roe and Bruemmer, 1981), pear (Bartley *et al.*, 1982) papaya (Lazan *et al.*, 1989), and peach (Pressey and Avants, 1978). Ripening mutants of tomato, such as *rin*, having little or no endo-PG activity, also display limited uronide solubilization and remain firm as ripening proceeds (Hobson, 1980; Tigchelaar *et al.*, 1978). In some fruit types, the relationship is less clear. For example, polyuronide solubilization and softening occurs in the absence of endo-PG activity in strawberry (Huber, 1984) and muskmelon (McCollum *et al.*, 1989). Significant textural changes also occur during ripening in apples and starfruit, yet both organs contain only exo-polygalacturonase activities (Ghazali and Leong, 1987; Knee and Bartley, 1981).

Current evidence has also suggested that the role of endo-PG in tomato fruit softening should be questioned. Although correlations do exist between endo-PG activity and fruit softening, changes in whole-fruit firmness as ripening ensues is rarely closely correlated with the magnitude of endo-PG increase (Brady *et al.*, 1985). However, uronide solubilization and endo-PG activity can be more closely associated with softening when tissue in which the enzyme is localized is chosen for firmness measurements (Ahrens and Huber, 1990). The mutant *dark green*, which remains firm when ripe, has twice the endo-PG activity but less soluble uronide released than does the normal wild-type (Tong and Gross, 1989). When the structural PG gene was inserted into the *rin* ripening mutant and subsequently expressed, polyuronides were solubilized, but softening did not occur (Giovanni *et al.*, 1989). Finally, when expression of the PG gene was markedly reduced by the incorporation of antisense RNA, transformed fruit softened to the same extent as did the normal control fruit (Smith *et al.*, 1988). Clearly, the interaction of endo-PG and the cell wall is a complex process and may not result in direct initiation of the textural changes associated with ripening.

Polygalacturonases, especially the exo-PG, may also be responsible for mechanical changes occurring during growth responses in the cell wall. Exo-PG was found in higher concentrations in rapidly growing hypocotyls of several monocots (Pressey and Avants, 1977). This spatial association with rapidly growing tissues suggests that this enzyme may be responsible cell-wall metabolism in growth responses. Alternatively, the polygalacturonases may also be important in transglycosylation reactions (Bolwell, 1988; Jarvis, 1984). Exo-PG has also been associated with abscission zones, where its appearance and activity are accelerated by ethylene (Riov, 1974). Growth or abscission could be permitted by the disconnection of covalent linkage points along the pectic chain by either enzyme.

II. THE LYASES

A. Introduction

Lyases that degrade pectin and D-galacturonan polymers and oligomers are enzymes that cleave the C–O glycosidic bond of α-D-$(1 \rightarrow 4)$galacturonans by the mechanism of β-elimination (Rexova-Benkova and Markovic, 1976). Products are generated with a 4,5-unsaturated galacturonosyl residue on the nonreducing end of the cleaved substrate. Unsaturated monomers released are quickly rearranged to form 4-deoxy-5-ketouronic acid (Albersheim *et al.*, 1960; Preiss and Ashwell, 1963).

Degradation of pectin or galacturonan by the lyases occurs in either a random (endo) or terminal (exo) fashion. With either action pattern, the unsaturated oligomeric products formed will absorb light at 235 nm. Products of β-elimination will also react with periodate to yield formylpyruvic acid. The latter product will react with thiobarbituric acid to give a red chromogenic product that absorbs light at 545 to 550 nm. Polygalacturonan does not react. Galacturonic acid will react, however, to yield a product that absorbs light at 510 nm. Thus, the products of lyase and polygalacturonase can be monitored simultaneously (Ayers *et al.*, 1966; Dean and Timberlake, 1989a).

B. Endo-Pectate Lyase

1. Substrates and Products of Hydrolysis

The optimal substrate of endo-pectate lyase (endo-PL, EC 4.2.2.2) is deesterified pectin or polygalacturonic acid (Karbassi and Luh, 1979; Konno and Yamasaki, 1982; Moran *et al.*, 1968a; Nasuno and Starr, 1967; Starr and Moran, 1962). Rate of substrate degradation decreases as degree of esterification increases. No activity is detected when polymethylpolygalacturonic acid is used as substrate. With some enzyme sources, however, optimal activity has been reported with substrates having a low degree of esterification (Rexova-Benkova and Markovic, 1976).

Substrate chain length affects enzyme affinity and rate of reaction. Polygalacturonic acid (DP, 134) was degraded at a higher rate than was the tetramer, trimer, or dimer of galacturonic acid (Moran *et al.*, 1968a). The enzyme has a much greater affinity for acid-soluble pectic acid (DP, 10–12) than has the successive series of octamer through trimer, respectively (Nagel and Anderson, 1965; Nagel and Wilson, 1970). As substrate chain length decreases, V_{max} decreases and K_m increases.

The products of endo-PL action, when polygalacturonic acid is used as substrate, are oligomers of various sizes and, in some cases, a small amount of monomer (Collmer and Bateman, 1982; Karbassi and Luh, 1979; Moran *et al.*, 1968a; Nasuno and Starr, 1967). As digestion times increase, larger oligomers disappear. Unsaturated digalacturonic acid is the predominant final end-product; however, the unsaturated trimer is the smallest end-product reported in a soil-borne *Bacillus aerobe* (Hasegawa and Nagel, 1966).

The action of endo-PL on saturated and unsaturated oligosaccharides is indicative of the final stages of polymer breakdown. The glycosidic bond adjacent to the reducing end of the oligomer (bond 1) is never attacked by the enzyme (Moran *et al.*, 1968a; Nagel and Anderson, 1965; Nasuno and Starr, 1967), and it has been suggested that the reducing group in some way prevents enzymatic attack of the adjacent bond. The breaking of bond 3 is preferred in a pentamer or tetramer, and occurs at a faster rate than that of bond 2. Cleavage of saturated pentamer yields predominantly saturated and unsaturated dimers and trimers. Bond 4 is cleaved at the beginning of the reaction, but other bonds are preferred as breakdown proceeds. Degradation of the unsaturated pentamer yields unsaturated dimer and unsaturated trimer only. Nagel and Hasegawa (1967) reported that 96% of saturated tetramer cleavage occurred at bond 3, whereas only 4% occurred at bond 2. Unsaturation of the terminal galacturonic acid group diminishes the activity of the enzyme. The presence of the unsaturated bond reduces the affinity of bond 3, but enhances it for bond 2 (Anderson and Nagel, 1964; Moran et al., 1968a; Nagel and Anderson, 1965; Nagel and Hasegawa, 1967; Nagel and Wilson, 1970). Saturated monomer and dimer, and unsaturated dimer and trimer are the products of saturated tetramer cleavage. Unsaturated tetramer yields only unsaturated dimer. In the case of the trimer, degradation is exclusively limited to bond 2. The rate of unsaturated trimer degradation is much less than the saturated trimer.

2. Enzyme Characterization

a. pH Optimum

The optimal pH for endo-PL activity has been reported to be between 8.0 and 9.5 (Ayers et al., 1966; Hasegawa and Nagel, 1966; Karbassi and Luh, 1979; Konno and Yamasaki, 1982; Moran et al.,1968a; Nagel and Anderson, 1965; Nagel and Hasegawa, 1967; Nasuno and Starr, 1967; Nagel and Wilson, 1970; Preiss and Ashwell, 1963; Starr and Moran, 1962). Within this pH range, a shift toward a lower optimum has been reported to occur with substrates of decreasing chain length (Nagel and Anderson, 1965).

b. Metal Requirement

The enzyme has a requirement for Ca^{2+}, the optimum being 1 mM (Kabassi and Luh, 1979; Moran et al., 1968a; Nasuno and Starr, 1967; Starr and Moran, 1962). In the presence of EDTA, significant inhibition occurs, which can be restored with Ca^{2+} additions. Ca^{2+} is thought to interact with the substrate to form a more-favorable substrate geometry for enzymatic attack (Nagel and Wilson, 1970). Equimolar Sr^{2+} additions could partially replace the Ca^{2+} requirement in some enzyme sources (Hasegawa and Nagel, 1966) but not in others (Nasuno and Starr, 1967). Ba^{2+} reduced endo-PL activity (Konno and Yamasaki, 1982; Nasuno and Starr, 1967) but had no effect with enzyme isolated from *Bacillus* (Hasegawa and Nagel, 1966). Hg^{2+}, Mg^{2+}, and Mn^{2+} at 1 mM were also inhibitory to enzyme activity, as were K^+ and Na^+ at higher concentrations (Konno and Yamasaki, 1982; Preiss and Ashwell, 1963).

c. Isolation and Purification

Isolation and partial purification of endo-PL has been achieved with minimal or no contamination from other pectin-degrading enzymes. The enzyme is usually concentrated by either precipitation or ultrafiltration and then separated with anion-exchange, cation-exchange, gel-filtration, and/or affinity chromatography. Further purification is achieved using nondenaturing gel-electrophoretic techniques (Collmer and Bateman, 1982; Dean and Timberlake, 1989b; Keen and Tamaki, 1986; Lei *et al.*, 1987, 1988; Thurn *et al.*, 1987).

d. Molecular Weight and pI

Column chromatography followed by isoelectric focusing has revealed multiple forms or isozymes of endo-PL, and this has been the case for enzyme isolated from *Bacillus polymyxa, Pseudomonas fluorescens, Erwinia carotovora*, and *E. chrysanthemi* (Keen and Tamaki, 1986; Lei *et al.*, 1988; Liao, 1989; Nagel and Wilson, 1970; Thurn *et al.*, 1987; Willis *et al.*, 1987). The isoelectric points of most of the isozymes range from 8.8 to 10.0 (Keen and Tamaki, 1986; Liao, 1989; Thurn *et al.*, 1987). One form of pectate lyase in *E. chysanthemi* and the enzyme from *Aspergillus nidulans* had an acidic pI of 4.2. In *Cytophaga johnsonae*, a pI of 6.7, and in *E. carotovora*, a pI of 6.6 was reported (Dean and Timberlake, 1989b; Liao, 1989; Thurn *et al.*, 1987; Willis *et al.*, 1987). Molecular weights of the enzyme and associated isozymes are 40,000 to 47,500 (Dean and Timberlake, 1989b; Keen and Tamaki, 1986; Lei *et al.*, 1987, 1988; Thurn *et al.*, 1987). The molecular weight of *C. johnsonae* endo-PL was reported as 35,000 (Liao, 1989). Slightly lower molecular weights have been deduced in processed protein arising from DNA sequencing (Lei *et al.*, 1987, 1988).

C. Oligogalacturonide Lyase

1. Substrates and Products of Hydrolysis

A lyase that degraded galacturonic oligomers was isolated from several bacterial sources. The optimal substrate was shown to be unsaturated dimer for the enzyme isolated from *E. carotovora* and *E. aroideae* (Hatanaka and Ozawa, 1970; Moran *et al.*, 1968b). Saturation of the substrate resulted in a threefold increase in K_m (Moran *et al.*, 1968b). Increasing chain length also decreased enzyme affinity. Very little activity occurred when acid-soluble pectic acid, polygalacturonate, or polymethylpolygalacturonate was utilized as substrate. The oligogalacturonide lyase (OG lyase, EC 4.2.2.6), isolated from a *Pseudomonas* source, was found to have different substrate preferences. Rate of degradation was unrelated to saturation, and the tetramer was preferentially cleaved over the dimer (Hatanaka and Ozawa, 1971).

The action pattern of oligomer cleavage indicates that the enzyme is largely capable of cleaving bond 1, but is not limited to that linkage. Unsaturated digalacturonic acid is cleaved to produce only 4-deoxy-5-ketouronic acid, whereas saturated dimer yields equal amounts of 4-deoxy-5-ketouronic acid and galacturonic acid (Hatanaka and Ozawa, 1970, 1971; Moran et al., 1968b). Larger substrates are attacked from the reducing end of the chain. Degradation of the saturated trimer yields 4-deoxy-5-ketouronic acid, unsaturated dimer, galacturonic acid, and saturated dimer. The presence of galacturonic acid and unsaturated dimer suggests that bond 2 is also susceptible to attack. However, only 4-deoxy-5-ketouronic acid is produced when larger substrates are degraded (Hatanaka and Ozawa, 1971).

2. Enzyme Characterization

a. pH Optimum and Metal Requirement

The pH optimum of the enzyme lies between 7.0 and 7.2, and no effect of substrate saturation is found (Hatanaka and Ozawa, 1970, 1971; Moran et al., 1968b). Increasing chain length resulted in decreased pH optima (Hatanaka and Ozawa, 1970). Ca^{2+} has no effect on activity of the enzyme isolated from Erwinia sources, but a 70–80% increase in activity was reported for the Pseudomonas enzyme with 0.5 mM Ca^{2+} (Hatanaka and Ozawa, 1971).

b. Isolation and Purification

Isolation and partial purification of OG lyase has been achieved by conventional techniques. Ammonium sulfate precipitation followed by anion-exchange chromatography has been used (Hatanaka and Ozawa, 1970, 1971; Moran et al., 1968b). Complete separation of endo-PL from OG lyase activity was reported with utilization of the appropriate chromatographic isolation techniques.

D. Exo-Pectate Lyase

1. Substrates and Products of Hydrolysis

Exo-pectate lyase (exo-PL, EC 4.2.2.9) has been isolated from only a few sources, and for this reason, is considered relatively rare (Rexova-Benkova and Markovic, 1976). Rate of substrate degradation is independent of chain length, but the trimer is the smallest substrate attacked. Affinity for the substrate decreases as degree of esterification increases (Macmillan and Vaughn, 1964; Macmillan et al., 1964). The substrate is attacked from the reducing end of the chain, and the only product of degradation is unsaturated digalacturonic acid (Macmillan and Vaughn, 1964; Macmillan et al., 1964; Okamoto et al., 1964).

2. Enzyme Characterization

a. pH Optimum and Metal Requirement

The optimal pH for enzymatic reaction with substrates of various chain lengths is between 8.0 and 8.5 (Macmillan and Vaughn, 1964; Okamoto et al., 1964). The enzyme appears to have a divalent cation requirement. For enzyme isolated from C. multifermentans, Ca^{2+} is required at 0.5 mM. Decreasing chain length in the presence of Ca^{2+}, however, results in an inhibition of enzyme activity. Ca^{2+} may bridge and interact with small substrates in a way that prevents substrate–enzyme binding (Macmillan and Vaughn, 1964). In other sources, Na^+, Co^{2+}, and Mn^{2+} have a slight stimulatory effect on enzyme activity, and a decrease in activity occurs with Cu^{2+} and Hg^{2+} (Hatanaka and Ozawa, 1972).

E. Endo-Pectin Lyase

1. Substrates and Products of Hydrolysis

Endo-pectin lyase (endo-PNL, EC 4.2.2.10) is the only enzyme known to degrade highly esterified pectins without the aid of additional pectic enzymes (Alana et al., 1989; Rexova-Benkova and Markovic, 1976). No exo-pectin lyase activity has been detected in sources so far examined. Polymethylpolygalacturonic acid is the preferred substrate for endo-PNL, and enzymatic activity decreases as degree of esterification decreases. Polygalacturonic acid is not degraded (Alana et al., 1989; Albersheim et al., 1960; Edstrom and Phaff, 1964a, 1964b). Enzymatic degradation proceeds more rapidly when the substrate contains blocks of esterified residues than when a random distribution of methyl esterification occurs (Edstrom and Phaff, 1964b; Rexova-Benkova and Markovic, 1976).

Products of various sizes are released as a result of endo-PNL activity. Intermediates between octa- and monogalacturonides were produced by *Rhizoctonia solani* with pectin as substrate (Marcus et al., 1986). Larger substrates are preferred by the enzyme. As substrate size decreases, however, rate of degradation decreases. Research with saturated polymethylated oligomers has indicated patterns of cleavage in the final stages of degradation. Unsaturated substrates have not been investigated. For enzyme from *Aspergillus fonsecaeus*, the saturated tetramer is the smallest substrate attacked. The third bond of oligomeric substrates is preferentially cleaved, but exhaustive treatment of an oligomer may yield products that suggest cleavage at bond 4, although at a much slower rate. Bonds 1 and 2 are not susceptible to attack, and therefore, the trimer is not degraded (Edstrom and Phaff, 1964b).

2. Enzyme Characterization

a. pH Optimum

The optimal pH for enzymatic activity with polymethylpolygalacturonic acid or pectin as substrate is between 4.9 and 6.5 (Baldwin and Pressey, 1989; Edstrom and Phaff, 1964a; Rexova-Benkova and Markovic, 1976). Decreasing esterification results in a slight decrease in pH optima (Rexova-Benkova and Markovic, 1976). Alkaline pH optima have been reported for fungal enzyme isolated from several infected plant tissues. Such different pH optima may reflect the ability of organisms to adapt and then further degrade the infected tissue despite its alkaline nature (Bugbee, 1990; Marcus et al., 1986; Wijesundera et al., 1984).

b. Metal Requirement

No metal requirement exists for enzymatic activity but, in most cases, Ca^{2+} at 0.5 to 1.0 mM has been shown to increase activity with pectin as substrate. Higher concentrations are inhibitory (Albersheim and Killias, 1962; Edstrom and Phaff, 1964a, 1964b; Wijesundera et al., 1984). No Ca^{2+} effect is seen with polymethylpolygalacturonic acid as substrate. Ca^{2+} is thought not simply to be a replacement for the methyl group in partially deesterified pectins, but rather to promote optimal interaction of substrate with enzyme (Edstrom and Phaff, 1964b). Mg^{2+} can partially replace Ca^{2+}, as can Na^+ with even less effect (Albersheim and Killias, 1962; Edstrom and Phaff, 1964a).

c. Isolation and Purification

Isolation and partial purification of endo-PNL has been achieved by the use of conventional techniques. As with endo-PL, the enzyme is often prepared from culture filtrates that contain specific substrates, precluding the need for lengthy clean-up procedures. Conventional column chromatography followed by high-pressure liquid chromatography on various ion-exchange columns have been successful in removing contaminating pectic enzymes from endo-PNL (Albersheim and Killias, 1962; Baldwin and Pressey, 1989; Edstrom and Phaff, 1964a; Marcus et al., 1986; Wijesundera et al., 1984). With these procedures, three isoenzymes have been identified in commercial pectinase preparations (Baldwin and Pressey, 1989), and two, in culture filtrates of Colletotrichum lindemuthianum (Wijesundera et al., 1984).

d. Molecular Weight and pI

Sephadex gel-filtration chromatography and SDS-PAGE have been used for assessment of molecular weight. The molecular weight of the enzyme and its different forms has been reported to be between 23,500 and 35,000 (Baldwin and Pressey, 1989; Bugbee, 1990; Wijesundera et al., 1984). The molecular weight of the enzyme prepared from Rhizoctonia solani was reported as 45,000 to 46,000 (Marcus *et al.*, 1986), perhaps as a result of dimerization. Isoelectric focusing revealed pI values between 8.1 and 10.1.

e. Inhibitors

Polyols have been found to inhibit endo-PNL activity in enzyme isolated from several filamentous fungi and yeasts. Inositol and mannitol were increasingly inhibitory at concentrations ranging from 0.5 to 10 mM. Glycerol is also effective in this range, with near-complete inhibition at 500 mM (Reyes *et al.*, 1984). Enzyme isolated from *Penicillium italicum* was very sensitive to glycerol inhibition. A 50% inhibition occurred with 6.5 mM glycerol (Alana *et al.*, 1989).

F. Production and Biology of the Lyases

The lyases that degrade pectin have been found exclusively in fungi and bacteria (Rexova-Benkova and Markovic, 1976). With exception of a single report in pea tissue (Albersheim and Killias, 1962), lyase activity has not been reported in higher plants. However, a possible additional exception has been reported in pollen of tomato anthers, in which protein sequences derived from anther-specific cDNAs have striking homology with *Erwinia* pectate lyase sequences (Wing *et al.*, 1989). Whether the homologous sequences actually result in lyase activity was not explored.

Most lyases are found in the extracellular spaces of the organisms that produce them. The enzymes are often produced and excreted in response to an induction event, such as the presence of suitable substrate (Collmer and Keen, 1986). Thus, endo-PL, exo-PL, and endo-PNL are considered to be extracellular enzymes (Collmer and Bateman, 1982; Collmer and Keen, 1986; Macmillan and Vaughn, 1964; Wijesundera *et al.*, 1984). OG lyase is thought to be intracellular, where its direct action on assimilated oligomeric substrates results in production of 4-deoxy-5-ketouronic acid. Further metabolism of this compound leads to the formation of pyruvate and 3-phosphoglyceraldehyde (Hugouvieux-Cotte-Pattat and Robert-Baudouy, 1985; Moran *et al.*, 1968b).

Endo-PLs produced by the *Erwinia* soft-rot bacteria are induced in culture and *in planta*. *E. chrysanthemi* and *E. carotovora* subspecies *carotovora* have been extensively studied. In culture, the rate of endo-PL production increases in the presence of isolated cell walls, polygalacturonan, or oligogalacturonides (Collmer and Keen, 1986; Tsuyumu, 1977). Endo-PL in these species is not directly induced by extracellular substrates. Induction of endo-PL begins with an extracellular digestion phase. Basal levels of digestion products, predominantly dimers, are imported into the intracellular space, where they are degraded by OG lyase. The various products of 4-deoxy-5-ketouronic acid metabolism form the inducers of lyase and other pectic enzyme synthesis.

Pectic enzyme–mediated cell death, in general, is thought to be initiated by the degradation of the pectic component of the cell wall. The weakened cell wall is unable to support the underlying protoplast, which then bursts. The initial stage of cell death can also involve release of substances from the cell wall, which may also cause death in certain hosts (Movahedi and Heale, 1990).

There have been conflicting reports about the role of the lyases in microbial pathogenicity of plants. Evidence for the role of pectate lyase in pathogenicity of *Erwinia* is accumulating. Deletion of the genes coding for the four isozymes of endo-PL in *E. chrysanthemi* (strain EC16) resulted in substantial loss of pathogenicity (Collmer and Keen, 1986). Incorporation of one or more endo-PL genes into *Escherichia coli* also confers a limited amount of pathogenicity to an organism that normally lacks pectic enzymes (Keen and Tamaki, 1986). A clear relationship between lyase production and pathogenicity in other organisms has not been established, however. *A. nidulans,* a saprophyte, produces levels of pectate lyase and polygalacturonase in culture comparable to the soft-rot organisms, yet has limited pathogenic potential (Dean and Timberlake, 1989a). Not all isozymes are necessary for pathogenicity, since the pathogenicity of *Xanthomomas campestris,* pv. *campestris* is unaffected by removal of one of three pectate lyase genes from the genome (Dow *et al.,* 1989).

Pectate lyase genes *(pel)* have been cloned into *E.coli* from *Erwinia.* Using this technique, five isozymes that represent two distinct gene clusters have been isolated from *E. chrysanthemi.* One gene cluster contains *pelB* and *pelC* genes, which share a high degree of homology. The other cluster contains genes *pelA, pelD,* and *pelE.* The *pelE* gene has been shown to have the highest macerating potential of *pel* genes examined (Thurn *et al.,* 1987). Three *pel* genes have been identified in *E. carotovora* subspecies *carotovora.* The three genes are highly homologous, with *pelA* and *pelB* exhibiting the highest homology (Lei *et al.,* 1988). The *pelA* and *pelB* genes from *E. carotovora* and *pelB* and *pelC* genes from *E. chrysanthemi* have considerable homology and may have been derived from the same ancestral gene sequence (Collmer and Keen, 1986).

References

Ahrens, M. J., and Huber, D. J. (1990). Physiology and firmness determination of ripening tomato fruit. *Physiol. Plant* **78,** 8–14.

Alana, A., Gabilondo, A., Hernando, F., Moragues, M. D., Dominguez, J. B., Llama, M. J., and Serra, J. L. (1989). Pectin lyase production by a *Pencillium italicum* strain. *Appl. Environ. Microbiol.* **55,** 1612–1616.

Albersheim, P., and Killias, U. (1962). Studies relating to the purification and properties of pectin transeliminase. *Arch. Biochem. Biophys.* **97,** 107–115.

Albersheim, P., Neukom, H., and Deuel, H. (1960). Uber die bildung von ungesattigten abbau-produkten durch ein pektinabbauendes enzym. *Helv. Chim. Acta* **43,** 1422–1426.

Ali, Z. M., and Brady, C. J. (1982). Purification and characterization of the polygalacturonases of tomato fruit. *Aust. J. Plant Physiol.* **9,** 155–169.

Anderson, M. M., and Nagel, C. W. (1964). Effect of the unsaturated bond on the degradation of the tetragalacturonic acids by a transeliminase. *Nature* **203,** 649.

Archer, S. A. (1979). Pectolytic enzymes and degradation of pectin associated with breakdown of sulphated strawberries. *J. Sci. Food Agr.* **30,** 692–703.

Archer, S. A., and Fielding, A. H. (1979). Polygalacturonase isoenzymes of fungi involved in the breakdown of sulphated strawberries. *J. Sci. Food Agr.* **30,** 711–723.

Ayers, W. A., Papavizas, G. C., and Diem, A. F. (1966). Polygalacturonate *trans*-eliminase and polygalacturonase production by *Rhizoctonia solani. Phytopathology* **56,** 1006–1011.

Baldwin, E. A., and Pressey, R. (1988a). Tomato polygalacturonase elicits ethylene production in tomato fruit. *J. Am. Soc. Hort. Sci.* **113,** 92–95.

Baldwin, E. A., and Pressey, R. (1988b). Treatment of tomatoes with an exo-enzyme increases ethylene and accelerates ripening. *Proc. Fla. State Hort. Soc.* **101,** 215–217.

Baldwin, E. A., and Pressey, R. (1989). Pectic enzymes in pectolyase. Separation, characterization, and induction of ethylene in fruits. *Plant Physiol.* **90,** 191–196.

Baldwin, E. A., and Pressey, R. (1990). Exopolygalacturonase elicits ethylene production in tomato. *Hort Science* **25,** 779–780.

Barash, I., and Eyal, Z. (1970). Properties of a polygalacturonase produced by *Geotrichum candidum. Phytopathology* **60,** 27–30.

Barash, I., Zilberman, E., and Marcus, L. (1984). Purification of *Geotrichum candidum* endopolygalacturonase from culture and from host tissue by affinity chromatography on cross-linked polypectate. *Physiol. Plant Pathol.* **25,** 161–169.

Barmore, C. R., and Brown, G. E. (1981). Polygalacturonase from citrus fruit infected with *Penicillium italicum. Phytopathology* **71,** 328–331.

Barmore, C. R., Snowden, S. E., and Brown, G. E. (1984). Endopolygalacturonase from Valencia oranges infected with *Diplodia natalensis. Phytopathology* **74,** 735–737.

Bartley, I. M., Knee, M., and Casimir, M.-A. (1982). Fruit Softening. I. Changes in cell-wall composition and endo-polygalacturonase in ripening pears. *J. Exp. Bot.* **33,** 1248–1255.

Biggs, M. S., and Handa, A. K. (1989). Temporal regulation of polygalacturonase gene expression in fruits of normal, mutant, and heterozygous tomato genotypes. *Plant Physiol.* **89,** 117–125.

Biggs, M. S., Harriman, R. W., and Handa, A. K. (1986). Changes in gene expression during tomato fruit ripening. *Plant Physiol.* **81,** 395–403.

Bishop, P. D., Pearce, G., Bryant, J. E., and Ryan, C. A. (1984). Isolation and characterization of the proteinase inhibitor–inducing factor from tomato leaves. Identity and activity of poly- and oligogalacturonide fragments. *J. Biol. Chem.* **259,** 13172–13177.

Bolwell, G. P. (1988). Synthesis of cell-wall components: Aspects of control. *Phytochemistry* **27,** 1235–1253.

Brady, C. J. (1987). Fruit ripening. *Annu. Rev. Plant Physiol.* **38,** 155–178.

Brady, C. J., McGlasson, W. B., Pearson, J. A., Meldrum, S. K., and Kopeliovitch, E. (1985). Interactions between the amount and molecular forms of polygalacturonase, calcium, and firmness in tomato fruit. *J. Am. Soc. Hort. Sci.* **110,** 254–258.

Brecht, J. K., and Huber, D. J. (1988). Products released from enzymatically active cell wall stimulate ethylene production and ripening in preclimacteric tomato (*Lycopersicon esculentum* Mill.) fruit. *Plant Physiol.* **88,** 1037–1041.

Bugbee, W. M. (1990). Purification and characteristics of pectin lyase from *Rhizoctonia solani. Physiol. Mol. Plant Pathol.* **36,** 15–25.

Cervone, F., De Lorenzo, G., Degra, L., and Salvi, G. (1987a). Elicitation of necrosis in *Vigna unguiculata* Walp. by homogeneous *Aspergillus niger* endo-polygalacturonase and by α-D-galacturonate oligomers. *Plant Physiol.* **85,** 626–630.

Cervone, F., De Lorenzo, G., Degra, K., Salvi, G., and Bergami, M. (1987b). Purification and characterization of a polygalacturonase-inhibiting protein from *Phaseolus vulgaris* L. *Plant Physiol.* **85,**631–637.

Cervone, F., Hahn, M. G., De Lorenzo, G., Darvill, A., and Albersheim, P. (1989). Host–pathogen interactions. XXXIII. A plant protein converts a fungal pathogenesis factor into an elicitor of plant defense responses. *Plant Physiol.* **90,** 542–548.

Cervone, F., De Lorenzo, G., Pressey, R., Darvill, A. G., and Albersheim, P. (1990). Can *Phaseolus* PGIP inhibit pectic enzymes from microbes and plants? *Phytochemistry* **29**, 447–449.

Chan, H. T., and Tam, S. Y. T. (1982). Partial separation and characterization of papaya endo- and exo-polygalacturonase. *J. Food Sci.* **47**, 1478–1483.

Collinge, D. B., and Slusarenko, A. J. (1987). Plant gene expression in response to pathogens. *Plant Mol. Biol.* **9**, 389–410.

Collmer, A., and Bateman, D. F. (1982). Regulation of extracellular pectate lyase in *Erwinia chrysanthemi:* Evidence that reaction products of pectate lyase and exo-poly-α-D-galacturosidase mediate induction on D-galacturonan. *Physiol. Plant Pathol.* **21**, 127–139.

Collmer, A., and Keen, N. T. (1986). The role of pectic enzymes in plant pathogenesis. *Annu. Rev. Phytopathol.* **24**, 383–409.

Collmer, A., Whalen, C. H., Beer, S. V., and Bateman, D. F. (1982). An exo-poly-α-D-galacturonosidase implicated in the regulation of extracellular pectate lyase production in *Erwinia chrysanthemi*. *J. Bacteriol.* **149**, 626–634.

Crookes, P. R., and Grierson, D. (1983). Ultrastructure of tomato fruit ripening and the role of polygalacturonase isoenzymes in cell-wall degradation. *Plant Physiol.* **72**, 1088–1093.

De Lorenzo, G., Ito, Y., D'Ovidio, R., Cervone, F., Albersheim, P., and Darvill, A. G. (1990). Host–pathogen interactions. XXXVII. Abilities of the polygalacturonase-inhibiting proteins from four cultivars of *Phaseolus vulgaris* to inhibit the endopolygalacturonases from three races of *Colletotrichum lindemuthianum*. *Physiol. Mol. Plant Pathol.* **36**, 421–435.

Dean, R. A., and Timberlake, W. E. (1989a). Production of cell wall–degrading enzymes by *Aspergillus nidulans:* A model system for fungal pathogenesis of plants. *Plant Cell* **1**, 265–273.

Dean, R. A., and Timberlake, W. E. (1989b). Regulation of the *Aspergillus nidulans* pectate lyase gene (*pelA*). *Plant Cell* **1**, 275–284.

DellaPenna, D., Kates, D. S., and Bennett, A. B. (1987). Polygalacturonase gene expression in Rutgers, *rin, nor,* and *Nr* tomato fruits. *Plant Physiol.* **85**, 502–507.

DellaPenna, D., Lincoln, J. E., Fischer, R. L., and Bennett, A. B. (1989). Transcriptional analysis of polygalacturonase and other ripening-associated genes in Rutgers, *rin, nor,* and *Nr* tomato fruit. *Plant Physiol.* **90**, 1372–1377.

Demain, A. L., and Phaff, H. J. (1954). Hydrolysis of the oligogalacturonides and pectic acid by yeast polygalacturonase. *J. Biol. Chem.* **210**, 381–393.

Dick, A. J., and Labavitch, J. M. (1989). Cell-wall metabolism in ripening fruit. IV. Characterization of the pectic polysaccharides solubilized during softening of Bartlett pear fruit. *Plant Physiol.* **89**, 1394–1400.

Dow, J. M., Milligan, D. E., Jamieson, L., Barber, C. E., and Daniels, M. J. (1989). Molecular cloning of a polygalacturonate lyase gene from *Xanthomonas campestris* pv. *campestris* and role of the gene product in pathogenicity. *Physiol. Mol. Plant Pathol.* **35**, 113–120.

Edstrom, R. D., and Phaff, H. J. (1964a). Purification and certain properties of pectin *trans*-eliminase from *Aspergillus fonsecaeus*. *J. Biol. Chem.* **239**, 2403–2408.

Edstrom, R. D., and Phaff, H. J. (1964b). Eliminative cleavage of pectin and of oligogalacturonide methyl esters by pectin *trans*-eliminase. *J. Biol. Chem.* **239**, 2409–2415.

Elkashif, M. E., and Huber, D. J. (1988). Electrolyte leakage, firmness, and scanning electron microscopic studies of watermelon fruit treated with ethylene. *J. Am. Soc. Hort. Sci.* **113**, 378–381.

Ghazali, H. M., and Leong, C. K. (1987). Polygalacturonase activity in starfruit. *Food Chem.* **24**, 147–157.

Giovanni, J. J., DellaPenna, D., Bennett, A. B., and Fischer, R. L. (1989). Expression of a chimeric polygalacturonase gene in transgenic *rin* (ripening inhibitor) tomato fruit results in polyuronide degradation but not fruit softening. *Plant Cell* **1**, 53–63.

Grierson, D., and Tucker, G. A. (1983). Timing of ethylene and polygalacturonase synthesis in relation to the control of tomato fruit ripening. *Planta* **157**, 174–179.

Grierson, D., Slater, A., Speirs, J., and Tucker, G. A. (1985). The appearance of polygalacturonase mRNA in tomatoes: One of a series of changes in gene expression during development and ripening. *Planta* **163**, 263–271.

Hart, H. E., Parish, M. E., Burns, J. K., and Wicker, L. (1990). Orange finisher pulp as a substrate for polygalacturonase production by *Rhizopus oryzae*. *J. Food Sci.* (in press).

Hasegawa, S., and Nagel, C. W. (1966). A new pectic acid transeliminase produced exocellularly by a *Bacillus*. *J. Food Sci.* **31**, 838–845.

Hatanaka, C., and Ozawa, J. (1970). An oligogalacturonate transeliminse from *Erwinia aroideae*. *Agr. Biol. Chem.* **34**, 1618–1624.

Hatanaka, C., and Ozawa, J. (1971). Pectolytic enzymes of exo-types. I. Oligogalacturonide transeliminase of a *Pseudomonas*. *Agr. Biol. Chem.* **35**, 1617–1624.

Hatanaka, C., and Ozawa, J. (1972). Exopectic acid transeliminase of an *Erwinia*. *Agr. Biol. Chem.* **36**, 2307–2313.

Hobson, G. E. (1980). Effect of the introduction of nonripening mutant genes on the composition and enzyme content of tomato fruit. *J. Sci. Food Agr.* **31**, 578–584.

Huber, D. J. (1983). Polyuronide degradation and hemicellulose modifications in ripening tomato fruit. *J. Am. Soc. Hort Sci.* **108**, 405–409.

Huber, D. J.(1984). Strawberry fruit softening: The potential roles of polyuronides and hemicelluloses. *J. Food Sci.* **49**, 1310–1315.

Hugouvieux-Cotte-Pattat, N., and Robert-Baudouy, J. (1985). Isolation of *kdgK-lac* and *kdgA-lac* gene fusions in the phytopathogenic bacterium *Erwinia chrysanthemi*. *J. Gen. Microbiol.* **131**, 1205–1211.

Jarvis, M. C. (1984). Structure and properties of pectin gels in plant cell walls. *Plant Cell Environ.* **7**, 153–164.

Jen, J. J., and Robinson, M. L. (1984). Pectolytic enzymes in sweet bell peppers (*Capsicum annuum* L.). *J. Food Sci.* **49**, 1085–1087.

Jin, D. F., and West, C. A. (1984). Characteristics of galacturonic acid oligomers as elicitors of casbene synthetase activity in castor bean seedlings. *Plant Physiol.* **74**, 989–992.

Karbassi, A., and Luh, B. S. (1979). Some characteristics of an endo-pectate lyase produced by a thermophilic *Bacillus* isolated from olives. *J. Food Sci.* **44**, 1156–1161.

Keen, N. T., and Tamaki, S. (1986). Structure of two pectate lyase genes from *Erwinia chrysanthemi* EC16 and their high-level expression in *Escherichia coli*. *J. Bacteriol.* **168**, 595–606.

Kim, J., Gross, K. C., and Solomos, T. (1987). Characterization of the stimulation of ethylene production by galactose in tomato (*Lycopersicon esculentum* Mill.) fruit. *Plant Physiol.* **85**, 804–807.

Kimura, H., Uchino, F., and Mizushima, S. (1973). Properties of a polygalacturonase produced by *Acrocylindrium*. *J. Gen Microbiol.* **74**, 127–137.

Knee, M., and Bartley, I. M. (1981). Composition and metabolism of cell-wall polysaccharides in ripening fruits. *In* "Recent Advances in the Biochemistry of Fruits and Vegetables" (J. Friend and M. J. C. Rhodes, eds.), pp. 133–148. Academic Press, London.

Knegt, E., Vermeer, E., and Bruinsma, J. (1988). Conversion of the polygalacturonase isoenzymes from ripening tomato fruits. *Physiol. Plant.* **72**, 108–114.

Konno, H., and Yamasaki, Y. (1982). Studies on the pectic substances of plant cell walls. III. Degradation of carrot root cell walls by endopectate lyase purified from *Erwinia aroideae*. *Plant Physiol.* **69**, 864–868.

Konno, H., Yamasaki, Y., and Katoh, K. (1989). Extracellular exo-polygalacturonase secreted from carrot cell cultures. Its purification and involvement in pectic polymer degradation. *Physiol. Plant.* **76**, 514–520.

Lazan, H., Mohd, Z., Liang, K. S., and Yee, K. L. (1989). Polygalacturonase activity and variation in ripening of papaya fruit with tissue depth and heat treatment. *Physiol. Plant.* **77**, 93–98.

Lee, S.-C., and West, C. A. (1981). Polygalacturonase from *Rhizopus stolonifer*, an elicitor of casbene synthetase activity in castor bean (*Ricinus communis* L.) seedlings. *Plant Physiol.* **67**, 633–639.

Lei, S.-P., Lin, H.-C., Wang, S.-S., Callaway, J., and Wilcox, G. (1987). Characterization of the *Erwinia carotovora pelB* gene and its product pectate lyase. *J.Bacteriol.* **169**, 4379–4383.

Lei, S.-P., Lin, H.-C., Wang, S.-S., and Wilcox, G. (1988). Characterization of the *Erwinia carotovora pelA* gene and its product pectate lyase A. *Gene* **62**, 159–164.

Liao, C.-H. (1989). Analysis of pectate lyases produced by soft rot bacteria associated with spoilage of vegetables. *Appl. Environ. Microbiol.* **55**, 1677–1683.

Liu, Y. K., and Luh, B. S. (1980). Quantitative aspects of pectic acid hydrolysis by endo-polygalacturonase from *Rhizopus arrhizus*. *J. Food Sci.* **45**, 601–604.

Macmillan, J. D., and Vaughn, R. H. (1964). Purification and properties of a polygalacturonic acid-*trans*-eliminase produced by *Clostridium multifermentans*. *Biochemistry* **3**, 564–572.

Macmillan, J. D., Phaff, H. J., and Vaughn, R. H. (1964). The pattern of action of an exopolygalacturonic acid-*trans*-eliminase from *Clostridium multifermentans*. *Biochemistry* **3**, 572–578.

Marcus, L., Barash, I., Sneh, B., Koltin, Y., and Finkler, A. (1986). Purification and characterization of pectolytic enzymes produced by virulent and hypovirulent isolates of *Rhizoctonia solani* Kuhn. *Physiol. Mol. Plant Pathol.* **29**, 325–336.

McCollum, T. G., Huber, D. J., and Cantliffe, D. J. (1989). Modification of polyuronides and hemicelluloses during muskmelon fruit softening. *Physiol. Plant.* **76**, 303–308.

Moran, F., Nasuno, S., and Starr, M. P. (1968a). Extracellular and intracellular polygalacturonic acid *trans*-eliminases of *Erwinia carotovora*. *Arch. Biochem. Biophys.* **123**, 298–306.

Moran, F., Nasuno, S., and Starr, M. P. (1968b). Oligogalacturonide *trans*-eliminase of *Erwinia carotovora*. *Arch. Biochem. Biophys.* **125**, 734–741.

Moshrefi, M., and Luh, B. S. (1983). Carbohydrate composition and electrophoretic properties of tomato polygalacturonase isoenzymes. *Eur. J. Biochem.* **135**, 511–514.

Movahedi, S., and Heale, J. B. (1990). The roles of aspartic proteinase and endo-pectin lyase enzymes in the primary stages of infection and pathogenesis of various host tissues by different isolates of *Botrytis cinerea* Pers ex. Pers. *Physiol. Mol. Plant Pathol.* **36**, 303–324.

Nagel, C. W., and Anderson, M. M. (1965). Action of a bacterial transeliminase on normal and unsaturated oligogalacturonic acids. *Arch. Biochem. Biophys.* **112**, 322–330.

Nagel, C. W., and Hasegawa, S. (1967). Kinetics and site of attack of oligogalacturonic acids by an endo-pectic acid transeliminase. *Arch. Biochem. Biophys.* **118**, 590–595.

Nagel, C. W., and Wilson, T. M. (1970). Pectic acid lyases of *Bacillus polymyxa*. *Appl. Microbiol.* **20**, 374–383.

Nasuno, S., and Starr, M. P. (1966). Polygalacturonase of *Erwinia carotovora*. *J. Biol. Chem.* **241**, 5298–5306.

Nasuno, S., and Starr, M. P. (1967). Polygalacturonic acid *trans*-eliminase of *Xanthomonas campestris*. *Biochem. J.* **104**, 178–185.

Okamoto, K., Hatanaka, C., and Ozawa, J. (1964). A saccharifying pectate *trans*-eliminase of *Erwinia aroideae*. *Agr. Biol. Chem.* **28**, 331–336.

Patel, D. S., and Phaff, H. J. (1960). Properties of purified tomato polygalacturonase. *Food Res.* **25**, 47–57.

Preiss, J., and Ashwell, G. (1963). Polygalacturonic acid metabolism in bacteria. I. Enzymatic formation of 4-deoxy-L-*threo*-5-hexoseulose uronic acid. *J. Biol. Chem.* **238**, 1571–1576.

Pressey, R. (1984). Purification and characterization of tomato polygalacturonase converter. *Eur. J. Biochem.* **144**, 217–221.

Pressey, R. (1986a). Polygalacturonases in higher plants. *In* "Chemistry and Function of Pectins" (M. L. Fishman and J. J. Jens, eds.), pp. 157–174. American Chemical Society Symposium Series No. 310, Washington, D. C.

Pressey, R. (1986b). Extraction and assay of tomato polygalacturonases. *HortScience* **21**, 490–492.

Pressey, R. (1986c). Changes in polygalacturonase isoenzymes and converter in tomatoes during ripening. *HortScience* **21**, 1183–1185.

Pressey, R. (1987). Exopolygalacturonase in tomato fruit. *Phytochemistry* **26**, 1867–1870.

Pressey, R. (1988). Reevaluation of the changes in polygalacturonases in tomatoes during ripening. *Planta* **174**, 39–43.

Pressey, R., and Avants, J. K. (1971). Effect of substrate size on the activity of tomato polygalacturonase. *J. Food Sci.* **36**, 486–489.

Pressey, R., and Avants, J. K. (1973a). Two forms of polygalacturonase in tomatoes. *Biochim. Biophys. Acta* **309**, 363–369.

Pressey, R., and Avants, J. K. (1973b). Separation and characterization of endopolygalacturonase and exopolygalacturonase from peaches. *Plant Physiol.* **52**, 252–256.

Pressey, R., and Avants, J. K. (1975a). Cucumber polygalacturonase. *J. Food Sci.* **40**, 937–939.

Pressey, R., and Avants, J. K. (1975b). Modes of action of carrot and peach exopolygalacturonases. *Phytochemistry* **14**, 957–961.

Pressey, R., and Avants, J. K. (1976). Pear polygalacturonases. *Phytochemistry* **15**, 1349–1351.

Pressey, R., and Avants, J. K. (1977). Occurrence and properties of polygalacturonase in *Avena* and other plants. *Plant Physiol.* **60**, 548–553.

Pressey, R., and Avants, J. K. (1978). Difference in polygalacturonase composition of clingstone and freestone peaches. *J. Food Sci.* **43**, 1415–1423.

Pressey, R., and Avants, J. K. (1982a). Pectic enzymes in 'long keeper' tomatoes. *HortScience* **17**, 398–400.

Pressey, R., and Avants, J. K. (1982b). Solubilization of cell walls by tomato polygalacturonases: Effects of pectinesterases. *J. Food Biochem.* **6**, 57–74.

Pressey, R., and Reger, B. J. (1989). Polygalacturonases in pollen from corn and other grasses. *Plant Sci.* **59**, 57–62.

Prusky, D., Gold, S., and Keen, N. T. (1989). Purification and characterization of an endopolygalacturonase produced by *Colletotrichum gleosporiodes*. *Physiol. Mol. Plant Pathol.* **35**, 121–133.

Rexova-Benkova, L., and Markovic, O. (1976). Pectic enzymes. *Adv. Carbohydr. Chem. Biochem.* **33**, 323–385.

Reyes, F., Martinez, M. J., and Lahoz, R. (1984). Characterization as glycerol of an inhibitor of pectin lyases from autolysing cultures of *Botrytis cinerea*. *Trans. Br. Mycol. Soc.* **82**, 689–696.

Riov, J. (1974). A polygalacturonase from citrus leaf explants. *Plant Physiol.* **53**, 312–316.

Robertsen, B. (1986). Elicitors of the production of lignin-like compounds in cucumber hypocotyls. *Physiol. Mol. Plant Pathol.* **28**, 137–148.

Robertsen, B. (1987). Endo-polygalacturonase from *Cladosporium cucumerinum* elicits lignification in cucumber hypocotyls. *Physiol. Mol. Plant Pathol.* **31**, 361–374.

Roe, B., and Bruemmer, J. H. (1981). Changes in pectic substances and enzymes during ripening and storage of Keitt mangos. *J. Food Sci.* **46**, 186–189.

Smith, C. J. S., Watson, C. F., Ray, J., Bird, C. R., Morris, P. C., Schuch, W., and Grierson, D. (1988). Antisense RNA inhibition of polygalacturonase gene expression in transgenic tomatoes. *Nature* **334**, 724–726.

Starr, M. P., and Moran, F. (1962). Eliminative split of pectic substances by phytopathogenic softrot bacteria. *Science* **135**, 920–921.

Tam, S. Y. (1983). A new calculation method for distinguishing endo- from exo-polygalacturonases. *J. Food. Sci.* **48**, 532–538.

Themmen, A. P. N., Tucker, G. A., and Grierson, D. (1982). Degradation of isolated cell walls by purified polygalacturonase *in vitro*. *Plant Physiol.* **69**, 122–124.

Thurn, K. K., Barras, F., Kegoya-Yoshino, Y., and Chatterjee, A. K. (1987). Pectate lyases of *Erwinia chrysanthemi: pelE*-like polypeptides and *pelE* homologous sequences in strains isolated from different plants. *Physiol. Mol. Plant Pathol.* **31**, 429–439.

Tigchelaar, E. C., McGlasson, W. B., and Buescher, R. W. (1978). Genetic regulation of tomato fruit ripening. *HortScience* **13**, 508–513.

Tong, C. B. S., and Gross, K. C. (1989). Ripening characteristics of a tomato mutant, *dark green*. *J. Am. Soc. Hort. Sci.* **114**, 635–638.

Tsuyumu, S. (1977). Inducer of pectic acid lyase in *Erwinia carotovora*. *Nature* **269**, 237–238.

Tucker, G. A., and Grierson, D. (1982). Synthesis of polygalacturonase during tomato fruit ripening. *Planta* **155**, 64–67.

Tucker, G. A., Robertson, N. G., and Grierson, D. (1980). Changes in polygalacturonase isoenzymes during the 'ripening' of normal and mutant tomato fruit. *Eur. J. Biochem.* **112**, 119–124.

Tucker, G. A., Robertson, N. G., and Grierson, D. (1981). The conversion of tomato-fruit polygalacturonase isoenzyme 2 to isoenzyme 1 *in vitro*. *Eur. J. Biochem.* **115**, 87–90.

Walker-Simmons, M., Jin, D., West, C. A., Hadwiger, L., and Ryan, C. A. (1984). Comparison of proteinase inhibitor–inducing activities and phytoalexin elicitor activities of a pure fungal endopolygalacturonase, pectic fragments, and chitosans. *Plant Physiol.* **76**, 833–836.

Walton, J. D., and Cervone, F. (1990). Endopolygalacturonase from the maize pathogen *Cochliobolus carbonum*. *Physiol. Mol. Plant Pathol.* **36**, 351–359.

Wijesundera, R. L. C., Bailey, J. A., and Byrde, R. J. W. (1984). Production of pectin lyase by *Colletotrichum lindemuthianum* in culture and in infected bean (*Phaseolus vulgaris*) tissue. *J. Gen. Microbiol.* **130**, 285–290.

Willis, J. W., Engwall, J. K., and Chatterjee, A. K. (1987). Cloning of genes for *Erwinia carotovora* subsp. *carotovora* pectolytic enzymes and further characterization of the polygalacturonases. *Phytopathology* **77**, 1199–1205.

Wing, R. A., Yamaguchi, J., Larabell, S. K., Ursin, V. M., and McCormick, S. (1989). Molecular and genetic characterization of two pollen-expressed genes that have sequence similarity to pectate lyases of the plant pathogen *Erwinia*. *Plant Mol. Biol.* **14**, 17–28.

CHAPTER 10

Analytical and Graphical Methods for Pectin

R. H. Walter

Cornell University
Department of Food Science
Geneva, New York

189

I. INTRODUCTION

The mostly linear, anionic α-1,4-D-galacturonans, collectively called pectin, that form food gels with water, sugar and acid or calcium, are simultaneously reducing carbohydrates, polyhydric alcohols, polyacids, and polyesters. They contain polar (carboxyl) and easily removable, nonpolar (methyl) groups, thereby displaying a degree of amphiphilicity that is important to their function. They are relatively stable at low pH, but are susceptible to alkali. Chemical and biological reagents may demethylate and depolymerize them. They are constituted by a multitude of macromolecular sizes. In aqueous media, this family of compounds responds to external stimuli in diverse ways. The myriad properties have engendered numerous analytical methods for pectin, understandably nonspecific and therefore often necessitating prior isolation of it.

In this chapter, a distinction is made between a dispersion and a solution. The former is a colloidal system with solute dimensions in the 1–10^3 nm diameter range. Solute in the latter has truly molecular dimensions. Dispersions are heterogeneous (two phases), as shown by the well-known Tyndall effect. Solutions are a monophase system.

With reference to food, water is the only solvent for pectin. Unfortunately, the comparatively high pectin-water affinity voids much of the classical polymer theory that discourses organic solvent systems with far less affinity for solute. This chapter is a summary of recent analytical methods for pectic substances, where possible, in the context of modern theory and elementary polymer methodology.

II. DETECTION OF PECTIN

The detection of pectin is based primarily on the chemistry of galacturonic acid. Pectin is usually first isolated from most of the congeneric substances occurring in crude plant extracts and formulated foods, then purified, before being subjected to confirmative tests involving decarboxylation and chromophore formation. Some tests are based on the incompatibility of pectin with organic solvents.

A. Precipitation

The alcohol insolubility of pectin has been developed into a test for traces of pectin in fruit juices. A positive reaction is indicated by the development of a stringy, gelatinous deposit (Miles Laboratories, Inc., 1984). The isolate may be distinguished from other acidic polysaccharides by precipitation with quaternary ammonium salts, e.g., cetylpyridinium bromide, whose complex with pectin has a much lower critical electrolyte concentration than many other polymers (Scott, 1965). Complexes between ethylenediamine and pectin containing dissimilar molecular weights, charge densities, and distribution patterns of carboxyl and amine groups in the chain are differentially soluble in the pH range 2–8 (Stutz and Deuel, 1955).

B. Chromophore Formation

In histological testing, pectin may be distinguished from surrounding nonpectin material by staining with ruthenium red (ammoniated ruthenium oxychloride). A positive test is evidenced by the typical pink color (pectin) on a gray background (lignin and cellulose) (Walter *et al.*, 1978). Mucilages and gums containing uronic acid, and any other host molecule with two negative charges in a specific geometry (Sterling, 1970), interfere with the reaction. The test, which necessitates a preceding or concurrent alkaline treatment to effect demethylation of high-methoxyl pectin, is most efficacious on low-methoxyl pectins.

Differential staining (in leaf sections) was accomplished with alkaline NH_2OH-$FeCl_3$ (Beam and Bornman, 1973). The reaction converts pectin carboxyl groups to hydroxamic acids, resulting in water-insoluble, red complexes in excess Fe(III). The test is specific for pectin (McReady and Reeve, 1955).

Pectin spots were purified by thin-layer chromatography and stained at the origin with $FeCl_3$-H_2SO_4. Gums and modified starches were neutral to this reagent (Martelli and Proserpio, 1973).

Carbazole (diphenyleneimine) is the basis of a test for pectin in dark fruit juices in which development of a red color denotes a positive reaction, and a pink-to-light-red color, a pectin concentration less than 0.02% (Miles Laboratories, Inc., 1984).

Many other reagents that form chromophores with hexuronic acids (Dische, 1962; Hough *et al.*, 1950) are applicable to pectin, but given the nonspecificity of most of them, independent confirmation is warranted. Thus, Ayers *et al.* (1952) characterized four spots from an enzyme hydrolyzate by spraying them with an acid indicator (bromphenol blue), then treating them with *p*-anisidine-HCl, a nearly universal, sugar reagent for color development (Hough *et al.*, 1950).

Aqueous pectin, heated in a strongly alkaline medium, transmutes to a yellow chromophore through a series of time-dependent, chemical events (β-elimination of water, depolymerization, enolization, and oxidation). These reactions are the basis of Moore's test for reducing carbohydrates. Simple sugars reach the end-point coloration in a much shorter time than does pectin, and are so distinguished from pectin.

C. Decarboxylation

A vigorous evolution of CO_2 and appearance of a pentose derivative, notably furfural, resulting from the action of strong mineral acids and heat on plant material, is claimed to be a good indication of the presence of uronic acids. If the temperature is 100°C or less, CO_2 generation from interfering hexoses is retarded. This decarboxylation of pectin is considered to be quantitatively more accurate than are colorimetric methods, because the procedure is less prone to errors. There are, however, the disadvantages of tediousness and the requirement of large sample size and complex apparatus (Marsh, 1966).

D. Ultraviolet Absorption

Reaction with H_2SO_4 causes neutral sugars to display an absorption maximum at 315 nm, whereas the absorption maximum for galacturonic acid is in the area of 298 nm. Berth (1988) arranged absorption data from this region of the spectrum into a quotient, obtaining $E_q = E_{298}/E_{315} = 1.20$ for pure galacturonic acid, and 0.42 for glucose, and through these numbers, monitored the increasing homogeneity of pectin in eluate from a gel column.

III. DEGREE OF ESTERIFICATION

By custom, the degree of esterification (DE) is an index of the extent to which carboxyl groups in the galacturonan chain exist as the methyl ester. DE may be reported as percentage of the total number of carboxyl groups, or as methoxyl content of total pectin. The respective, theoretical maximum for each is 100% and 16.32% (Doesburg, 1965). Lower DE is generally brought about by enzyme and acid hydrolysis. Alkali accomplishes the same results, but chain scission is a major, secondary reaction. Deesterification in dilute hydrochloric acid has been reported to be pseudo-first order, not influenced by pectin origin (Kirtchev *et al.*, 1989).

Many of the functional applications of pectin are directly or indirectly related to methyl ester content. Consequently, DE measurements have become routine.

A. Titrimetry

DE may be obtained directly by titrating a pectin dispersion before and after saponification. The first volume of titrant, NaOH, for example, will be equivalent to the unesterified carboxyl groups, and the second volume following saponification will be equivalent to the total number of carboxyl groups. Back-titration of the excess base and insertion of the relevant data in the neutralization equation reduce the exercise to a final number. The degree of amidation may be calculated similarly after inclusion of a distillation step (National Academy of Sciences, 1981). Mizote *et al.* (1975) described titration with poly-N,N-dimethyldiallyl-ammonium chloride, and claimed a rate acceleration to the quantitative endpoint at 0°C above that occurring with NaOH alone. Alkaline demethylation in dimethyl sulfoxide (DMSO) (80–90% by vol with water) has been reported to be unusually rapid (Vinson *et al.*, 1966). In another instance with pectin, DMSO did not show any advantage over NaOH (Walter *et al.*, 1983).

B. Methanol Assay

Methanol is an end-product of the deesterification of pectin. It is directly measurable from a hydrolyzate by gas chromatography (Walter *et al.*, 1983). This technique is probably the simplest means of obtaining DE. With or without the use of a precolumn, the procedure involves few steps. The data for the calculation are area measurements taken off a chromatogram (Fig. 1). Alternatively, methanol may be oxidized chemically (Wood and Siddiqui, 1971) or enzymatically

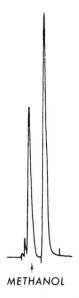

METHANOL

Figure 1 Resolution of methanol from a pectin hydrolyzate by gas chromatography on 0.2% Carbowax 1500 on Carbopak (80–100 mesh) and a flame ionization detector.

(Klavons and Bennett, 1986), and the DE calculated. An estimate of the degree of methylation may also ensue from a series of chemical steps that terminates in thiosulfate reduction (Laver and Wolfrom, 1962).

C. Copper Reduction

Copper, reduced and precipitated by pectin, was determined before and after saponification of the pectin. This reaction was developed into a method for DE that is particularly suited to plant extracts in which copper-complexing substances are absent. Starch will interfere with the reaction if it contains phosphate. This interference leads to a low DE (Keijbets and Pilnik, 1974).

IV. METHYLATION

It may sometimes be advantageous to increase the methoxyl content of pectin, as for example, in structural elucidation research. Diazomethane is a common pectin-methylating reagent (Hough and Theobald, 1963), although its use for this purpose may be accompanied by depolymerization (Smit and Bryant, 1969), especially in the presence of phosphate (Inoue et al., 1987). Water may catalyze methylation of the hydroxyl groups (Hough and Jones, 1952). These shortcomings might be mitigated by methylating at very low temperatures (Pippen et al., 1953; Berth et al., 1982) in a mostly hydrophobic medium (Kiyohara et al., 1988). The methylation reaction does not proceed to completion, but remethylation does facilitate incorporation of a very high percentage of methoxyl groups.

V. ELECTRICAL METHODS

Pectin migrates in an electrical field toward the anode at a rate in proportion to its charge density. Kwak (1978) observed that polypectate conformed with Manning's limiting law for conductance of rod-like polyanions only at relatively low charge densities. Aspinall and Cottrell (1970) applied the migration principle to the testing of homogeneity of pectin fractions. Pechanek et al. (1982) developed an electrophoretic method for the quantitative and qualitative determination of gelling and thickening agents including pectin. By differential-pulse polarography, Reisenhofer et al. (1984) determined that pectate had a higher copper-binding affinity than did many other polyelectrolytes. These transport methods of analysis offer an economy of time without sacrificing precision and accuracy; the necessary apparatuses must, however, be available.

VI. SPECTROSCOPIC METHODS

Some methods of pectin analysis exploit the very intense coloration developed with chromogenic, carbohydrate reagents. Other methods rely on the polyionic character of pectin, and still others, on its colloidal behavior.

A. Colorimetry and Spectrophotometry

These methods first require that the sugar monomer, specifically uronic acid, be chemically dehydrated, and the anhydrosugar species be combined with a chromogen (Dische, 1962; Wedlock *et al.*, 1984). Alternative procedures utilize the ability of glucose to reduce copper in an alkaline medium, and the reduced copper to develop color with arsenomolybdate, for example (Nelson, 1944). This last reaction is the basis of a reducing end-group method (EGM) of determining pectin molecular weights (Milner and Avigad, 1967).

Alcian blue, a cationic Cu(II)-phthalocyanine dye, precipitates polyanions from solution, and an analytical method of determining acidic polysaccharides (e.g., pectin) relies on the proportionality between the amount of the dye removed and the polyanion concentration (Ramus, 1977).

The quantitative determination of isolated pectins is most commonly made by an improved carbazole (Bitter and Muir, 1962) or a *meta*-hydroxydiphenyl (Blumenkrantz and Asboe-Hansen, 1973) assay of the pectin equivalent of 4 to 40 μg galacturonic acid. The precision of the methods is high. Greater sensitivity and specificity are attributed to the *meta*-hydroxydiphenyl method. The two chromophores are stable for 16 to 24 hr. The determinations are subject to interference from other carbohydrates. The carbazole method apparently suffers from the additional limitation of giving high results in the presence of phenolic impurities (Robertson, 1981). Using *meta*-hydroxydiphenyl, Kintner and Van Buren (1982) estimated that the absorbance due to 100 parts of nonuronide carbohydrates was equivalent to that from one part of galacturonic acid. The addition of borate ions to the carbazole–H_2SO_4 reagent reduces the interference and provides other advantages (Bitter and Muir, 1962). Urea also reduces the interference (Alacheva *et al.*, 1973). The *meta*-hydroxydiphenyl method is claimed to have the advantage of speed of execution. Automated carbazole analyses were performed at a rate of 10 samples per hour (Thibault and Robin, 1975), while the automated *meta*-hydroxydiphenyl analysis rate was 15 samples per hour (Thibault, 1979).

Another reagent used in pectin analysis is 2-thiobarbituric acid (Albersheim *et al.*, 1960). The color reaction is preceded by decarboxylation and dehydration of galacturonic acid. The rates of decarboxylation and dehydration are relatively faster than for a number of possibly interfering carbohydrates, and may consequently be used to determine pectin in the presence of other uronide compounds. The method's sensitivity depends on the sequence of reagent additions (Wedlock *et al.*, 1984).

B. Circular-Dichroism Spectroscopy

This method has been adapted to studying intra- and inter-molecular associations of pectin (Plashchina *et al.*, 1978), and pectin interactions with counterions (Gidley *et al.*, 1979; Kohn and Sticzay, 1977; Thibault and Rinaudo, 1986). According to Johnson (1985), one of its weaknesses is the current state of underdevelopment of the theory relative to conformation. Nevertheless, Bystricky *et al.* (1979) interpreted circular-dichroism (CD) data from D-galacturonan to mean the polymer's existence in a close-to-helical structure in solution. Rees and Wright (1971) concluded from CD calculations that multiple helix formation by α-1,4-galacturonan is impossible. On the strength of CD, light-scattering, and rheometric evidence, Chou and Kokini (1988) concluded that the rheological behavior of pectin solutions was dominated by hydrogen bonding and electrostatic interactions, that hydrophobic interactions did not play a significant role, and that pectins in solution existed as random coils.

C. Infrared Spectroscopy

This method has occasionally been used in the quantitative analysis of pectin, but given the very sensitive spectrophotometric methods available, it is preferably adapted to compositional and structural studies involving, for example, the carboxyl function and associated hydrogen bonding in the range of the CO stretching vibrations (Bellamy, 1954; Rao, 1963; Tipson, 1967). To the extent that introduction and withdrawal of methyl groups, into and out of the galacturonan, disrupt intra- and intermolecular bonding, shifts in the absorption band positions can be correlated with structural modifications. The method is claimed to be capable of differentiating between secondary and tertiary pectin structures (Filippov, 1978).

When dilute pectin dispersions are completely dried on a clean metal or glass surface, they make thin, transparent, easily removable films suitable for mounting on infrared spectrometer film holders. These films may contain as much as 18% water, even though they appear to be dry (Palmer and Hartzog, 1945). For rigorous analytical study of polysaccharide films, the bands corresponding to absorbed water can be identified by treatment with deuterium oxide, which will replace the $-OH$ stretching and bending modes with the corresponding $-OD$ modes in the spectrum (Passaglia and Marchessault, 1965). The study of film mounts led Filippov *et al.* (1988) to conclude that pectin orients itself into dimeric, nonionized carboxyl groups, during film formation.

For carbohydrates and their derivatives, the 1000–1250 cm^{-1} interval is not very useful; the 700–1000 cm^{-1} interval has proved to be more informative (Whiffen, 1957). However, according to Filippov and Shamshurina (1972), bands in the 1000–2000 cm^{-1} region are independent of pectin source and production practices, and may therefore be used to identify pectin. Filippova and Shkolenko (1968) extended the useful range from 700 cm^{-1} to 3600 cm^{-1}. Kawabata and Sawayama (1976) analyzed commercial and laboratory-isolated pectins in KBr pellets between 3600 cm^{-1} and 800 cm^{-1}. They obtained DE values comparable to those obtained from titrimetry. Infrared spectra of carbohydrate derivatives including pectin, galacturonic acid, polygalacturonic acid, and their methyl esters and amides have been published (Kuhn, 1950; Solms *et al.*, 1954; Tipson, 1967; Kawabata and Sawayama, 1976). Figure 2 illustrates similarities in the infrared spectrum of some polysaccharides, which make them almost indistinguishable from each other. Observe that the BAKING (high DE) shows one, and LM 18CB (low DE) shows two carboxylate absorption bands in the 5.9–6.3 μm (1700–1580 cm^{-1}) segment of the spectrum. The best that may be hoped for in routine, infrared analysis is a comparison of the spectrum of unknown pectin with the spectra of known standards.

D. Nuclear Magnetic Resonance Spectroscopy

This method offers a means of directly measuring structure, conformation, and concentration. ^{13}C-NMR signals relating to some polysaccharide structures have been published (Gorin, 1981). The interpretation of the structure of large molecules is complicated by ambiguities in their NMR spectrum (Bock, 1985), but Keenan *et al.* (1985) nevertheless established that rhamnosyl residues in sugarbeet pectin were a backbone and not a side-chain component. Irwin *et al.* (1985) used NMR spectroscopy and discovered that polyuronide content and degree of methylation remained constant during ripening of apple tissue.

E. Light-Scattering Photometry

Solute with at least one dimension in the colloidal domain, 1-1000 nm, is capable of scattering incident light in conformance with the Debye equation (Allcock and Lampe, 1984; Billmeyer, Jr., 1984; Everett, 1988). Pectin molecules are within this limit, and pectin molecular weights obtained by light scattering are of the order of 10^4 to 10^8 daltons; more often than not, the upper limit is the highest obtainable by any method. The light-scattering molecular weight is a

MICRONS

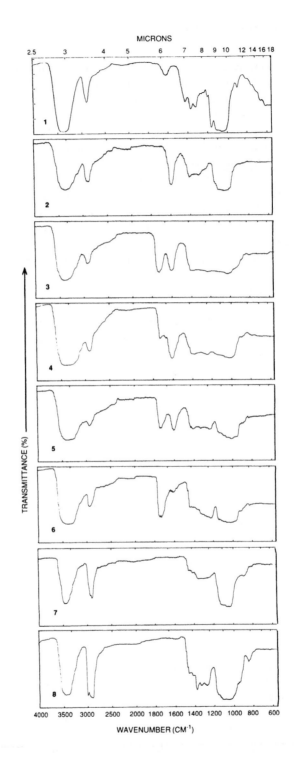

TRANSMITTANCE (%)

WAVENUMBER (CM⁻¹)

weight-average property ($\overline{M}_w$). Smith (1976) compared pectin molecular weights by light scattering to those by analytical centrifugation and found fourfold higher numbers for the former over the latter.

Other useful information obtained from light-scattering data and the corresponding Zimm plots, following concentration (C) and angular (θ) extrapolations to 0 g/ml and 0°, respectively, is the radius of gyration, $\langle s^2 \rangle^{\frac{1}{2}}$, (from the slope of the C = 0 line) and the second virial coefficient, A_2, (from the slope of the C = 0 line) and the second virial coefficient, A_2, (from the slope of the θ = 0 line). $\overline{M}_w$ is computed from the value at the intersection of these two extrapolated lines on the ordinate axis. A foreknowledge of certain constants may enable $\overline{M}_w$ to be calculated from measurement of one concentration only (Kamata and Nakahara, 1973). The $\langle s^2 \rangle^{\frac{1}{2}}$ gives information about chain flexibility. A_2 gives information about the hydrodynamic interactions between polymer molecules or micelles, and by corollary, about pectin-solvent compatibility.

Pectin shows an occasional tendency to yield negative Zimm plots (Plashchina et al., 1985). Such plots are likely due to the non-random, macromolecular organization induced by charge (Stacey, 1956). Electrostatic charge on a co-polymer of methacrylic acid and 2-dimethylaminoethyl methacrylate, in the absence of an electrolyte in the solvent, has been implicated in negative Zimm plots (Stacey, 1956). Polydispersity has been implicated in curved Zimm plots for polymethyl methacrylate (Billmeyer and de Than, 1955).

Aided by light-scattering photometry, Axelos et al. (1987) established that the low-methoxyl pectin studied was a nondraining, flexible coil. Kawabata and Sawayama (1977) described the conformation of laboratory and commercial pectins as a random coil. In the literature, other descriptions vary from a rigid, extended conformation to one in which rhamnose bestows some degree of flexibility (Morris, 1986). Owens et al. (1946) claimed that their results were indicative of a rigid, rod-like structure in aqueous salt solutions. In Smith's (1976) opinion, a single conformational model could not account for shapes in solution of all the pectins studied.

Pectin-chain flexibility has been shown to pass through a maximum, as DE increased from 43% to 58% (Plashchina et al., 1985). In their study, $\overline{M}_w$ was found to be independent of DE. Berth et al. (1982) claimed that A_2 and $\langle s^2 \rangle^{\frac{1}{2}}$ were independent of DE, and that conformation of pectin molecules is therefore unaffected by DE (in salt solution). $\overline{M}_w$, however, increased with DE, but was independent of ionic strength, whereas $\langle s^2 \rangle^{\frac{1}{2}}$ was inversely dependent on ionic strength. A_2 may be negative, zero, or positive (Sorochan et al., 1971; Berth et al., 1982; Plashchina et al., 1985; Jordan and Brant, 1978) in electrolyte solution.

←——

Figure 2   The infrared spectrum of a thin (3–9 μm film of (1) cellulose; (2) caboxymethylhydroxyethyl cellulose; (3) Kelcoloid HVF (propyleneglycol alginiate); (4) Keltrol *syn*. Xanthan gum [3-(manno-1,4-glucuronic acid-1,2-mannosidyl-1-cellulose]; (5) LM18CB pectin; (6) Baking pectin; (7) hydroxyethylcellulose; (8) hydroxypropyl cellulose.

VII. CHROMATOGRAPHIC METHODS

A. Ion-Exchange Chromatography

Pectic substances, given their negative charge, are particularly amenable to isolation and fractionation on cationic adsorbents conventionally held in the geometry of a column, permitting the neutral fraction to pass through without interaction in the usually acidic solvent. With increasing ionic strength and pH, the polyanions elute in order of their relative charge density, from the lowest to the highest. The resolving power of this technique was demonstrated by Knee (1970), who identified two components from a pectinic acid sample that had hitherto been shown by electrophoresis to contain a single species. Higher buffer concentrations are needed to elute lower DE pectins (Van Deventer-Schriemer *et al.*, 1976).

The high affinity of pectin carboxyl groups for Cu(II) ions was taken advantage of in an ion-exchange chromatographic determination of pectin content and DE of potato tissue, without first isolating the pectin. The analysis was preceded by a prior direct exchange of the chemically bound copper for H^+ (Keijbets and Pilnik, 1974).

B. Gel Chromatography

This method is also called gel-permeation chromatography and size-exclusion chromatography, because its fundamental principle is the retention and migration of fractions of a heterogeneous, polymeric solute at different rates through a gel matrix, predominantly on the basis of size. One molecular weight as high as 10^7 (O'Beirne *et al.*, 1981) has been fractionated. More accurately, resolution is a logarithmic function of the product of the hydrodynamic volume (v) and the intrinsic viscosity ([η]). When plotted against elution volume, this product (v[η]) delineates a universal calibration locus that is independent of a polymer's morphology and chemistry (Grubisic *et al.*, 1967). The calibration line for sunflower pectin differed modestly from that of some related polysaccharides (Anger and Berth, 1985). Elution does not always occur in increasing order of molecular weights; for example, Berth (1988) showed that the method did not differentiate between lower-molecular-weight extended coils and compact, higher-molecular-weight fractions in pectin. Neither does gel chromatography always separate related monomeric acids or saturated and unsaturated oligomers (Voragen *et al.*, 1982). The method nevertheless facilitates the study of heterogeneous pectin and pectin degradation products (Grubisic *et al.*, 1967; Voragen *et al.*, 1986). Davis *et al.* (1980), suggested that elution behavior during gel chromatography is influenced by reversible aggregation as well as by the chain length of individual molecules.

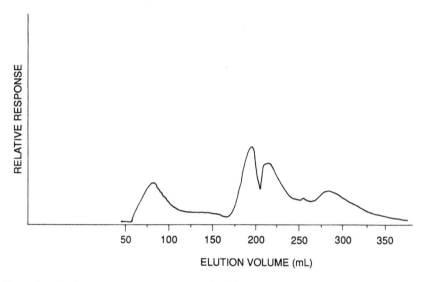

Figure 3   Gel fraction of a commercial pectin dispersed in water (no electrolyte) on Biogel P100, using a Bio-Rad high-sensitivity conductivity monitor (Bio-Rad Chemical Division, Richmond, CA).

Gel chromatography of commercial pectins results in overlapping band elution profiles rather than in discreet segments (Fig. 3), but a narrow, size distribution may still be separated, if small volumes (*e.g.,* 5 ml) are collected in order of their position in the elution series.

C. High-Performance Liquid Chromatography

This method, abbreviated to HPLC, describes the special condition of gel- or ion-exchange chromatography at elevated solvent flow-pressures, whereby a column's resolution is enhanced many times above that obtained at atmospheric pressure. Hence, the method is synonymously called high-pressure liquid chromatography. With reference to ion-exchange HPLC, the polyanions having the greatest charge density are again the last to elute. Hicks *et al.* (1985) reversed the normal sequence and separated uronic acids and other related compounds on a cationic exchange resin. In this mode, the anionic solute was the first to emerge from the column. The authors recommended the reversed sequence as a useful, high-resolution alternative to traditional gas chromatography.

In another adaptation of HPLC, Fishman *et al.* (1984) demonstrated the variability of the radius of gyration of dispersed pectin as a function of intrinsic and extrinsic properties. Fishman *et al.* (1986a) assumed a rod-like structure and calculated number-average lengths and the degree of polymerization (DP) of citrus pectin from the radii of gyration data obtained from HPLC.

Speed of pectin analysis by HPLC relative to atmospheric (low-pressure) chromatography is a major advantage (Barth, 1980; Giangiacomo *et al.*, 1982). Within only minutes, Schols *et al.* (1989) followed the pattern of methoxyl distribution along the galacturonan chain in an enzyme hydrolyzate.

D. Thin-Layer Chromatography

In this exceptional procedure, liquid–liquid partition of solute occurs two-dimensionally on a surface. Solvent flow is due to capillarity, and the resulting absence of a measurable void volume minimizes or altogether eliminates solute diffusion. As a result, resolution is enhanced. These are fundamental differences with column chromatography. The extraordinarily sharp separations obtained with thin-layer chromatography (TLC) make possible much smaller sample volumes (microliters) and concentrations (1–5 micrograms) than those required for column chromatography; pectin purity can be evaluated by TLC with only trace concentrations. A water–methanol elution solvent tends to leave pectin at the point of deposition, where it may be detected by $FeCl_3$ in H_2SO_4 (Martelli and Proserpio, 1973). Anisaldehyde (*p*-methoxybenzaldehyde) in H_2SO_4 is a broad-spectrum, oligosaccharide, TLC, detection reagent (Weill and Hanke, 1962) applicable also to pectin.

TLC can be executed in a high-pressure (HP) mode. The resolving power of the combined high pressure and capillary flow is shown by the results of Rombouts and Thibault (1986). From the HPTLC of a 0.1 ml sample of a beet-pulp hydrolyzate, they calculated that the content of water-soluble ferulic acid in purified beet pectin was 0.1% of the dry matter.

E. Gas Chromatography–Mass Spectrometry

In gas chromatography (GC), resolution of a mixture of relatively volatile compounds is predicated on the principle of differential rates of diffusion and transport in an inert gas stream. Pectin can be degraded to impart relative volatility through chemical derivatization of the degradation products. Thus, the enzyme hydrolyzates of citrus and apple pectin were analyzed by their content of silyl ethers (Wiley and Tavakoli, 1969). The potential for pectin decarboxylation in an acid medium may require a protective reduction to alcohol of the uronic acid groups before silylation (Taylor and Conrad, 1972; Lau *et al.*, 1985). Alkaline hydrolysis is milder, yields methanol at a faster rate than does acid hydrolysis (Kim, 1978), and is therefore the preferred reaction for GC determinations of DE by methanol analysis (Forni *et al.*, 1984; McFeeters and Armstrong, 1984).

The quantitative determination of galacturonic acid in pectin by GC agreed well with the results from carbazole (Petrzika and Linow, 1985). The *in vitro* hydrolysis may be omitted, and the sample, pyrolyzed directly into a gas chromatograph. The resulting pyrograms will contain information that may be statistically evaluated and correlated with the degree of methylation (Barford *et al.*, 1986).

The silylated derivatives of pectin may be analyzed by GC-mass spectrometry (Lau *et al.*, 1985; Petrzika and Linow, 1986). The reproducibility of the respective fragmentation patterns in the mass spectrum was a factor in identifying food thickeners including pectin by this combination (Sjoberg and Pyysalo, 1985).

VIII. MEMBRANE OSMOMETRY

Membrane osmometry, applied to dilute pectin dispersions, is an extension of the kinetic theory of gases, wherein solvent (water) diffuses across a pectin-retaining semipermeable membrane and reaches an equilibrium state in which the pressure on both sides of the membrane is equal. The osmotic pressure (π in atmospheres), replacing the gas pressure in the gas equation, is proportional to the number of moles of pectin. The molecular weight thus obtained is a number-average value ($\overline{M}_n$). Osmometers do not require calibration, since $\overline{M}_n$ is an absolute, colligative property. Accuracy of the method depends on the completeness of molecular separation in the dispersion, a pectin state that is difficult to achieve.

The practical range of $\overline{M}_n$ measurements by osmometry is 5000–500,000 daltons (Ulrich, 1974). Under certain (unspecified) conditions, it may be extended to 1,000,000 daltons (Burge, 1977) beyond which π becomes too small to be accurately measured (Tanford, 1961).

If one assumes a pectin molecular weight of 10,000, a 0.5% aqueous dispersion has a 0.005 molar concentration—too little even at this small number to elevate and lower the boiling and freezing points of water substantially, but large enough to influence crystallization. A host of other physical characteristics important to food quality is influenced by small amounts of pectin (Nash, 1960).

Under MW 5000 daltons, vapor-phase osmometry complements membrane osmometry, but relative to pectin, the information would be of doubtful value, inasmuch as it would pertain exclusively to oligomers that do not possess the necessary functionality important to food.

The simplest osmometer is a single- or dual-cell device, mounted with one or two capillaries. The solvent rises in the single or the sensing capillary under the influence of π, and is measured at equilibrium as the height differential (in cm) between the initial and final solvent levels (single cell), or between liquid columns in the reference and sensing capillaries (dual cell). Towle (1972) described the operation of the Ten-yuan static osmometer. This device is simple to construct, is relatively insensitive to temperature fluctuations, and attains equilibrium in a comparatively short interval. Equilibration times that were previously 24–48 hr are reduced to minutes. This is dynamic osmometry, in which π is counterbalanced and approximated by a pressure-sensing servo-mechanism before equilibration.

Membrane osmometry is attended with a number of potential and actual problems. The measurement is customarily performed in a dilute electrolyte solution in which dissociation of the pectinic acid leads to a total concentration that relates π to the combined H^+ and polyanion molarity. The nondiffusibility of the polyanion is a source of error, in that the Donnan effect of the negative charge on pectin is to accumulate cations (e.g., Na^+ in a NaCl solvent) on the side of the polyanion, the end result being an exaggerated π and a consequently low $\overline{M}_n$. Methylation to the highest DE possible (Berth et al., 1980, 1982) will minimize the Donnan effect. The nitro, acetyl (Henglein and Schneider, 1936), and propionyl (Owens et al., 1948) derivatives may serve the same purpose. Nitration is drastic, and therefore presents the possibility of altering physical properties and causing degradation of pectin. Acylation can be performed under very mild conditions, ultimately with far less structural degradation (Owens et al., 1948).

The membrane is a potential source of determinate error. Mindful of the heterogeneity of pectin, one should be selected with a low enough permeability rate to preclude passage of average-size pectin molecules. The escape of solvent has been observed in this laboratory from the periphery of a supply-house, regenerated, circular cellulose membrane placed between and interfacing with an open reference and an open sensing cell. The cells were filled with water, covered in the osmometer jar in a humid atmosphere, and observation was made of the outward migrating water between the cell surfaces.

Aging of pectin dispersions is yet another source of error. During the equivalent of an equilibration interval required for static osmometers, viscosity data showed that dispersions changed their hydrodynamic status with time (Walter and Sherman, 1983). To the extent that the size, number, and distribution of dispersed particles in a fixed-weight concentration (c) change, π and $\overline{M}_n$ change accordingly, as the colloidal units become numerically less and larger in size, or vice versa.

π is a concentration-dependent property. However, for dilute pectin, the reduced osmotic pressure, (π/c), when plotted against c, yields a straight line, the intercept (infinite dilution) of which is inversely proportional to $\overline{M}_n$. The slope (β) of π/c versus c gives A_2 (the second virial coefficient). This β yields information about solute–solvent interaction (Cowie, 1973), permitting inferences to be drawn about dispersion conditions. For example, $\beta = 0$ indicates the theta condition where theoretically the solvent–solvent, solvent–polymer and polymer–polymer interactions are equal. $\beta = 0$ also images a lack of flexibility of the galacturonide chains (Owens et al., 1946). A positive β (positive A_2) denotes an affinity between pectin and solvent to which the magnitude of A_2 is scaled. A negative β (negative A_2) reflects solvent incompatibility under the influence of external stimuli, and presages a tendency for the pectin to desolvate and precipitate.

The theoretical and mathematical origins of A_2 derived from light-scattering photometry [in units of $cm^3/g^{-2}/mole$ (Frank and Mark, 1955) or $cm^3 \times g^{-1}$ (Allcock and Lampe, 1981)] and from osmometry (also in $cm^3 \cdot g^{-1}$) incorporate concentration-dependence principles. Not surprisingly therefore, A_2 was similar for the two independent methods by which a polystyrene fraction was collaboratively tested (Frank and Mark, 1955).

IX. VISCOMETRY

Deformation and flow measurements are a common approach to the physical characterization of colloids. The interpretation of viscosity data caused Smith (1976) to doubt that the pectins studied existed in random coil conformation.

For pectin dispersions, the useful viscosity property is resistance of the pectin solute to flow, relative to water containing an electrolyte. The purpose of the

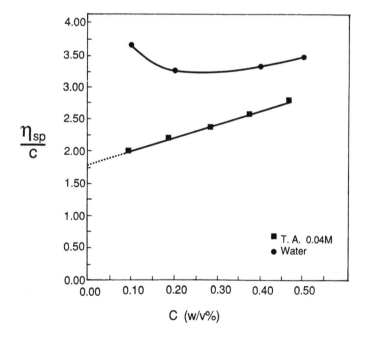

Figure 4 Viscosity number (reduced specfiic viscosity) as a function of concentration, c, of a dispersion of pectin in water (no electrolyte) and in 0.04 M tartaric acid. Note the steep negative slope in dilute concentrations, attributed to electroviscosity.

electrolyte is to depress ionization of the pectin carboxyl groups. Pectin then behaves as a neutral polymer (Pals and Hermans, 1952), instead of behaving as a polyelectrolyte exhibiting a large electroviscous effect (Fig. 4).

Viscometers operate on many basic principles. The capillary and rotational designs have been more frequently applied to pectin.

A. Capillary Viscometry

In capillary viscometry, flow is initiated downward through a vertical, capillary glass-column at a constant pressure and temperature. The data are treated to yield ultimately the intrinsic viscosity (syn. limiting viscosity number), $[\eta]$. The unit of $[\eta]$ is expressed in volume/weight, usually milliliter/gram or deciliter/gram (Carpenter and Westerman, 1975). Huggins (1942) related $[\eta]$ to solute concentration, and by means of the derived equation, a novel approach to studying pectin functionality, centered on the Huggins interaction coefficient and its ramifications (Eirich and Riseman, 1949), was recommended (Walter and Sherman, 1983; Walter and Sherman, 1984). Smidsrod and Haug (1971) devised a method involving $[\eta]$ and ionic strength to test comparative stiffness of polyelectrolyte chains in response to salt.

B. The Mark–Houwink Equation

$[\eta]$ is related to molecular weight through the Mark–Houwink equation, $[\eta] = KM^a$, where M is a viscosity-average molecular weight ($\overline{M}_v$) (Billmeyer, 1984). K is the intercept, and exponent a is the slope of the log–log plot. $\overline{M}_v$ is generally closer to $\overline{M}_w$ than to $\overline{M}_n$. The equation is valid to about 10^5 (Anger and Berth, 1986). The exponent is variable between 0.5 for tightly coiled polymers and 1.7 for rigidly extended molecules. Different solvents, polymers, and temperatures affect a (Daniels et al., 1956, 1970). It was found to be sensitive to solution clarification techniques, and more directly, to the microgel fraction that was removable by purification (Hourdet and Muller, 1987). Smidsrod (1970) equated a high numerical value for it with an extended alginate molecule that was arguably an indication of a very rigid chain conformation.

The difficulty in any determination of molecular weight with the use of the Mark–Houwink equation is the reliance of the computation on K and a. These constants are seldom known beforehand, and when obtained from similar systems, they differ considerably (Table I). Unlike synthetic polymers, there is no homologous series of pectin with known molecular weights that would definitively establish a, K, and M. Moreover, the dependence of a and K on $\overline{M}_v$ of some pectins (Deckers et al., 1986) may further compound the difficulty in universally applying the equation.

Table I Mark–Houwink constants for pectin

Pectin	Solvent	K	a	Reference
DE = 38%	0.10*M* NaCl	—	0.62	Axelos (1987)
DE = 10–40%	0.2*M* NaCl	—	0.85, 1.3	Hourdet and Muller (1987)
DE = 32–95%	0.09*M* NaCl	9.55×10^{-2}	0.73	Anger and Berth (1986)
High DE	Acetate buffer pH 3.7	0.49×10^{-3}	0.79	Deckers *et al.* (1986)
DE = 70%	—	2.16×10^{-2}	0.79	Berth *et al.* (1977)
Laboratory commercial	0.1*M* NaCl	—	0.64	Kawabata and Sawayama (1977)
L.M., H.M.	0.1*M* NaCl	—	0.80, 1.50	Smith (1976)
HM	1% sodium hexametaphosphate pH 4.5	4.7×10^{-5}	1	Christenson (1954)

C. Rotational Viscometry

This method requires larger samples than capillary viscometry does, and is more suitable at higher concentrations. Using this technique, Kawabata (1977) examined the time, pH, shear, concentration, and temperature effects on pectin solutions, and Walter *et al.* (1985, 1986) characterized apple- and grape-pomace extracts containing a high percentage of pectin.

X. MOLECULARITY AND DISPERSITY PARAMETERS

A parameter is a mathematical auxiliary variable, used in common to define two dependent variables, each being separately a dependent function of the auxiliary variable. In pectin research, as in many other disciplines involving physical phenomena, the word has acquired a new meaning, to indicate an index of or limit on physical occurrences. Jowet (1988) takes issue with the broadened definition. Notwithstanding the mathematician's preemption of the term, contemporary usage is hereinafter adopted in discussing some biopolymeric principles of pectin.

A. Aggregation

Aggregates of pectin are formed spontaneously in an aqueous dispersion (Sorochan *et al.*, 1971; Davis *et al.*, 1980; Fishman *et al.*, 1983, 1984; Fishman and Pepper, 1985), particularly in an acidic medium (Sawayama *et al.*, 1988). Aggregation is believed to be a nonequilibrium process resulting in chain dimensions that are larger than those estimated from osmometry data in a completely molecular dispersion (Jordan and Brant 1978). Considering this tendency to aggregate, molecular dispersion is seldom the prevailing physical state of pectin in an aqueous medium. Many analysts routinely filter pectin dispersions through micropore filters as a final step in removing the so-called microgel, before light-scattering photometry. In this laboratory, we have found that 0.45 micron Acrodiscs (Gelman Sciences, Inc., Ann Arbor, MI) remove approximately 5% of an ultracentrifuged sample. Whether or not this agglomerate is an artifact or an intrinsic fraction of the pectin resulting from a spontaneous process has not been answered. Under what circumstances the colloidal assembly should or should not be removed is a subjective and arbitrary decision. $\overline{M}_n$ may be acquired from a pectin dispersion with the greatest certainty only by the reducing EGM, mindful of its limitations nevertheless, whereby one specific reactive site per molecule is in effect counted.

Aggregation is heat reversible (Fishman and Pepper, 1985), with particle size increasing between pH 2.6 and 3.2, and decreasing above pH 3.2, as a function of increasing temperature (Gubenkova et al., 1988).

Pectin molecules in apple tissue have been observed under the electron microscope (by staining with ruthenium red) to form aggregates that were dispersed after the charge density was increased by enzyme hydrolysis (Hanke and Northcote, 1975). Fishman et al. (1986b) calculated the axial ratio (ratio of length to width) of pectin aggregates to be 120 to 200. Pectin aggregates have been regarded as random coils (Berth et al., 1977).

B. Molecular Weights and the Degree of Polymerization

The three molecular-size parameters previously discussed are $\overline{M}_w$, $\overline{M}_n$, and $\overline{M}_v$. Barring aggregation, $\overline{M}_w > \overline{M}_v > \overline{M}_n$. Another molecular weight, $\overline{M}_z$, is obtainable from ultracentrifugation data (see Smith, 1976), but it is not a frequently reported parameter in pectin research. The DP is calculated from $\overline{M}_p^1$ and one half the molecular weight of the repeating dinner ($\overline{M}_{0.5D}$) (Fig. 5), so that DP = $(\overline{M})/\overline{M}_{0.5D}$, where the denominator is a weight-average number that depends on the methyl ester content[2]. The quotient is accurate only for pectins containing a very minimum amount of nonuronide material.

Macromolecular size can be visualized readily from DP. If (according to Saverborn, 1945) the length of pectinic acid molecules, varying widely with the method of measurement, is from 730 A° to 4500 A°, and a monomer length of 5.15 A° (Kertesz, 1951) is assumed, the DP of pectin is 142–874. The contribution of methyl groups to molecular weights consisting of chain lengths of this magnitude is minor.

---

[1] $\overline{M}_p$, Average molecular weight of pectin.

[2] Based on one-half the molecular weight of $C_{12}H_{16}O_{12}(CH_3)$ (DE = 0%), $C_{12}H_{15}O_{12}(CH_3)$ (DE = 50%), or $C_{12}H_{14}O_{12}(CH_3)_2$ (DE = 100%), M_p = 176.183 or 190, respectively. See Fig. 5.

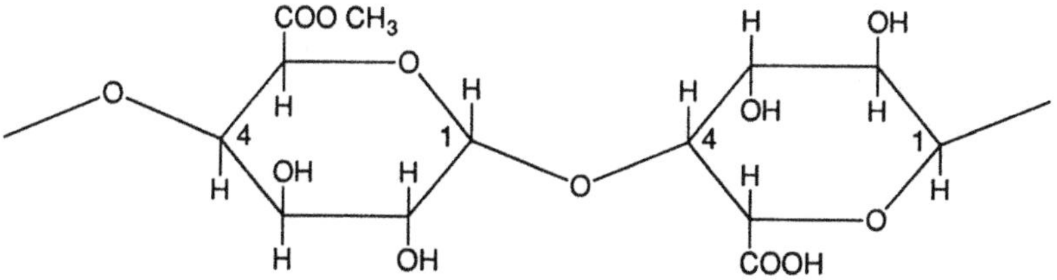

Figure 5   Repeating dimer in pectin, having DE = 50%.

$\overline{M}_w$, originating from a summation of light-scattering data, is a squared property of each mass (m_i), whereas, $\overline{M}_n$, based on osmometry, considers an average of the product of m_i and its respective particulate concentration (k) (Billmeyer, 1984). Hence, $\overline{M}_w = f(m_i{}^2)$ and $\overline{M}_n = f(km_i)$. For a constant weight of pectin, it is seen that longer chains and larger aggregates (heavier m_i) will have a greater impact on $\overline{M}_w$ than will shorter chains and smaller aggregates (lighter m_i). The opposite (shorter chains and smaller aggregates) has a greater impact on $\overline{M}_n$. The reducing EGM, predicated solely on chemical composition, gives $\overline{M}_n$ that is theoretically insensitive to changes in mass and distribution. The reducing EGM results do vary, however, and different methods for different concentrations might become necessary (Voragen *et al.*, 1971).

C. Heterogeneity and the Degree of Polymerization

Polydispersity conventionally refers to the state of colloidal matter existing in a broad spectrum of molecular sizes. In other words, the system is heterogeneous. In a monodisperse system the molecules exist in approximately one size (Everett, 1988). The DP of pectin is an index of the length of a galacturonan.

Elias (1977) makes a distinction between polydispersity and polymolecularity. In his view, solute consisting of many different molecular weights (different DP) is polymolecular, and polydispersity is "the dispersion of properties of any particles." By this distinction, the narrower the band of isolated pectin (Fig. 3), *i.e.*, the less polymolecular a fraction is, the more constant will be the DP. Elias' definition of polydispersity suggests various patterns of behavior of a given concentration of (colloidal) solute wrought by path-dependent responses to external stimuli.

Everett's definition of polydispersity (Everett, 1988) may be reconciled with Elias' definition of polymolecularity (Elias, 1977), if it is understood that in varying, nonequilibrium stages of aggregation a monodisperse pectin (by Everett's definition), identical to a molecularly homogeneous pectin (by Elias' definition) may exhibit different properties or behavior under different circumstances (Elias' polydispersity). Alternatively stated, a constant molecularity (same chemical composition, DP and $\overline{M}$) may lead to different polydispersities, as, for example, in hysteresis, when hydrocolloidal dispersions are dried, and the xerogel is rehydrated at different rates.

The pectins of commerce are polymolecular isolates principally from citrus and apples.

In the sense that pectin may contain monomers of sugars in the galacturonan chain, it is a heteropolymer. This particular kind of heterogeneity affects the accuracy of the DP of pectin, because the higher the percentage of nongalacturonan inclusions, the greater is the fraction of weight excluded from the specific pectin assay. The same applies to molecular weight determinations.

XI. MECHANICAL METHODS FOR PECTIN GELS

Pectin jellies have traditionally been characterized by the sag method of determining gel firmness. The specifications require that the jelly be cooked in an open vessel, poured into cups, and left to set for 24 hr before testing (IFT, 1959). Subsequently, the jelly cup is inverted on a glass plate, and the tendency to collapse is measured against a calibrated rod. Rao *et al.* (1989) tested gel strength with a Voland-Stevens Texture Analyzer (VSTA) and an Instron Universal Testing Machine. For these tests, the open vessel in the IFT procedure was replaced by a virtually closed, 6-liter distillation flask in a refluxing system (Fig. 6). The flask allowed better control of the variables, and hence better precision of the measurements. Also, on other occasions, the IFT formula was reduced to one tenth the specified quantities, and the jelly was prepared and left to set in a much smaller, custom-made, wide-neck flask (450 ml, ST 55/50, 16.5 cm high) (Fig. 6) that accommodated the VSTA and Instron measuring plungers directly in the flask without first interrupting the gel. While the ridgelimeter measures gel strength only, the texture profiles of the VSTA and Instron methods provide information on hardness, fracturability, and cohesiveness.

XII. NOVEL METHODS OF PHYSICAL CHARACTERIZATION

Greenwood (1952) was among the first to discern the urgency of defining polysaccharide materials on the basis of their high-polymer characteristics. The burgeoning use of pectin in fabricated food makes this need more compelling, inasmuch as former empirical parameters are obsolete. For example, as it was pointed out in Chapter 3, percentage of sag has no relevance to pectin for yogurt. Modern methods of characterization should have their fundamentals in contemporary polymer theory, because pectin is a biopolymer, in many instances, behaving similarly to synthetic polymers, albeit modified by its high affinity for

Figure 6 Pectin jelly made by refluxing the IFT formula in a 6-liter and a 450-ml flask.

water. This section is an attempt to show how pectin may be characterized more scientifically than has previously been the case.

A. The Energetics of Pectin Jelly Sols

The dominant property of dispersed pectins in food is their ability to gel with sugar and acid or calcium. The role of jelly ingredients has been elucidated in this text and elsewhere (Olsen, 1934; Doesburg, 1965). Sugar, in addition to having the presumed purpose of deactivating water, may be assigned a volume-exclusion role in the water–pectin–acid system. Gel firmness and softness are a matter of Brownian movement in the matrix. The higher the sugar content, the less unoccupied space there is for kinetic motion, and the firmer is the gel (lower percentage sag). Conversely, the lower the sugar content, the more unoccupied volume there is, and the softer is the gel (greater percentage sag).

High-methoxyl gels are physically bonded, are structurally reversible (as may be shown by dialysis), but are, however, not ordinarily thermally reversible (Walter and Sherman, 1986). Low-methoxyl gels are chemically bonded (through calcium bridges) and are thermally reversible. The two gels consist of polymer chains with a microstructure in short, stiff, and long flexible conformations, interrupted by junction zones (Mitchell and Blanshard, 1979; Rees, 1970, 1972). They appear to fall in Russo's F2 and L2 gel categories (Russo, 1987). The common feature of all these pectin gels is their formation by a loss of heat from a sol. The process (Fig. 7) begins by a transfer of kinetic energy, $\Delta H(+)$, from water to the ground-state pectin. The hydrated pectin reacts endothermically $[\Delta E(+)]$ with sugar in the manner of an activated complex at the boiling temperature (105°C), where light-scattering centers disappear from the heated sol before the addition of acid. Upon cooling, the sol gradually loses an increment of the accumulated energy, $\Delta E'(-)$, leaving $\Delta H'(-)$, so that $\Delta H'(-) = \Delta E(+)$

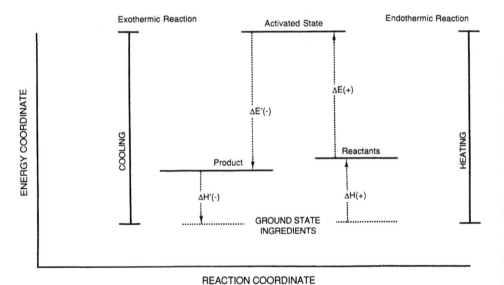

Figure 7   The energetics of pectin solation and gelation.

Table II Reduction in percentage sag (percentage RS)[a] of high-methoxyl (OMe) pectin jellies with time

Pectin jelly	pH	% OMe	% G.A[b]	Time (hr)				% RS[a]
				4	7	24	30	
Baking	2.33	8.3	98	25.8	25.3	24.7	24.5	5.0
BBRS	2.53	8.4	—	25.7	25.3	24.4	24.2	5.1
DDSS	2.52	6.3	86	25.9	25.4	24.4	24.0	7.3
DDextSS	2.70	7.6	81	26.3	25.4	24.6	24.3	7.6

[a]Average of 3 measurements.
[b]Percentage galacturonic acid.
[c](Percentage RS at 4 hr − percentage RS at 30 hr)/percentage RS at 4 hr × 100.

$+ \Delta H(+) - \Delta E'(-)$. Syneresis may be interpreted as a separation of liquid from an aging gel, in an attempt by the metastable pectin network containing $\Delta H'(-)$ to return to the ground state. $\Delta E'(-)$ is the apparent activation energy of viscous flow for pectin jelly sols, obtainable from a log of viscosity *versus* reciprocal absolute temperature plot (Walter and Sherman, 1981).

The pectin sol–gel transition from viscosity to elasticity is complete within 24 hr after incipient gelation. Approximately 75–85% of the attainable gel strength is reached within the first 30 min (Beveridge and Timbers, 1989). These shifts are the basis of the sag test for jelly grade (IFT, 1959). At the end of the 24-hr interval, as much as a 7.6% sag reduction is possible (Table II).

B. The Gelation Temperature

The loss of $\Delta E'(-)$ occurs over a range of temperatures. In some instances, a definite inflection ($T_{gel.}$), reminiscent of a glass transition temperature, precedes the incipient reticulum (Walter and Sherman, 1981; Walter and Sherman, 1986). This $T_{gel.}$ can be ascertained from a plot of viscosity or viscosity differentials against reciprocal, absolute temperatures. Under specified conditions, $T_{gel.}$ is an intrinsic property, at times independent of cooling rates (Walter and Sherman, 1981), and at times, inversely dependent on cooling rates (Hinton, 1950). Once cooling has surpassed the $T_{gel.}$, flow of high-methoxyl pectin jelly sols is not reversible (Walter and Sherman, 1981), although they may soften at a critical temperature, and may even harden with more heating (Beveridge and Timbers, 1989).

Another method of determining $T_{gel.}$ relies on the changing slope pattern of convection and conduction cooling on a temperature–time curve (Fig. 8). The first segment, convection cooling, begets a steeper, negative slope antecedent to the practically zero slope of the second segment, conduction cooling. The

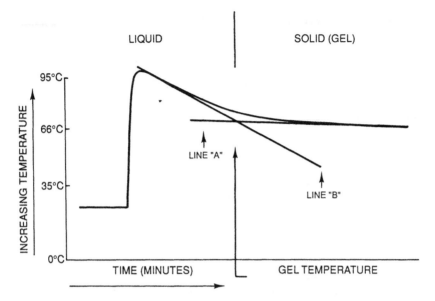

Figure 8 Determiantion of the gelation temperature of pectin jelly by intersecting tangents to convection and conduction cooling curves. Reproduced with permission of Hercules, Inc., Wilmington, DE.

tangents to these two segments intersect at T_{gel}. The conduction-cooling segment does not distinguish between the asymptote of the temperature gradient (without the gel), as the system approaches equilibrium with its surroundings, and the asymptote resulting from the approach of the conduction-cooling gel temperature to the water-bath temperature (30°C).

C. Partial Molal Volume

The volume fraction is an extensive property used in the process of characterizing synthetic polymers. Biopolymers present a practical difficulty, and as a result, partial volumes of biopolymer solutes have rarely been measured (Eisenberg, 1976). A procedure for deriving this quantity for a dispersed pectin was recently reported (Walter and Matias, 1989).

Pectin has the ability to stabilize a foam at an air–water interface and, through agitation, incorporate a large amount of air in aqueous systems. These are functional attributes in foods, but they encumber accurate and precise measurements like density, specific volumes, partial molal quantities, etc., before calculation of partial molal volumes. Volume fractions were computed from the partial molal volumes. A higher level of methylation was found to result in a lower, positive entropy change (Table III). The significance of this may best be understood by comparison with crystallization processes, in which there are negative energy contributions (Bueche, 1962; Cowie, 1973). During cooling, a pectin jelly sol imparts a large negative enthalpy to the viscous mass that more than compensates for a positive entropy. The net effect is macromolecular structuration of pectin. As Table III indicates, a water dispersion of a high-methoxyl pectin is prone to greater organization than is a low-methoxyl pectin. These preliminary data have implications for hydrophobic bonding in high-DE pectin jellies, since the smaller ΔS for BAKING and DDextSS pectins suggests their greater tendency to organize themselves into a reticulum, while the higher ΔS for LM22CG and LM12CB (the low-DE pectins) indicates a smaller tendency to do so.

The advantage of being able to characterize pectin through volume-fraction computations is that chemical composition will be less a factor, and state and path-dependent equations will more aptly define the hydrodynamic condition, and thus obviate the many assumptions, variations, and discrepancies in current empirical measurements.

D. Hydrophilicity

When xerogels are immersed in water, they undergo volume expansion (ΔV). This ΔV was recently measured against weight–weight or weight–volume concentrations (Walter and Matias, 1989; Walter and Talomie, 1990), and the slope of the linear regression line (using molar or molal concentration) was treated with the necessary constants to give a swelling parameter defined as the hydrophilicity (Walter and Talomie, 1990). High-methoxyl pectin showed almost twice the water-binding capacity of low-methoxyl pectin. The inference is that

Table III Configurational entropy (ΔS) of a constant molar concentration of pectin in 0.04 M tartaric acid

Pectin	DE%	MW[a] $\times 10^4$	$V_p{}^a \times 10^4$	X[a] $\times 10^{-3}$	$\Delta S^b \times 10^{-1}$ (Joule Kelvin^{-1})
Baking[b]	68–71	3.02	2.97	5.14	2.37
		2.85	1.95	8.26	3.81
DDextSS[b]	61–64	3.28	4.48	8.02	3.72
			3.05	8.00	5.49
LM22CG	45–53	3.36	4.40	13.8	6.43
LM12CB	27–35	3.41	7.69	16.0	7.46

[a]MW, molecular weight; DE, degree of esterification; V_p, partial molal volume of pectin; X, volume fraction of pectin.
[b]Data from two separate experiments.

hydrophilicity decreases with decreasing methylester content of pectins. Pectin hydrophilicity was many times lower than that of carboxymethylcellulose and guar gum (Walter and Talomie, 1990).

E. Aggregation Number

On the assumption that osmometry gave a number-average molecular weight of pectin aggregates, Fishman *et al.*, (1984) rationalized into an aggregation number the $\overline{M}_n$ from osmometry with the $\overline{M}_n$ from end-group titration. If aggregation is indeed a non-equilibrium event, any $\overline{M}_n$ obtained therefrom is transitory, and, accordingly, the aggregation number is inexact. Instead, the $\overline{M}_w$ from light-scattering data, has been rationalized with the $\overline{M}_n$ from the reducing EGM to give an aggregation number, explained as the relative amount of unaggregated pectin molecules entering into a relatively stable, colloidal association (micelle) in an aqueous medium (Walter and Matias, 1991). In light-scattering photometry, solute motion is in dynamic equilibrium in a narrow beam of light impinging on a unit volume of solution or dispersion. Use of a light-scattering $\overline{M}_w$ in the numerator and an unambiguous chemical quantity in the denominator ensures that this parameter is a stable equilibrium index of aggregation of single pectin molecules.

The quotient $(\overline{M}_w/\overline{M}_n)$ has frequently been described as a measure of molecular-weight distribution, but exactly how it elaborates this circumstance has not been explained. This interpretation wrongly presupposes complete molecular separation. The ratio is claimed to tell very little about polydispersity (Vold and Vold, 1983). It is reasonable to hold that properly assembled quotients might be more informative of aggregation phenomena than as indices of molecularity and dispersity. For example, a larger *vis-à-vis* a smaller aggregation number foretells a higher predisposition of a dispersed solute to flocculate and precipitate (Doty, 1947). It certainly connotes greater solute–solute affinity, and by corollary, less solute–solvent interaction.

F. Coordinate Orientation

The ultimate goal of investigations into pectin behavior is to facilitate a specific choice of pectin for a variety of industrial and commercial applications. Attempts have been made to systematize the selection of hydrocolloids for food and beverages (Trudso, 1989), but selection is still largely a trial-and-error undertaking.

Severs (1962) commented that more polar macromolecules have a higher activation energy than less polar molecules, that increasing temperature had a

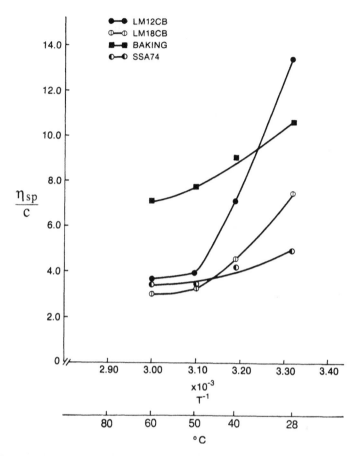

Figure 9 Orientation of pectin dispersions in a coordinate plane circumscribed by viscosity numbers and reciprocal, absolute temperatures.

more drastic effect on polymers having a higher activation energy, and that the energy required for a viscosity change is less at high temperatures than at low temperatures. It so happens that the impact of the numerous variables on pectin's hydrodynamic response to heat can be integrated into a flow profile, circumscribed by reciprocal, absolute temperatures and a viscosity index that is more unique to a pectin than to a pectin class (Walter and Sherman, 1984) (Fig. 9).

An example of how coordinate orientation may inform, is as follows: if a manufacturer wishes to use pectin to impart a high viscosity to a liquid formula, the choice should be LM12CB (Fig. 9). If a high viscosity in another product is to be maintained at an elevated temperature, the choice should be BAKING. For a food use requiring enhanced solids content simultaneous with a relatively low viscosity, SSA74 should be the pectin of choice.

G. Cooling-Rate Index

Hot pectin–jelly sols in the viscous (pregelation[3]) interval are amenable to measurement . A time–temperature cooling gradient was substituted parametrically with its viscosity–time counterpart, and the latter was integrated into a quotient. Theoretically, the cooling-rate quotient is ≥ 1 (Walter and Sherman, 1990). This approach substitutes the subjectivity in the end-point determinations of conventional, setting-time measurements (Hinton, 1950) with mathematical indexing. The algebraic logic dictates that rapid-set pectins should give a large index, and slow-set pectins should give a small index.

XIII. SUMMARY

Numerous objective methods exist to detect and measure pectin, but variations of pectin form and behavior have led to inconsistencies in the relevant literature. Many of the traditional procedures do not have an origin in fundamental theory. Contemporary polymer science has, however, offered an opportunity for pectin (and other hydrocolloids) to begin or continue to be characterized by classical equations, whereby chemical composition, for example, has less of an influence than state- and path-dependent functions.

References

Alacheva, D. T., Parshikova, L. P., Edel'man, G. A., Andreev, V. V., and Filippov, M. P. (1973). Colorimetric determination of uronides in pectic substances. Chem. Abstr. 80, 58455v (1974).

Albersheim, P., Neukom, H., and Deuel, H. (1960). Splitting of pectin chain molecules in neutral solution. *Arch. Biochem. Biophys.* **90,** 46–51.

Allcock, H. R., and Lampe, F. W. (1981). "Contemporary Polymer Chemistry." Prentice-Hall, Englewood Cliffs, N.J.

Anger, H., and Berth, G. (1985). Gel-permeation chromatography of sunflower pectin. *Carbohydr. Polym.* **5,** 241–250.

Anger, H., and Berth, G. (1986). Gel-permeation chromatography and the Mark–Houwink relation for pectins with different degrees of esterification. *Carbohydr. Polym.* **6,.** 193–202.

Aspinall, G. O., and Cottrell, I. W. (1970). Lemon-peel pectin. II. Isolation of homogeneous pectins and examination of some associated polysaccharides. *Can. J. Chem.* **48,** 1283–1289.

Axelos, M. A. V., Lefebvre, J., and Thibault, J. F. (1987). Conformation of a low-methoxyl citrus pectin in aqueous solution. *Food Hydrocolloids* **5/6,** 569–570.

Ayres, A., Dingle, J., Phipps, A., Reid, W. W., and Solomons, G. L. (1952). Enzyme degradation of pectic acid and the complex nature of polygalacturonase. *Nature* **170,** 834–836.

Barford, R. A., Magidman, P., Phillips, J. G., and Fishman, M. L. (1986). Estimation of degree of methylation of pectin by pyrolysis–gas chromatography. *Anal. Chem.* **58,** 2576–2578.

Barth, H. G. (1980). High-performance gel-permeation chromatography of pectins. *J. Liq. Chromatogr.* **3,** 1481–1496.

---

[3]Pregelation: meaning before the gel develops.

Bean, J. E., and Bornman, C. H. (1973). Staining of pectic substances as extracts and in the transfusion tracheids of *Welwitschia*. *J. S. Afr. Bot.* **39,** 23–33.

Bellamy, L. J. (1954). "The Infrared Spectra of Complex Molecules." Wiley, New York.

Berth, G. (1988). Studies on the heterogeneity of citrus pectin by gel-permeation chromatography on Sepharose 2B/Sepharose 4B. *Carbohydr. Polym.* **8,** 105–117.

Berth, G., Anger, H., Plashchina, I. G., Braudo, E. E., and Tolstoguzov, V. B. (1982). Structural study of the solutions of acidic polysaccharides. II. Study of some thermodynamic properties of the dilute pectin solutions with different degrees of esterification. *Carbohydr. Polym.* **2,** 1–8.

Berth, G., Anger, H., and Lexow, D. (1980). Determination of the molecular weights of pectins by means of osmometry in aqueous solutions. *Food Sci. Technol. Abstr.* **12,** 8A 581.

Berth, G., Anger, H., and Linow, F. (1977). Light-scattering and viscosimetric studies for molecular-weight determination of pectins in aqueous solutions. *Chem. Abstr.* **88,** 188477e.

Beveridge, T., and Timbers, G. E. (1989). Small-amplitude oscillatory testing (SAOT) application to pectin. *J. Texture Stud.* **20,** 317–324.

Billmeyer, F. W., Jr. (1984). "Textbook of Polymer Science." 3rd Ed. Wiley, New York.

Billmeyer, F. W., Jr., and de Than, C. B. (1955). Dissymmetry of molecular light scattering in polymethyl methacrylates. *J. Am. Chem. Soc.* **77,** 4763–4767.

Bitter, T., and Muir, H. M. (1962). A modified uronic acid carbazole reaction. *Anal. Biochem.* **4,** 330–334.

Blumenkrantz, N., and Asboe-Hansen, G. (1973). New method for quantitative determination of uronic acids. *Anal. Biochem.* **54,** 484–489.

Bock, J. L. (1985). Recent developments in biochemistry. Nuclear magnetic resonance spectroscopy. *In* "Methods of Biochemical Analysis" (D.Glick, ed.) Vol. 31, pp. 259–315. Wiley, New York.

Bueche, F. (1962). "Physical Properties of Polymers." Interscience, New York.

Burge, D. E. (1977). Molecular-weight determination by osmometry. *American Laboratory* **9,** 41–51.

Bystricky, S., Kohn, R., and Sticzay, T. (1979). Effect of polymerization of oligogalacturonates and D-galacturonans on their circular dichroic spectra. *Chem. Abstr.* **90,** 152515t (1979).

Carpenter, D. K., and Westerman, L. (1975). Viscometric methods of studying molecular weight and molecular weight distribution. *In* "Polymer Molecular Weights," Part II. (P. E. Slade, Jr., ed.), p. 398. Marcel Dekker, New York.

Chou, T. C., and Kokini, J. L. (1988). Role of hydrophobic interaction, hydrogen-bonding interaction, electrostatic interaction on the rheological and spectroscopic properties of apple, tomato, and citrus pectins. Abstract 308, Annual Meeting (June), Institute of Food Technologists, New Orleans.

Christenson, P. E. (1954). Methods of grading pectin in relation to the molecular weight (intrinsic viscosity) of pectin. *Food Res.* **19,** 163–172.

Cowie, J. M. G. (1973). "Polymers: Chemistry and Physics of Modern Materials." Billing & Sons, London.

Daniels, F., Mathews, J. H., Williams, J. W., Bender, P., and Alberty, R. A., (1956). "Experimental Physical Chemistry," 5th Ed. McGraw-Hill, New York.

Daniels, F., Williams, J. W., Bender, P., Alberty, R. A., Cornwell, C. D., and Harriman, J. E. (1970). "Experimental Physical Chemistry," 7th Ed. McGraw-Hill, New York.

Davis, M. A. F., Gidley, M. J., Morris, E. R., Powell, D. A., and Rees, D. A. (1980). Intermolecular association in pectin solutions. *Int. J. Biol. Macromol.* **2,** 330–332.

Deckers, H. A., Olieman, C., Rombouts, F. M., and Pilnik, W. (1986). Calibration and application of high-performance size-exclusion columns for molecular-weight distribution of pectins. *Carbohydr. Polym.* **6,** 361–378.

Dische, Z. (1962). Color reactions of hexuronic acids. *Meth. Carbohydr. Chem.* **I,** 497–501.

Doesburg, J. J. (1965). "Pectic Substances in Fresh and Preserved Fruits and Vegetables." I.B.V.T. Communication Nr. 25, Wageningen, The Netherlands.

Doty, P., Wagner, H., and Singer, S. (1947). The association of polymer molecules in dilute solution. *J. Phys. Colloid Chem.* **51**, 32–57.

Eirich, F., and Riseman, J. (1949). Some remarks on the first interaction coefficient of the viscosity-concentration equation. *J. Polym. Sci.* **4**, 417–434.

Eisenberg, H. (1976) "Biological Macromolecules and Polyelectrolytes in Solution." Clarendon Press, Oxford.

Elias, H. (1977). "Macromolecules. I. Structure and Properties." (Trans. from German by J. W. Stafford). Plenum, New York.

Everett, D. H. (1988). "Basic Principles of Colloid Science." Royal Society of Chemistry, London.

Filippov, M. P. (1978). "Infrakrasnye Spektry Pektinovyh Veschestv (Infrared Spectra of Pectic Substances)." (G. V. Lazur'evsky, ed.). Shtiinza Publishers, Kishinev, USSR. Book Review. *Carbohydr. Res.* **80**, C25–C27.

Filippov, M. P., Komissarenko, M. S., and Kohn, R. (1988). Investigation of ion exchange on films of pectic substances by infrared spectroscopy. *Carbohydr. Polym.* **8**, 131–135.

Filippov, M. P., and Shamshurina, S. A. (1972). Comparative study of pectin substances by IR spectroscopy. *Food Sci. and Technol. Abstr.* **6**, 7A260 (1974).

Filippova, T. V., and Shkolenko, G. A. (1968). Infrared spectra of pectins in sugar beets and grapes. *Chem. Abstr.* **71**, 46688z (1969).

Fishman, M. L., and Pepper, L. (1985). Reconciliation of the differences in number-average molecular weights (M_n) by end-group titration and osmometry for low-methoxy pectins. *In* "New Dev. Ind. Polysaccharides", Proceedings, Conference on New Developments in Industrial Polysaccharides, pp. 159–166. Gordon & Breach, New York.

Fishman, M. L., Pfeffer, P. E., Barford, R. A., and Doner, L. W. (1984). Studies of pectin solution properties by high-performance size-exclusion chromatography. *J. Agric. Food Chem.* **32**, 372–378.

Fishman, M. L., Pepper, L., Damert, W. C., Phillips, J. G., and Barford, R. A. (1986a). A critical reexamination of molecular weight and dimensions for citrus pectins. *In* "Chemistry and Function of Pectins" (M. L. Fishman and Joseph J. Jen, eds.), pp. 22–37, Amer. Chem. Soc., Monograph Series 310, Washington, D. C.

Fishman, M. L., Pepper, L., and Pfeffer, P. E. (1984). Dilute solution properties of pectin. *Polym. Mater. Sci. Eng.* **51**, 561–565.

Fishman, M. L., Pepper, L., and Pfeffer, P. E. (1986b). Dilute solution properties of pectin. *In* "Water Soluble Polymers; Beauty With Performance" (J. E. Glass, ed.), pp. 57–70. Amer. Chem. Soc. Advan. Chem. Series 213, Washington, D.C.

Fishman, M. L., Pepper, L., Pfeffer, P. E., and Barford, R. A. (1983). Pectin aggregation as measured by number-average molecular weights. Amer. Chem. Soc., Agr. and Food Div. 186th Meeting. Abstracts of Papers, Washington, D.C.

Forni, E. Rizzolo, A., and Gargano, A. (1984). Gas-chromatographic determination of the methoxy number of pectins. *Chem. Abstr.* **101**, 128984v (1984).

Frank, H. P., and Mark, H. F. (1955). Report on molecular-weight measurements of standard polystyrene samples. II. International Union of Pure and Applied Chemistry. *J. Polym. Sci.* **17**, 1–20.

Garmon, R. G. (1975). End-group determinations. *In* "Polymer Molecular Weights," Part I. (P. E. Slade, Jr., ed.), pp. 31–78. Marcel Dekker, New York.

Giangiacomo, R., Polesello, A., and Marin, F. (1982). Quantitative determination of galacturonic acid in commercial pectins by HPLC. *Food Sci. and Technol. Abstr.* **15**, 2G123 (1983).

Gidley, M. J., Morris, E. R., Murray, E. J., Powell, D. A., and Rees, D. A. (1979). Spectroscopic and stoichiometric characterization of the calcium-mediated association of pectate chains in gels and in the solid state. *J. Chem. Soc., Chem. Commun.* 990–992.

Gorin, P. A. (1981). Carbon-13 nuclear magnetic resonance spectroscopy of polysaccharides. *Adv. Carbohydr. Chem. Biochem.* **38**, 13–104.

Greenwood, C. T. (1952). The size and shape of some polysaccharide molecules. *Adv. Carbohydr. Chem.* **7**, 289–318.

Grubisic, Z., Rempp, P., and Benoit, H. (1967). A universal calibration for gel-permeation chromatography. *Polym. Lett.* **5**, 753–759.

Gubenkova, A. A., Somov, A. A., and Shenson, V. A. (1988). Physicochemical properties of pectin and pectin-based solutions and jellies, *Food Sci. and Technol. Abstr.* **22**, (1990). No. 3, 3T18.

Hanke, D. E., and Northcote, D. H. (1975). Molecular visualization of pectin and DNA by ruthenium red. *Biopolymers* **14**, 1–17.

Henglein, F. A., and Schneider, G. (1936). Esterification of pectin substances. *Chem. Abstr.* **30**, 3780[9] (1935).

Hicks, K. B., Lim, P. C., and Haas, M. J. (1985). Analysis of uronic and aldonic acids, their lactones, and related compounds by high-performance liquid chromatography on cation-exchange resins. *J. Chromatogr.* **319**, 159–171.

Hinton, C. L. (1950). The setting temperature of pectin jellies. *J. Soc. Food Agr.* **1**, 300–307.

Hough, L., and Jones, J. N. K. (1952). Methylation of carbohydrate with diazomethane. *Chem. Ind. (London)*, 380.

Hough, L., Jones, J. K. N., and Wadman, W. H. (1950). Quantitative analysis of mixtures of sugars by the method of partition chromatography. Part V. Improved methods for the separation and detection of the sugars and their methylated derivatives. *J. Chem. Soc.*, 1702–1706.

Hough, L., and Theobald, R. S. (1963). Methylation with diazomethane. *In* "Methods in Carbohydrate Chemistry," II, pp. 162–166, Academic Press, New York.

Hourdet, D., and Muller, G. (1987). Solution properties of pectin polysaccharides. I. Aqueous size exclusion chromatography of Flax pectins. *Carbohydr. Polym.* **7**, 301–312.

Huggins, M. L. (1942). The viscosity of dilute solutions of long-chain molecules. 4. Dependence on concentration. *J. Am. Chem. Soc.* **64**, 2716-2718.

IFT. (1959). Pectin standardization. Final Report of the IFT Committee. *Food Technol.* **13**, 496–500.

Inoue, Y., Yamamoto, A., and Nagasawa, K. (1987). Depolymerization of pectin with diazomethane in the presence of a small proportion of phosphate buffer. *Carbohydr. Res.* **161**, 75–90.

Irwin, P. L., Gerasimowicz, W. V., Pfeffer, P. E., and Fishman, M. L. (1985). H-^{13}C polarization transfer studies of uronic acid polymer systems. *J. Agric. Food Chem.* **33**, 1197–1201.

Johnson, W., Jr. (1985). Circular dichroism and its empirical application to bipolymers. *In* "Methods of Biochemical Analysis" (D. Glick, ed.), pp. 61–163. Wiley, New York.

Jordan, R. C., and Brant, D. A. (1978). An investigation of pectin and pectic acid in dilute aqueous solution. *Biopolymers* **17**, 2885–2895.

Jowet, R. (1988). Editorial. *J. Food. Eng.* **7**, 1–3.

Kamata, T., and Nakahara, H. (1973). One concentration method in light scattering. *J. Coll. Interface Sci.* **43**, 89–96.

Kawabata, A. (1977). Studies on chemical and physical properties of pectic substances in fruits. *Memoirs Tokyo Univ. Agr.* **19**,115–160.

Kawabata, A., and Sawayama, S. (1976). Infrared methods for fruit pectic substances. *Food Sci. and Technol. Abstr.* **9**, 2J126 (1977).

Kawabata, A., and Sawayama, S. (1977). Molecular weights and molecular size of pectins. *J. Agr. Chem. Soc. Japan* **51**, 15–21.

Keenan, M. H., Belton, P. S., Matthew, J. A., and Howson, S. J. (1985). A ^{13}C-n.m.r. study of sugar-beet pectin. *Carbohydr. Res.* **138**, 168–170.

Keijbets, M. J. H., and Pilnik, W. (1974). Some problems in the analysis of pectin in potato tuber tissue. *Potato Res.* **17**, 169–177.

Kertesz, Z. I. (1951). "The Pectic Substances." Interscience, New York.

Kim, W. J., Smit, C. J. B., and Rao, V. N. M. (1978). Demethylation of pectin using acid and ammonia. *J. Food Sci.* **43**, 74–78.

Kintner, P. I., and Van Buren, J. P. (1982). Carbohydrate interference and its correction in pectin analysis using the *m*-hydroxydiphenyl method. *J. Food Sci.* **47**, 756–759.

Kirtchev, N., Panchev, I., and Kratchanov, C. (1989). Kinetics of acid-catalysed deesterification of pectin in a heterogeneous medium. *Int. J. Food Sci. Technol.* **24,** 479–486.

Kiyohara, H., Cyong, J., and Yamada, H. (1988). Structure and anticomplementary activity of pectic polysaccharides isolated from the root of *Angelica acutiloba* Kitagawa. *Carbohydr. Res.* **182,** 259–275.

Klavons, J. A., and Bennett, R. D. (1986). Determination of methanol using alcohol oxidase and its application to methyl ester content of pectins. *J. Agric. Food Chem.* **34,** 597–599.

Knee, M. (1970). The separation of pectic polymers from apple fruit by chromatography on diethylaminoethyl-cellulose. *J. Exp. Bot.* **21,** 651–662.

Kohn, R., and Sticzay, T. (1977). Circular dichroism and the cation binding to polyuronates. *Chem. Abstr.* **87,** 201964n (1977).

Kuhn, L. P. (1950). Infrared spectra of carbohydrates. *Anal. Chem.* **22,** 276–283.

Kwak, J. C. T., Murphy, G. F., and Spiro, E. J. (1978). The equivalent conductivity of aqueous solutions of alkali metal salts of a number of ionic polysaccharides. *Chem. Abstr.* **89,** 129825 Q (1978).

Lau, J. M., McNeil, M., Darvill, A. G., and Albersheim, P., (1985). Structure of the backbone of rhamnogalacturonan I, a pectic polysaccharide in the primary cell walls of plants. *Carbohydr Res.* **137,** 111–125.

Laver, M., and Wolfrom, M. L. (1962). Methoxyl and ethoxyl determination. *In* "Methods in Carbohydrate Chemistry I." (R. L. Whistler *et al.* eds.), pp. 454–461. Academic Press, New York.

Marsh, C. A. (1966). Chemistry of D-glucuronic acid and its glycosides. *In* "Glucuronic Acid: Free and Combined" (G. J. Dutton, ed.), pp. 3–136. Academic Press, New York.

Martelli, A., and Proserpio, G. (1973). A rapid TLC method for the identification of alginates and pectins in foods and cosmetics. *Chem. Abstr.* **84,** 162973f (1976).

McFeeters, R. F., and Armstrong, S. A. (1984). Measurement of pectin methylation in plant cell walls. *Anal. Biochem.* **139,** 212–217.

McReady, R. M., and Reeve, R. M. (1955). Test for pectin based on reaction of hydroxamic acids with ferric ion. *J. Agric. Food Chem.* **3,** 260–262.

Miles Laboratories, Inc. (1984). "Pectin and Starch Tests for Fruit Juice. Product Information. Processing." Biotech Products Division, Elkhart, IN.

Milner, Y., and Avigad, G. (1967). A copper reagent for the determination of hexuronic acids and certain ketohexoses. *Carbohydr. Res.* **4,** 359–361.

Mitchell, J. R., and Blanshard, J. M. V. (1979). On the nature of the relationship between the structure and rheology of food gels. *In* "Food Texture and Rheology" (P. Sherman, ed.), pp. 425–435. Academic Press, New York.

Mizote, A., Odagiri, H., Toei, K., and Tanaka, K. (1975). Determination of residues of carboxylic acids (mainly galacturonic acid) and their degree of esterification in industrial pectins by colloid titration with Cat-Floc. *Analyst* **100,** 822–826.

Morris, V. J. (1986). Gelation of polysaccharides. *In* "Functional Properties of Food Macromolecules" (J. R. Mitchell and D. A. Ledward, eds.), p. 144*ff.* Elsevier Applied Science, New York.

Nash, N. H. (1960). Functional aspects of hydrocolloids in controlling crystal structure in foods. *In* "Physical Functions of Hydrocolloids" pp 45–58. Adv. Chem. Series, no. 25. Am. Chem. Soc., Washington, D.C.

National Academy of Sciences (1981). "Food Chemicals Codex," 3rd Ed. Washington, D.C.

Nelson, N. (1944). Photometric adaptation of the Somogyi method for the determination of glucose. *J. Biol. Chem.* **153,** 375–380.

O'Beirne, D., Van Buren, J. P., and Mattick, L. R. (1981). Two distinct pectin fractions from senescent Ida Red apples extracted using nondegradative methods. *J. Food Sci.* **47,** 173–176.

Olsen, A. G. (1934). Pectin studies. III. General theory of pectin jelly formation. *J. Phys. Chem.* **38,** 919–930.

Owens, H. S., Lotzkar, H., Schultz, T. H., and Maclay, W. D. (1946). Shape and size of pectinic acid molecules deduced from viscometric measurements. *J. Am. Chem. Soc.* **68**, 1628–1632.

Owens, H. S., Miers, J. C., and Maclay, W. D. (1948). Distribution of molecular weights of pectin propionates. *J. Colloid Sci.* **3**, 277–291.

Palmer, K. J., and Hartzog, M. B. (1945). An x-ray diffraction investigation of sodium pectate. *J. Am. Chem. Soc.* **67**, 2122–2127.

Pals, D. T. F., and Hermans, J. J. (1952). Sodium salts of pectin and of carboxy methyl cellulose in aqueous sodium chloride. *Recueil des Travaux Chimiques Des Pays-Bas* **71**, 433–467.

Passaglia, E., and Marchessault, R. H. (1965). Preparation and testing of films. *In* "Methods in Carbohydrate Chemistry V," (R. L. Whistler *et al.* eds.), p. 222. Academic Press, New York.

Pechanek, U., Blaicher, G., Pfannhauser, W., and Woidich, H. (1982). Electrophoretic method for qualitative and quantitative analysis of gelling and thickening agents. *J. Assoc. Offic. Anal. Chem.* **65**, 745–752.

Petrzika, M., and Linow, F. (1985). Quantitative gas chromatographic determination of monomeric pectin constituents. *Food Sci. and Technol. Abstr.* **18**, 2A40 (1986).

Petrzika, M., and Linow, F. (1986). MS of trimethylsilyl oxime derivatives of monosaccharides and galacturonic acid. *Food Sci. and Technol. Abstr.* 9A12 (1986).

Pippen, E. L., Schultz, T. H., and Owens, H. S. (1953). Effect of degree of esterification on viscosity and gelation behavior of pectin. *J. Colloid Sci.* **8**, 97–104.

Plashchina, I. G., Braudo, E. E., and Tolstoguzov. (1978). Circular dichroism studies of pectin solutions. *Carbohydr. Res.* **60**, 1–8.

Plashchina, I. G., Semenova, M. G., Braudo, E. E., and Tolstoguzov, V. B. (1985). Structural studies of the solutions of anionic polysaccharides. IV. Study of pectin solutions by light-scattering. *Carbohydr. Polym.* **5**, 159–179.

Ramus, J. (1977). Alcian blue: A quantitative aqueous assay for algal acid and sulfated polysaccharides. *J. Phycol.* **13**, 345–348.

Rao, M. A., Cooley, H. J., Walter, R. H., and Downing, D. L. (1989). Evaluation of texture of pectin jellies with the Voland-Stevens Texture Analyzer. *J. Texture Stud.* **20**, 87–95.

Rao, C. N. R. (1963). "Chemical Applications of Infrared Spectroscopy." Academic Press, New York.

Rees, D. A. (1970). Structure, conformation and mechanism in the formation of polysaccharide gels and networks. *In* "Advances in Carbohydrate Chemistry and Biochemistry," **24**, pp. 267–332. Academic Press, New York.

Rees, D. A. (1972). Polysaccharide gels: A molecular view. *Chem. Ind. (London)* **19**, 630–636.

Rees, D. A., and Wright, A. W. (1971). Polysaccharide conformation. Part VII. Model building computations for α-1,4 galacturonan and the kinking function of L-rhamnose residues in pectic substances. *J. Chem. Soc.,* Section B, Part II, 1366–1372.

Reisenhofer, E., Cesaro, A., Delben, F., Manzini, G., and Paoletti, S. (1984). Copper(II) binding by natural ionic polysaccharides. Part II. Polarographic data. *Chem. Abstr.* **102**, 149687 (1985).

Robertson, G. L. (1981). The determination of pectic substances in citrus juices and grapes by two spectrophotometric methods. *J. Food Biochem.* **5**, 139–143.

Rombouts, F. M., and Thibault, J. F. (1986). Feruloylated pectic substances from sugar-beet pulp. *Carbohydr. Res.* **154**, 177–188.

Russo, P. S. (1987). A perspective on reversible gels and related systems. *In* "Reversible Polymeric Gels and Related Systems" (P. S. Russo, ed.), pp. 5–9. Amer. Chem. Soc. Symposium Series No. 350. Washington, D.C.

Saverborn, S. (1945). "A Contribution to the Knowledge of the Acid Polyuronides." Almqvist and Wiksells, Uppsala, Sweden. Cited in Kertesz, Z., "The Pectic Substances," p. 59. Interscience, New York.

Sawayama, S., Kawabata, A., Nakahara, H., and Kamata, T. (1988). A light-scattering study on the effects of pH on pectin aggregation in aqueous solution. *Food Hydrocoll.* **2**, 31–37.

Schols, H. A., Reitsma, J. C. E., Voragen, A. G. J., and Pilnik, W. (1989). High-performance ion-exchange chromatography of pectins. *Food Hydrocoll.* **3**, 115–121.

Scott, J. E. (1965). Fractionation by precipitation with quaternary ammonium salts. *In* "Methods in Carbohydrate Chemistry V," pp. 38–44. Academic Press, New York.

Severs, E. T. (1962). "Rheology of Polymers." Reinhold, New York.

Sjoberg, A.M., and Pyysalo, H. (1985). Identification of food thickeners by monitoring of their pyrolytic products. *J. Chromatogr.* **319**, 90–98.

Smidsrod, O. (1970). Solution properties of alginate. *Carbohydr. Res.* **13**, 359–372.

Smidsrod, O., and Haug, A. (1971). Estimation of the relative stiffness of the molecular chain in polyelectrolytes from measurements of viscosity at different ionic strengths. *Biopolymers* **10**, 1213–1227.

Smit, C. J. B., and Bryant, E. F. (1969). Changes in molecular weight of pectin during methylation with diazomethane. *J. Food Sci.* **34**, 191–193.

Smith, J. (1976). "The Molecular Weight of Pectin." Ph.D. Thesis, University of Leeds, England.

Solms, J. Denzler, A., and Deuel, H. (1954). Uber polygalakturonsaureamide. *Helv. Chim. Acta* **37**, 2153–2160.

Sorochan, V. D., Dzizenko, A. K., Bodin, N. S., and Ovodov, Y. S. (1971). Light-scattering studies of pectic substances in aqueous solution. *Carbohydr. Res.* **20**, 243–249.

Stacey, K. A. (1956). "Light Scattering in Physical Chemistry." Butterworths, London.

Sterling, C. (1970). Crystal-structure of ruthenium red and stereochemistry of its pectic stain. *Am. J. Bot.* **57**, 172–175.

Stutz, E., and Deuel, H. (1955). Polyampholyte mit verschiedener ladungsverteilung (Polyampholyte with different charge distribution). *Helv. Chim. Acta* **38**, 1757–1763. English summary.

Tanford, C. (1961). "Physical Chemistry of Macromolecules." Wiley, New York.

Taylor, R. L., and Conrad, H. E. (1972). Stoichiometric depolymerization of polyuronides and glycosaminoglycuronans to monosaccharides following reduction of their carbodiimide-activated carboxyl groups. *Biochemistry* **11**, 1383–1388.

Thibault, J. F. (1979). Automisation du dosage des substances pectiques par la methode au meta-hydroxydiphenyl. *Lebensm.-Wiss. Technol.* **12**, 247–251.

Thibault, J. F., and Rinaudo, M. (1986). Interactions of counterions with pectins studied by potentiometry and circular dichroism. *In* "Chemistry and Function of Pectins" (M. L. Fishman and J. J. Jen, eds.), pp. 61–72. Am. Chem. Soc. Series Symposium, 310. Washington, D.C.

Thibault, J. F., and Robin, J. P. (1975). Automation of uronic acid determination by the carbazole method. *Chem. Abstr.* **84**, 42080r (1976).

Tipson, R. S. (1967). Infrared spectroscopy of carbohydrates. 1967-O-290-632. U.S. Government Printing Office, Washington, D.C.

Towle, G.A. (1972). Determination of molecular weights by osmometry. *In* "Methods in Carbohydrate Chemistry" (R. L. Whistler et al., eds.), Vol. VI, pp. 510–512. Academic Press, New York.

Trudso, J. E. (1989). Hydrocolloids: Selection and application. *Food and Beverage Technol. Internat., USA*, 155–162.

Ulrich, R. D. (1974). Membrane osmometry. *In* "Polymer Molecular Weights" (P. E. Slade, Jr., ed.), pp. 9–30, Marcel Dekker, New York.

Van Deventer-Schriemer, Wytske, H., and Pilnik, W. (1976). Fractionation of pectins in relation to their degree of esterification. *Lebensm-Wiss. Technol.* **9**, 42–44.

Vinson, J. A., Fritz, J. S., and Kingsbury, C. A. (1966). Quantitative determination of esters by saponification in dimethyl sulfoxide. *Talanta* **123**, 1673–1677.

Vold, R. D., and Vold, M. J. (1983). "Colloid and Interface Chemistry." Addison-Wesley, New York.

Voragen, A. G. J., Rombouts, F. M., Hooydonk, M. J., and Pilnik, W. (1971). Comparison of endgroup methods for the measurement of pectin degradation. *Lebensm.-Wiss. Technol.* **4**, 7–11.

Voragen, A. G. Schols, H. A., De Vries, J. A., and Pilnik, W. (1982). High-performance liquid chromatographic analysis of uronic acids and oligogalacturonic acids. *J. Chromatogr.* **244,** 327–337.

Voragen, A. G. J., Schols, H. A., and Pilnik, W. (1986). Determination of the degree of methylation and acetylation of pectins. *Food Hydrocoll.* **1,** 65–70.

Walter, R. H., and Matias, H. L. (1989). Volume fraction of a dispersed pectin. *Food Hydrocoll.* **3,** 205–208.

Walter, R. H. and Matias, H. L. (1991). Pectin aggregation number by light scattering and reducing end-group analysis. Carbohydrate Polym. **15,** 33–40.

Walter, R. H., Rao, M. A., Cooley, H. J., and Sherman, R. M. (1985). The isolation and characterization of a hydrocolloidal fraction from grape pomace. *Am. J. Enol. Vitic.* **36,** 271–274.

Walter, R. H., Rao, M. A., Cooley, H. J., and Sherman, R. M. (1986). Characterization of hydrocolloidal extracts from apple pomace. *Lebensm.-Wiss. Technol.* **19,** 253–257.

Walter, R. H., and Sherman, R. M. (1981). Apparent activation energy of viscous flow in pectin jellies. *J. Food Sci.* **46,** 1223–1225.

Walter, R. H., and Sherman, R. M. (1983). The induced stabilization of aqueous pectin dispersions by ethanol. *J. Food. Sci.* **48,** 1235–1241.

Walter, R. H., and Sherman, R. M. (1984). Flow profiles of aqueous dispersed pectins. *J. Food Sci.* **49,** 67–69.

Walter, R. H., and Sherman, R. M. (1986). Rheology of high-methoxyl pectin jelly sols prepared above and below the gelation temperature. *Lebensm.-Wiss. Technol.* **19,** 95–100.

Walter, R. H., and Sherman, R. M. (1990). Derivation of a cooling-rate quotient for high-methoxyl pectin jelly sols. *Colloids and Surfaces* **44,** 1–16.

Walter, R. H., Sherman, R. M., and Lee, C. Y. (1983). A comparison of methods for polyuronide methoxyl determination. *J. Food Sci.* **48,** 1006–1007.

Walter, R. H. and Talomie, T. G. (1990). Quantitative definition of polysaccharide hydrophilicity. Food Hydrocoll. **4,** 197–203.

Walter, R. H., Van Buren, J. P., and Sherman, R. M. (1978). Dispersion of pectin and cellulose in dimethylsulfoxide. *J. Food Sci.* **43,** 1882.

Wedlock, D. J., Phillips, G. O., and Bachmann, M. (1984). A new colourimetric assay for pectins. *In* "Gums and Stabilizers for the Food Industry. 2. Applications of Hydrocolloids" (G. O. Phillips, ed.), pp. 529–533. Proceedings 2nd Int. Conf. Pergamon, Oxford.

Weill, C. E., and Hanke, P. (1962). Thin-layer chromatography of malto-oligosaccharides. *Anal. Chem.* **34,** 1736–1737.

Whiffen, D. H. (1957). Applications of infrared spectroscopy. *Chem. Ind.* February, 129–131.

Wiley, R. C., and Tavakoli, M. (1969). Trimethylsilyl derivatives of commercial pectins. *Food Technol.* **23,** 565–566.

Wood, P. J., and Siddiqui, I. R. (1971). Determination of methanol and its application to measurement of pectin ester content and pectin methyl esterase activity. *Anal. Biochem.* **39,** 418–428.

CHAPTER 11

Rheology of Pectin Dispersions and Gels

M. A. V. Axelos and J. Lefebvre

Institut National de la Recherche Agronomique
Laboratoire de Physico-Chimie des Macromolécules
Nantes Cedex 03, France

C-G. Qiu and M. A. Rao

Department of Food Science and Technology
Cornell University
Geneva, New York

The ability to form gels is the most important property of pectins. A substantial portion of the pectin produced in the world is used in making fruit jellies and products such as jams, preserves, and marmalades that require complete or partial gelation (May, 1990). A number of ways of standardizing the jelly-forming ability of pectins have been devised. For example, the term jelly grade was defined in 1926 as the sugar-carrying power of dry, powdered pectin (Kertesz, 1951). However, in order to understand the role of pectins in the formation of gels and gel characteristics, knowledge of rheological behavior of pectin dispersions and of pectin jellies during sol–gel transition will be very useful.

In this chapter, the rheological characteristics of pectin solutions and jellies are described, and the results of studies that have yielded fundamental rheological parameters and textural parameters are emphasized. Related publications on this subject include a review of instrumental evaluation of pectin jelly strength

(Crandall and Wicker, 1986), and of pectin and other biopolymer gels by Mitchell (1976). Reviews dealing with the general subject of rheology of fluid and solid foods and rheological measurements can be found in Mitchell (1984), Ross-Murphy (1984), M. A. Rao (1986).

I. RHEOLOGY OF PECTIN DISPERSIONS

Early attempts at evaluating pectin grades from viscosity measurements of pectin dispersions were deemed to be useful in preliminary grading to ascertain the amount of pectin to be used in test jellies (Kertesz, 1951). The non-Newtonian, especially the shear-thinning, nature of concentrated pectin dispersions was evident in many early studies. Aqueous pectin dispersions show the typical constant viscosity at low shear rates (zero shear viscosity, η_0), followed by a power law region at moderate shear rates (Fig. 1), while it is difficult owing to the limitations of instruments, a constant viscosity can also be obtained at high shear rates (infinite shear viscosity, η_∞). Aqueous pectin dispersions follow the same curve of dimensionless shear rate, defined as ($\dot\gamma/\dot\gamma_{0.1}$), against dimensionless viscosity, defined as (η/η_0), as many biopolymer solutions (Morris *et al.*, 1981). The coordinates of this curve were tabulated by Morris *et al.* (1981), so that the curve can be used to calculate the viscosity of pectin dispersions from knowledge

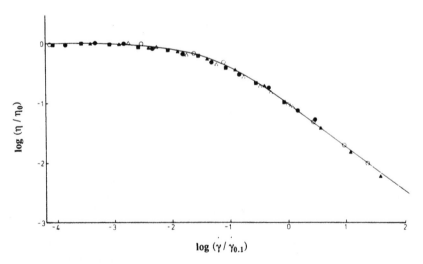

Figure 1   Plot of log (η/η_0) versus log ($\dot\gamma \cdot \dot\gamma_{0.1}$) applicable to concentrated solutions of disordered polysaccharides including pectin solutions; open triangles: guar gum; closed triangles: lambda carrageenan; closed circle: locust bean gum; closed square: high mannuronate alignate; and open circle, hyaluronate (Source: Morris *et al.*, 1981).

of $\dot{\gamma}_{0.1}$. The important point here is that aqueous pectin dispersions behave in a manner similar to that of a number of other biopolymers in terms of shear rate–apparent viscosity behavior.

In addition, the effect of concentration on zero-shear viscosity of pectin dispersions can be expressed in a plot of log $[(\eta_0 - \eta_s)/\eta_s]$ against log $C[\eta]$ (Morris *et al.*, 1981), where the intrinsic viscosity $[\eta]$ can be obtained from dilute-solution viscosity data as:

$$[\eta] = \lim_{C \to 0} (\eta_{sp}/C) \tag{1}$$

where $\eta_{sp} = [(\eta_0 - \eta_s)/\eta_s]$. We note that there are several ways of determining the intrinsic viscosity from dilute solution data including the use of Equation (1) (Tanglertpaibul and Rao, 1987). Because $[\eta]$ indicates the hydrodynamic volume of the molecule and is related to the molecular weight and to the radius of gyration, it reflects important molecular characteristics of a biopolymer. A plot of log $C[\eta]$ against log $[(\eta_0 - \eta_s)/\eta_s]$ of pectins from several sources (Chou and Kokini, 1987) is shown in Fig. 2. The data shown can be divided into two regions: (1) a region of dilute dispersions where the viscosity dependence on concentration follows a 1.4 power, and (2) a region of concentrated dispersions

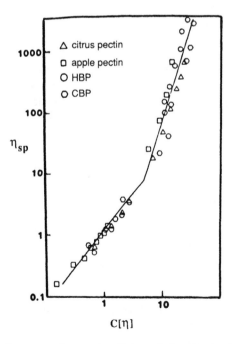

Figure 2   Plot of log $[\eta_0 - \eta_s)/\eta_s]$ versus log $C[\eta]$; pectin from hot-break tomato paste (HBP), (CBP), pectin from cold-break tomato paste. (Source: Chou and Kokini, 1987).

where the viscosity dependence on concentration follows a 3.3 power. The product $C[\eta]$, called the *coil-overlap parameter*, is dimensionless and indicates the volume occupied by the polymer molecule in the solution.

II. STRUCTURAL CONSIDERATION OF PECTIN GELS

Pectin gels may be considered to be the result of the association of long stretches of polymer chain into conformationally ordered junction zones that are stabilized by arrays of noncovalent interactions. The polymer chain is 1,4-linked α-D-galacturonate, which may be partially methyl esterified. The level of esterification affects the gelling properties of commercial pectins, and the regular buckled-ribbon sequences that associate noncovalently form the effective ordered junction zones (Thom *et al.*, 1982). Further, because dissociation of such junction zones requires more energy than is available from Brownian motion, the cross-linked network structure formed has long-term stability (Morris, 1984). Factors that can limit the ordered associations include deviation from regularity of esterified and nonesterified sequences, kinks, or intrusion of 1,2-linked-L-rhamnose in the main chain, and neutral sugar side-chains (Thom *et al.*, 1982).

With high-methoxyl pectins, gel strength increases with ester content up to about 70%, but further increases in esterification result in decrease of gel strength, so that at 95%, esterification gelation does not take place. Acid content is also important for gel formation in that it must be high enough to counteract the negative charge on unesterified galacturonate residues. At the optimal ester content of 70%, increasing the pH results in relatively weak gels, possibly due to electrostatic repulsion between chains. However, it is important to realize that ester groups make a positive contribution to chain interactions, while electrostatic repulsions are not desirable for gel formation. The general view of high-methoxyl pectin gelation is that junction zones containing large aggregates of esterified chain regions are formed, and the formation can be aided by lowering activity and by suppression of charges on pectin chains (Thom *et al.*, 1982).

Pectins with low levels of esterification form firm gels with Ca^{2+} ions; there is evidence that galacturonate blocks and not methyl galacturonate sequences are important for Ca^{2+} binding. Gel strength is a function of network formation due to intermolecular associations, and it increases with decrease in the number of methyl-esterified sequences (Thom *et al.*, 1982). The occurrence of a low level of randomly spaced esterified galacturonate residues in a poly-D-galacturonate sequence is more disruptive than is the occurrence of such residues in blocks. In the absence of counterions, methyl-esterified residues may become part of the structure when each bound Ca^{2+} is associated with at least one carboxyl groups; when Ca^{2+} ions are present in excess, aggregation occurs irrespective of the sequence of esterified and galacturonate blocks. However, the resulting gel structure is weaker and less stable than that of primary junction zones (Thom *et al.*, 1982).

III. RHEOLOGICAL CHARACTERIZATION OF PECTIN GELS

Pectin gels possess both the viscous properties of liquids and the elastic properties of solids, i.e., they are viscoelastic materials. Therefore, their rheological properties are expressed in terms of elastic moduli and Newtonian viscosities. We note that modulus is defined as stress:strain, and it can be determined under different time-dependent strain measurements. The moduli can be determined in experiments involving extension, compression, and shear. However, because pectin gels are viscoelastic, the strains will depend on both the applied stress and on the time of application of the stress. Most studies on viscoelastic materials are conducted using small magnitudes of strains. Further, when the ratio of stress to strain is independent on the strain, the behavior is linear viscoelastic, and one can assume that the Boltzman superposition principle is applicable.

Viscoelastic behavior of pectin gels has been studied by means of dynamic-shear, creep-compliance, and stress-relaxation techniques. The application of each of these techniques to pectin gels is considered in the following sections. It is customary to employ different symbols for the various rheological parameters in different types of deformation: shear, bulk, or simple extension. Following Ferry (1980) and Mitchell (1984), the symbols employed for the three types of deformation are given in Table I.

A. Dynamic Viscoelastic Properties of Pectin Gels

Dynamic measurements consist in applying a harmonic oscillation (strain or stress) to the sample; if the material behaves linearly, the resulting response (stress or strain respectively) is itself harmonic with the same frequency. The elastic and viscous components of the viscoelastic behavior are then easily obtained from the amplitudes ratio and the phase lag of the two signals. Results of dynamic experiments are often analyzed in terms of the complex modulus G^*:

$$G^* = G' + i\, G'' \tag{2}$$

Table I Viscoelastic parameters from shear, simple extension, and bulk compression

Parameter	Shear	Simple extension	Bulk compression
Stress-relaxation modulus	$G(t)$	$E(t)$	$K(t)$
Creep compliance	$J(t)$	$D(t)$	$B(t)$
Storage modulus	$G'(\omega)$	$E'(\omega)$	$K'(\omega)$
Loss modulus	$G''(\omega)$	$E''(\omega)$	$K''(\omega)$
Complex modulus	$G^*(\omega)$	$E^*(\omega)$	$K^*(\omega)$
Dynamic viscosity	$\eta'(\omega)$	$\eta_e'(\omega)$	$\eta_v'(\omega)$

The storage modulus G', corresponding to the stress in phase with the strain, is a measure of the energy stored in the material; for a gel, it is directly related to the cross-link density of the network. The loss modulus G'', corresponding to the 90° out-of-phase stress amplitude, is a measure of the energy dissipated as heat (Ferry, 1980). The reciprocal of the angular frequency ω defines the time scale of the dynamic measurements. By scanning the frequency, the time scale of the mechanical response of the material can be explored. The shape of the variations of G' and G'' with ω (mechanical spectrum) allows a qualitative determination of the nature of the material (Ferry, 1980).

It is emphasized that the mathematical analysis of data from dynamic experiments is usually based on the assumption that the viscoelastic response is linear; this means that the amplitudes ratio and the phase lag of stress and strain must be independent of the amplitude of the oscillation applied. The assumption is valid for many materials when strain amplitude is kept sufficiently small. For the study of low-methoxyl pectins–calcium gels, we have used a controlled-stress rotational-shear rheometer (Carri-Med Rheometer, Dorking, Surrey, UK) with a cone-plate device working in the frequency range 10^{-3}–10 Hz. The cone-plate geometry ensures uniform shear in the sample, small inertia and small sample volume, and easy handling. Mechanical spectra recorded for calcium–pectin gels generally show that G' and G'' are independent of frequency and that G' is larger than G'' indicating the predominant character of these gels (Fig. 3).

Gross *et al* (1982) studied gels containing low-methoxyl pectins that were prepared by NH_4OH-alcohol demethylation. The gels were cast cylinders: 1.5 cm height × 4.5 cm diameter and aged 18–24 hr. The dynamic modulus and phase angle, as well as the storage modulus, E', and the loss modulus, E'', were determined as a function of frequency, 100–240 Hz. From plots of storage and loss moduli, resonance dispersion was observed at a frequency of 220 Hz, i.e.,

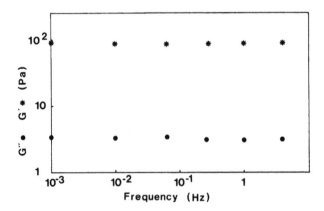

Figure 3 Mechanical spectrum obtained at the end of the kinetics on low-methoxyl pectin-calcium gels. C = 6.7 g/l, R = 2 $(Ca^{2+})/(COO^-)$ = 0.47, T = 20 °C, pH = 7 without added salt.

the loss modulus showed a minimum; simultaneously the storage modulus showed a maximum. The dynamic modulus calculated at 220 Hz was the most sensitive for distinguishing physical differences between the 14 gels. The role of chemical composition on E^* was explained best by the equation ($R = 0.96$) that accounted for 92% of the total variation in E^* between the 14 gels:

$$E^* = (3.4 \times 10^6) \text{ % esterification} - (1.98 \times 10^6) \text{ % AGA} \qquad (3)$$

Percentage of esterification accounted for 96% of the explained variation, while percentage of AGA accounted for the remaining 4%. In contrast, 65% of the variation in phase angle, θ_{220}, was accounted by % esterification and molecular weight:

$$\theta_{220} = (-1.32) \text{ % esterification} - (4.4 \times 10^{-4}) \text{ molecular weight} \qquad (4)$$

B. Creep-Compliance of Pectin Gels

In a creep experiment, an undeformed sample is suddenly subjected to a constant shearing stress, σ_c. The strain will increase with time and approach a steady state where the strain rate is constant. The data from this experiment are analyzed in terms of creep-compliance, defined by the relationship:

$$J(t, \sigma_c) = \gamma/\sigma_c \qquad (5)$$

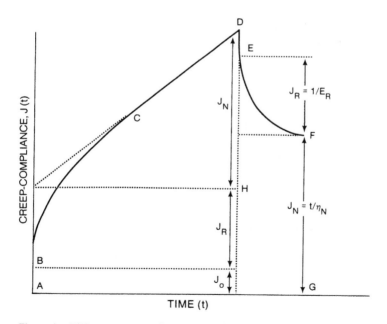

Figure 4   Different segments of a creep-compliance curve. (From Sherman, 1970.)

A typical creep-compliance curve (Fig. 4) can be divided into three principal regions (Sherman, 1970):

1. a region (A–B) of instantaneous compliance J_0 in which the bonds between the different structural units are stretched elastically. In this region, if the stress is removed, the structure of the sample will recover completely. We note that

$$J_0 = 1/E_0 \tag{6}$$

   where, E_0 is the instantaneous elastic modulus.

2. Region B–C corresponds to a time-dependent retarded elastic region with a compliance, J_R. In this region, the bonds break and reform, but all of them do not break and reform at the same rate. The equation for this part, using mean values for the parameters, is

$$J_R = J_m \left[1 - \exp \left(-t/\tau_m \right) \right] \tag{7}$$

   In Equation (7), J_m is the mean compliance of all the bonds and τ_m is the mean retardation time; τ_m equals $J_m \, \eta_m$, where η_m is the mean viscosity associated with elasticity. One can replace the mean quantities with the spectrum of retarded elastic moduli (E_i) and the viscosities (η_i):

$$J_R = J_m \left\{ 1 - \exp \left[- t/(J_i \eta_i) \right] \right\} \tag{8}$$

   where $J_i = 1/E_i$.

3. C–D is a linear region of Newtonian compliance in which the units, as a result of rupture of the bonds, flow past one another. The compliance, J_N, and the viscosity are related by

$$J_N = t/\eta_N \tag{9}$$

In addition, when the stress is suddenly removed at D, the pattern consists of an elastic recovery (D–E), followed by a retarded elastic recovery (E–F). Because bonds between structural units are broken in region C–D, a part of the structure is not recovered. After the applied stress is removed, the recovered strain approaches a maximal value, S_r, called recoverable shear (Dealy, 1982):

$$S_r = \sigma_c J_e^0 \tag{10}$$

where σ_c is the applied stress and J_e^0 is the value of the straight-line portion of the creep-compliance extrapolated to zero time. Further, J_e^0 is equal to the sum of J_r and J_0. A shear modulus, G_r, can be defined as the reciprocal of the steady-state compliance, J_e^0:

$$\sigma_c = G_r S_r \tag{11}$$

This equation is called *Hooke's law of shear*.

1. Role of Temperature

Kawabata (1977) studied the creep behavior of 65% sucrose jellies containing 1.5%, 2%, and 2.5% crude and purified commercial high-methoxyl pectins over the temperature range 10–50°C, and of low-methoxyl pectin gels containing calcium chloride and milk as gelling agents over the temperature range 10–40°C with a parallel-plate viscoelastometer. Using different reference temperatures, master creep curves were constructed for the creep data at each test temperature of the jelly samples; the energy of activation of the shift factor, a_T, was found to be 124.8, 130.2, and 273.4 kJ/mol for the three pectin concentrations, respectively. Creep behavior of pectin gels was also reported by Watson (1966), Dahme (1985), and Plaschina et al. (1979, 1982).

2. Role of Fructose

A jelly sample made with a washed medium rapid-set pectin and 100% sucrose had a large magnitude of instantaneous elastic modulus (E_0) and a large magnitude of viscosity (Table II), but the region of retarded compliance was not extensive (Fig. 5) (Qiu et al., 1990); the torsional creep data were obtained with a controlled-stress rheometer. Substitution of fructose for sucrose in the proportion 50% sucrose to 50% fructose resulted in increase in the retarded compliance region (Fig. 5). One clear indication of the effect of including fructose as a cosolute was the increase in the area under the creep-compliance curve. For the creep data recorded on the jelly samples between 0 and 3.0 min, the increase in area due to 50% fructose substitution was 17%, and for 75% substitution, it was 53%. The change in area under the creep-compliance curve appears to be a quantitative tool in studying the roles of different sugars on rheological properties.

Table II Creep-compliance parameters of pectin jellies

Sample	$E_0 \times 10^{-3}$ (dyne/cm^2)	$E_1 \times 10^{-3}$ (dyne/cm^2)	τ_1 (sec)	$\eta_1 \times 10^{-5}$ (sec, min)	$\eta \times 10^{-6}$ (Pa.s)
Slow-set apple	7.02	2.75	107	3.0	2.1
Rapid-set citrus	2.67	7.36	37	2.7	9.4
Sugar-free rapid set	0.48	1.27	8	0.1	3.0
Medium-rapid set (100% sucrose)	4.39	112.4	20	22.9	31
Medium-rapid set (50% sucrose + 50% fructose)	4.50	21.6	25	5.4	31
Medium-rapid set (25% sucrose + 75% fructose)	3.36	15.4	31	4.8	12

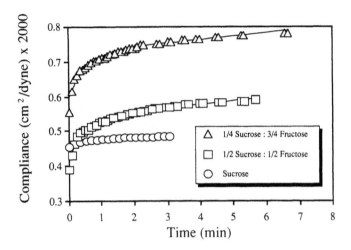

Figure 5   Experimental and predicted creep-compliance of 65% sugar jellies using medium rapid set pectin: (1) with all sucrose, (2) with 1/2 sucrose and 1/2 fructose, and (3) with 1/4 sucrose and 3/4 fructose (Source: Qiu *et al.*, 1990).

In terms of the effect of substituting fructose for sucrose on the viscoelastic parameters, the instantaneous elastic moduli changed slightly, but there was a large reduction in the retarded elastic modulus and a slight increase in the retardation time constant (Table II). Further increase in the proportion of fructose, i.e., total sugar in the proportion 25% sucrose to 75% fructose, accentuated the increase in the magnitude of retarded compliance. There was a further reduction in the retarded elastic modulus and a slight increase in the retardation time constant. It can be concluded that fructose acted as a plasticizer, i.e., it softened the jelly and increased the retarded compliance region that manifested in large changes in creep parameters (Table II). The results of the role of fructose substitution indicate that, in addition to lowering the setting temperature (May, 1990), significant structural changes take place in the gels, and that these changes can be quantified in a creep test. Simple inferences indicate that the molecular weight or the presence of starch oligomers are not sufficient to explain the observed phenomena. It appears that changes in the number and the nature of junction zones need to be examined.

3. Creep–Shear Modulus (G_r)

As stated earlier, the creep recovery or elastic recoil after the removal of the applied stress can also provide useful information on viscoelastic materials. One recovery curve for a pectin gel is illustrated in Fig. 6. First, the presence of recoverable strain exhibited by the pectin gels means that they did not behave

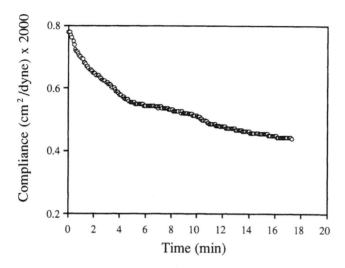

Figure 6   Recovery portion of a creep curve (Source: Qiu *et al.*, 1990).

as cross-linked polymers, but behaved as uncross-linked polymers (Ferry, 1980). Denoting the time at which the stress was removed as t_2, the strain during recovery, γ_r is given by the expression:

$$\gamma_r(t) = \sigma_c [J_e^0 + t/\eta - J(t - t_2)] \tag{12}$$

where J_e^0 is the steady-state compliance. From the recoverable strain (γ_r), the creep–shear modulus (G_r) that is equal to the reciprocal of the steady-state compliance was calculated from Equation (11) (Dealy, 1982). Equation (11) is noteworthy because it implies that the modulus G_r is not a strong function of shear rate. The modulus G_r is of interest also because it is related to the zero shear limit first normal stress difference function ($\psi_{1,0}$) and also to the ratio of elastic to viscous forces in simple shear flow (Dealy, 1982). Magnitudes of the relaxation modulus G_r, calculated from recoverable strain, followed the same trends of instantaneous elastic modulus (E_0).

C. Stress Relaxation

In a stress-relaxation experiment, the decay of stress as a function of time is recorded at a fixed magnitude of deformation. The experiments can be performed with an instrument such as the Instron Universal Testing machine. However, in order to minimize errors due to friction between the test sample and the surfaces of the platens in contact with the sample, either the sample must be bonded to the platens or the contact surfaces must be well lubricated. In many studies, lubrication of the contact surfaces was employed because it is easier than bonding the sample to the platens. To further explain salient features of the stress-relaxation technique and the types of results that can be obtained, we consider a study of the rheological behavior of pectin–sugar gels made from a commercial slow-set pectin as a function of concentration, pH, and sugar content, and aged for 48 hr (Comby *et al.*, 1986). A typical stress–time curve is presented in Fig. 7. During the loading period (time, $t < t^*$), a linear relationship between force and displacement was observed, and a compression modulus (E_0) was calculated from the relationship:

$$E_0 = \sigma^*/\varepsilon^* = (L_0/A_0)\,(F^*/\Delta l) \tag{13}$$

where F^* is the force corresponding to the deformation ε^*, and A_0 is the initial cross-section of the gel; $\varepsilon^* = (\Delta l/L_0)$, where Δl is the displacement, and L_0 is the initial height of the gel sample; $F^*/\Delta l$ is the slope of the loading curve. From the relaxation curves, the Young's modulus $E(t)$ can be obtained, after taking into consideration the relaxation phenomenon during the loading period. One such procedure is the Zappas approximation reported by Lin and Aklonis (1980):

$$E(t - t^*/2) = \sigma(t)/\varepsilon^* \tag{14}$$

This method can be used to estimate $E(t)$ very accurately using data from $t = 2t^*$; but from $t = 1.3\,t^*$, the approximation is good. The relaxation curves were characterized by an inflection point between 20 and 60 sec. Assuming that the inflection point corresponds to the classic plateau zone, the value of the modulus at this point (E_p) may be viewed as a pseudo-elastic modulus and interpreted according to the theory of gel elasticity that was proposed for gels with low

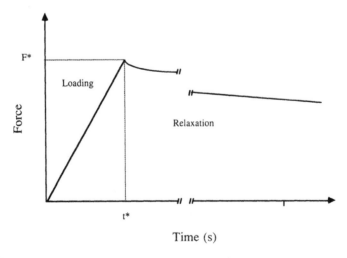

Figure 7   A typical stress relaxation response curve due to the imposition of a constant strain; t^* is the time at which the imposition of a fixed strain is completed. (From Comby *et al.*, 1986.)

concentrations (C < 0.4%) (Oakenfull, 1984). One important determination was that the molecular weight of the junction zones was about one quarter that of the number average molecular weight, i.e., that each molecule can be involved in only two or three junction zones.

In addition, Comby *et al.* (1986) derived the parameters of a two-element Maxwell model:

$$E(t) = E_1 \exp(-t/\tau_1) + E_2 \exp(-t/\tau_2) \tag{15}$$

where τ_i is the relaxation time. Magnitudes of E_0 calculated using Eq. (13), E_p, calculated from the pseudoelastic plateau of the relaxation curves, and E_1, increased with increase in pectin concentration. However, E_2 and τ_2 were not much affected up to pectin concentration (C) of 1.7%, suggesting that in turn, the relaxation processes of chain segments between cross-linking zones were not affected. However, when C was increased from 1.7% to 1.8% there was a sharp increase in τ_1, reflecting a decrease in chain mobility. The effect of pH on stress-relaxation curves was examined for 1.7% pectin gels: E_p and E_0 increased steadily when pH decreased from 3.2 to 2.6, but remained constant below pH of 2.6. Increasing sucrose content of the 1.7% pectin gels from 60.3 to 72.3% resulted in an increase in the elastic modulus from 6,750 to 28,300 Pa.

D. Static Measurement of Modulus

Saunders and Ward (1954) showed that a sample of gel placed in a U-tube together with a capillary containing an indexing fluid can be used to determine the rigidity modulus of weak gels. A version of their apparatus in which the indexing fluid is in a horizontal position is shown in Fig. 8. Assuming that the gel is entirely elastic, the shear modulus, G, can be calculated from the equation:

$$G = PR^4 8a^2 hL \tag{16}$$

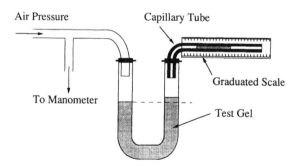

Figure 8   Modified Saunders–Ward U-tube apparatus.

where P is the effective pressure (i.e., the applied pressure corrected for the back pressure of the indexing liquid), R and a are the radii of the wide and narrow tubes, h is the displacement of the indexing liquid, and L is the length of the gel. Because the strain varies from a maximal value of $4a^2h/R^3$ at the tube wall to zero at the tube center, the volume average strain is obtained as $8a^2h/3R^3$. The stress is usually applied for 30 sec, and the displacement of the indexing liquid is recorded as a function of time. However, many food gels show a time-dependent strain (Stainsby *et al.*, 1984). For gels that exhibit syneresis, the test sample as a whole slips along the tube when pressure is applied. Attempts to minimize slip of the sample include introducing dimples into the wall of the wide tube (Komatsu and Sherman, 1974) and coating the tube with a single layer of antibumping granules about 1 mm in diameter (Stainsby *et al.*, 1984). The latter technique, using tubes 6.25 mm and 0.478 mm in radii, was found to be suitable for determining the rigidity moduli of starch, amylose, κ-carrageenan, and egg-white gels. Of further interest here is that with the modified apparatus, time-dependent strain was not observed for the studied gels, so that it may not be possible to obtain creep data with this technique.

Oakenfull *et al.* (1989) presented a method for determining the absolute shear modulus (μ) of gels from compression tests in which the force, F, the strain or relative deformation (δ/L) are measured with a cylindrical plunger with radius r, on samples in cylindrical containers of radius R, as illustrated in Fig. 9. Assuming that the gel is an incompressible elastic solid, the following relationships were derived:

$$\mu = f \times Y_e \tag{17}$$

where f is a parameter whose magnitude depends on the geometrical ratios r:R and R:L, as shown in Table III, and

$$Y_e = (F/\pi\ r^2)/(\delta/L) \tag{18}$$

A correction must be applied to the measured force by subtracting the buoyancy effect, F_h:

$$F_h = \pi\ \rho\ g\ \delta\ r^2\ R^2/(R^2 - r^2) \tag{19}$$

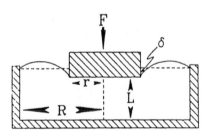

Figure 9 Apparatus for rigidity modulus from compression tests (Source: Oakenfull *et al.*, 1989).

Table III Factors for converting apparent Young's modulus into absolute shear modulus for different sizes of dish and probe

r/R	L/R = 0.1	L/R = 0.2	L/R = 0.5	L/R = 1.0
0.05	0.157	0.0752	0.0381	0.0186
0.10	0.208	0.104	0.0634	0.0331
0.20	0.185	0.102	0.0912	0.0522
0.40	0.0549	0.0632	0.0930	0.0574
0.50	0.0264	0.0490	0.0803	0.0485
0.60	0.0138	0.0384	0.0622	0.0359

Although the technique was illustrated for gelatin gels, it should be applicable to pectin gels also. We note here that because magnitudes of the force, F, were very small, they were measured with a sensitive electronic balance.

Ring and Stainsby (1985) determined the storage modulus, G', of food gels using a pulse shearometer, in which the shear wave generated by a disc was transmitted through an elastic sample. The storage modulus was calculated from the density of the sample, and the velocity, the attenuation, and wavelength of the shear wave.

E. Measurement of Break Strength

Another important rheological parameter of pectin jellies is the yield or break strength. However, many instruments (e.g., pektinometers) measure the yield strength in arbitrary units (Rao et al., 1989). In contrast, the vane method proposed by Dzuy and Boger (1983) iscapable of yielding magnitudes of yield in fundamental units. Each pectin–sugar sol was poured into a jacketed vessel (7.14 cm diameter, 11.7 cm height) that had a six-bladed impeller (4 cm diameter, 6 cm height) in place. The impeller was attached to a viscometer (Haake RV2, Fison Instruments, Saddle Brook, N.J.), and magnitude of torque was recorded as a function of time at 0.4 rpm. The maximal value of torque corresponding to the yield strength was employed in calculating the yield stress (Qiu et al., 1990).

$$T_m = \frac{\pi D_v^3}{2} \left(\frac{H}{D_v} + \frac{1}{3} \right) \sigma_v \tag{20}$$

Magnitudes of yield stress of several pectins are listed in Table IV; they show the well-known trend that citrus pectins are friable and that slow-set pectins are stronger. The important point here is that, with the vane method, one can obtain magnitudes of yield stress of pectin jellies in fundamental units. However, yield strength alone cannot provide insights into the rheological behavior of pectin jellies, and one must employ a fundamental rheological test for such a purpose.

Table IV Magnitudes of yield stress (dyne/cm^2) of
pectin jellies using the Vane method

Pectin identification	Yield stress (dyne/cm^2)
Slow-set apple	1548
Medium-rapid set	1916
Baker's rapid set	1672
Slow-set citrus	1080
Extra-slow set	1851
Rapid-set citrus	618

IV. RHEOLOGY OF SOL–GEL TRANSITION

As a consequence of the formation of a continuous network structure, the sol–gel transition results in a change in the rheological behavior of the system from an essentially viscous to a markedly viscoelastic one; the elastic character of the gel increases with the number and stability of cross-links. Thus, rheology provides proper means of monitoring the gelation process and getting an insight into gel structure. When used for these purposes, the rheological method must satisfy several conditions: (1) it must be nondestructive and must not interfere with gel formation; (2) the time involved in the measurements should be short relative to the characteristic times of the process; and (3) the results should preferably be expressible in fundamental terms in order to be related to the structure of the network. Dynamic rheological experiments meet these requirements.

In addition, the Peltier effect temperature control of the Carri-Med Rheometer allowed the sample to be subjected to quick temperature changes, and to the formation of gel *in situ* by quenching. For a given set of experimental conditions, the magnitudes of G′ and G″ were recorded as a function of aging time after quenching the sample to the working temperature. When the gel reached a metastable state, the mechanical spectrum was plotted. During the experiments, the strain amplitude was kept around 0.05, a magnitude well within the linear viscoelastic range for these gels, determined by us.

A. Effect of Composition on Kinetics

When a solution of low-methoxyl pectins containing calcium ions, prepared and set between cone and plate geometry of the Carri-Med Rheometer at 70°C, is quenched at the working temperature, the gel forms very quickly. G′ becomes

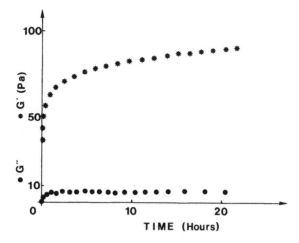

Figure 10 Evolution of the storage and loss modulus with time. C = 6.7 g/l, R = 2 (Ca²⁺)/ (COO⁻) = 0.47, T = 20 °C, pH = 7 without added salt.

much higher than G'' very soon, and it shows a rapid increase during the first hour of aging (Fig. 10). After a certain time, t*, the variation of G' becomes a linear function of the logarithm of the aging time (Fig. 11). This time, t*, depends on the frequency, ω, and on the ratio, R = 2(Ca²⁺)/(COO⁻). The slope of the linear relation, which can be taken as the rate of aging, increases with the calcium content. For very long times, G' does not follow this linear relationship, but tends toward an asymptotic value.

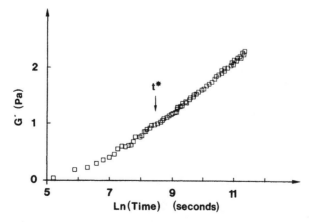

Figure 11 Storage modulus plotted against the logarithm of aging time. C = 14.9 g/l, R = 2 (Ca²⁺)/(COO⁻) = 0.11, T = 20 °C, pH = 7 without added salt.

B. Mechanical Spectra

Mechanical spectra recorded in the final stage of the aging process show that
G' and G'' are independent of the frequency in the range explored, and that the
magnitude of G' is more than ten times that of G''. This observation points to a
predominantly elastic and solid-like behavior of low-methoxyl pectin–calcium
gels, with a rather permanent character of the junction zones, no rearrangement
being evidenced in the network structure on the time scale 150 to 0.02 sec. We
may assume that $G'(\omega) \simeq Ge$, where Ge is the equilibrium shear modulus
($t \rightarrow \infty$ limit of the shear-relaxation modulus).

   The dependence of Ge on pectin concentration, C, for a given ratio, R, is
plotted on Fig. 12. The curve can be fitted by an equation of the form:

$$Ge \simeq (C - C_0)^\alpha \tag{21}$$

The gelation threshold concentration, C_0, depends on the chain conformation;
$C_0[\eta]$ was found to be around 0.8, showing that C_0 is very close to the critical
concentration C^* that denotes the transition from the dilute to the semidilute
regime of pectin dispersions. We found that $C^*[\eta] \simeq 0.7$ (Axelos *et al.*, 1989).

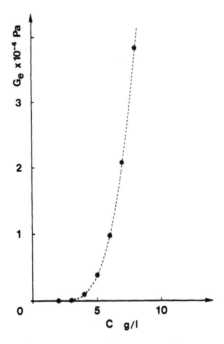

Figure 12   *Equilibrium shear modulus plotted against pectin concentration for a contant R =
$2(Ca^{2+})/(COO^-) = 0.88$, T = 20 °C pH = 7 in NaCl 0.1 M. Dashed line, ---- corresponds to the
fit by Equation (21) with $C_0 = 1.9$ g/l and $\alpha = 3.1$.

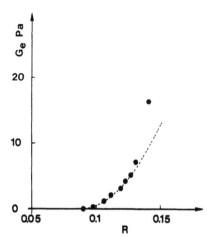

Figure 13   *Equilibrium shear modulus plotted against $R = 2(Ca^{2+})/(COO^-)$ for a pectin concentration $C = 14.9$ g/l, $T = 20\ °C$, pH = 7 without added salt. Dashed line, ---- corresponds to the fit by Equation (22) with $R_0 = 0.086$ and $\beta = 1.9$.

Thus, gelation cannot occur in the dilute regime. The exponent α was found to be 3.0 ± 0.2 in good agreement with the classical theory of network formation (Gordon and Ross-Murphy, 1975).

The effect of calcium concentration at a given pectin concentration is displayed on Fig. 13. At low calcium levels just above the sol–gel transition, the elastic modulus increases according to the relationship:

$$Ge \simeq (R - R_0)^\beta \qquad (22)$$

where R_0 is the limiting ratio below which the gel does not form, and β is the critical exponent of the sol–gel transition. At high enough ionic strength, β was found to be equal to 1.9 ± 0.2. This result points to the fact that the framework of percolation theory seems to be adequate to describe the behavior near the gelation threshold in the case of biopolymers networks such as pectin gels. For such polymers, gelation occurs through the formation of extended junction zones, stabilized by cooperative low-energy interactions, and not through covalent cross-linking (Axelos and Kolb, 1990). Above a certain value of R, Ge showed an upward departure from the preceding equation (Fig. 13).

C. Effect of Temperature

When the temperature increased, G' decreased sharply (Fig. 14), but the general features of the kinetics and of the mechanical spectrum were kept intact. The activation energy of mechanical relaxation E_a and the standard breakdown

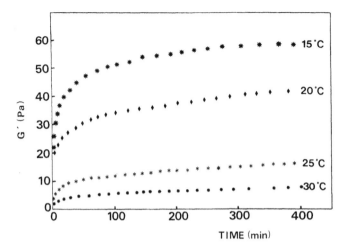

Figure 14   Kinetics of the gel formation at different temperatures. C = 5.13 g/l, R = 2 (Ca^{2+})/ (COO$^-$) = 1, pH = 7 in NaCl 0.1 M. (From Durand *et al.*, 1990.)

enthalpy of the junction zones, $\Delta H°$, were calculated from the shift factors (horizontal and vertical, respectively) for the superimposition of G' versus time curves in the temperature range 5–50°C. We obtained E_a = 50 kJ/mole and $\Delta H°$ = 100 kJ/mole for the 4.27 g/liter gel with R = 1.1. In Table V, this $\Delta H°$ value is compared to the results obtained by Braudo *et al.* (1984). The $\Delta H°$ value for low-methoxyl pectin is intermediate between that for gelatin (300 kJ/mole) and that for high-methoxyl pectin gels (20 kJ/mole). The high value for gelatin gel is to be ascribed to the conformational transition implicit in its formation: triple-helix formation entails a large loss of entropy, which has to be compensated for by a high ΔH value in order to ensure the stability of junction zone. Gelation of high-methoxyl pectins does not need any change in the chain conformation (Oakenfull and Scott, 1984), and thus is driven by a small enthalpy

Table V   Standard breakdown enthalpy of the network cross-links

Gel composition	Polymer concentration g/L	$\Delta H°$ kJ/mole
High-methoxyl pectin, sucrose (70%, w/w) water, pH 3.0	2.5	26 ± 1[a]
Gelatin water, pH 4.7	5.0	295 ± 13[a]
Low-methoxyl pectin, R = 2 (Ca^{2+})/(COO$^-$) = 1.1, water, pH 7.0	4.3	100 ± 10[b]

[a]From Braudo *et al.* (1984).
[b]From Durand *et al.* (1990).

variation. Gelation of low-methoxyl pectins induced by calcium complexation does not involve substantial conformational change at the macromolecular level (Durand *et al.* 1990); the relatively high ΔH value reflects the cooperativity of the process generating an effective junction zone.

The setting characteristics of two 67% sugar jellies containing 0.40, 0.60, 0.85, and 1.0% of two commercial high-methoxyl pectins were studied by Beveridge and Timbers (1989) by means of small-amplitude oscillatory tests. Gel strength, measured as increase in amplitude during cooling, was linearly correlated with temperature. Further, the slope, and intercept of the linear relation were related to the pectin concentration up to 0.85%, but 1% pectin could not be distinguished from 0.85%. On reheating, the gels softened initially, but softened again as the temperature exceeded about 70°C.

V. TEXTURE-PROFILE ANALYSIS OF PECTIN GELS

Texture-profile analysis (TPA) is an empirical technique that has found extensive use in the characterization of the texture of many foods. The double-penetration test is an objective test that imitates the biting action of the teeth and provides force–deformation data required for the evaluation of the seven texture parameters: fracturability, hardness, cohesiveness, adhesiveness, springiness, gumminess, and chewiness (Bourne, 1982). Rao *et al.* (1989) studied the feasibility of determining the TPA parameters, shown in Fig. 15 (Bourne, 1982) with a Voland-Stevens Texture Analyzer (VSTA) (Voland Corp., Hawthorne, NY). The VSTA worked satisfactorily on the downstroke with respect to the crosshead speed, but the depth of penetration could not be controlled, because the depth of penetration was not based on measured distance; it appears to be based

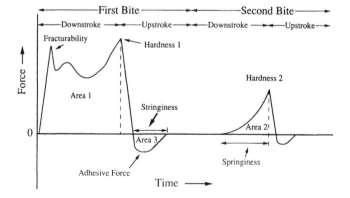

Figure 15   Texture profile analysis parameters (adapted from Bourne, 1982).

Table VI Magnitudes of texture-profile analysis parameters using a 1.27-cm diameter plunger

		[a]Voland		[b]Instron		
		Aver.	SD	Aver.	SD	% Difference
Fracturability	(N)	0.216	0.022	0.192	0.034	11.06
Hardness 1	(N)	0.140	0.009	0.134	0.014	4.46
Adhesiveness	(N)	—	—	0.028	0.006	—
Hardness 2	(N)	0.096	0.007	0.095	0.007	1.61
Stringiness	(mm)	—	—	14.714	1.330	—
Springiness	(mm)	15.333	0.116	15.054	0.132	1.83
AREA1	(N*cm)	0.180	0.012	0.153	0.015	14.98
AREA2	(N*cm)	0.091	0.005	0.063	0.003	31.56
AREA3	(N*cm)	—	—	−0.023	0.008	—
Cohesiveness	(−)	0.508	0.028	0.408	0.023	19.75
Chewiness	(N*cm)	1.090	0.074	0.819	0.060	24.84
Gumminess	(N)	0.071	0.005	0.055	0.004	23.35

[a]Magnitudes are averages of six measurements.
[b]Magnitudes are averages of four measurements.

on the resistance sensed by the load cell. A precise magnitude of adhesive force could not be obtained. the net result is that area 3 of the TPA curve could not be determined accurately. Also, there was a built-in delay at the end of the downstroke and before the beginning of the upstroke. In summary, reliable values of the TPA parameters of fracturability, hardness 1, hardness 2, springiness, area 1, and area 2 were obtained with the VSTA.

The magnitudes of TPA parameters using 1.27-cm diameter plunger on the VSTA and the Instron, as well as their standard deviations, are given in Table VI. The adhesive force, springiness, and area 3 with the VSTA were not included for the aforementioned reasons. In general, the TPA parameters with the VSTA were higher than those with the Instron. The percentage difference of the values ranged from 1.61% for hardness to 31.6% for area 2. The standard deviations of the TPA with the two instruments parameters were of the same order of magnitude.

It appears that the VSTA can provide reliable values of TPA data that are based on the force–time curves recorded during the downstrokes of the plexiglas plunger. Although TPA parameters are empirical in nature, they usually provide good correlation with sensory data. In addition, by providing more than one index of quality, they have an advantage over single-point quality control methods, such as the sag method and those based on break strength.

References

Axelos, M. A. V., and Kolb, M. (1990). Crosslinked biopolymers: Experimental evidence for scalar percolation theory. *Phys. Rev. Lett.* **64**, 1457–1460.

Axelos, M. A. V., Thibault, J. F., and Lefebvre, J. (1989). Structure of citrus pectins and viscometric study of their solution properties. *Int. J. Biol. Macromol.* **11,** 186–191.

Beveridge, T., and Timbers, G. E. (1989). Small amplitude oscillatory testing (SAOT) application of pectin gelation. *J. Texture Stud.* **20,** 317–324.

Bourne, M. C. (1982). "Food Texture and Viscosity: Concept and Measurement." Academic Press, New York.

Braudo, E. E., Plaschina, I. G., and Tolstoguzov, V. B. (1984). Structural characterisation of thermoreversible anionic polysaccharide gels by their elastoviscous properties. *Carbohydr. Polym.* **4,** 23–48.

Chou, T. C., and Kokini, J. L. (1987). Rheological properties and conformation of tomato paste pectins, citrus and apple pectins. *J. Food Sci.* **52,** 1658–1664.

Comby, S., Doublier, J. L., and Lefebvre, J. (1986). Stress-relaxation study of high-methoxyl pectin gels. *In* "Gums and Stabilisers for the Food Industry 3 (Proceedings of the 3rd International Conference, Wrexham, Wales)" (G. O. Phillips, D. J. Wedlock, and P. A. Williams, eds.), pp. 203–212. Elsevier Science Publishers, New York.

Crandall, P. G., and Wicker, L. (1986). Pectin internal gel strength: Theory, measurement, and methodology. *In* "Chemistry and Function of Pectins" (M. L. Fishman and J. J. Jen, eds.), pp. 88–101. American Chemical Society, Washington, D.C.

Dahme, A. (1985). Characterization of linear and nonlinear deformation behavior of a weak standard gel made from high-methoxyl citrus pectin. *J. Texture Stud.* **16,** 227-239.

Dealy, J. M. (1982). "Rheometers for Molten Polymers." Van Nostrand Reinhold, New York.

Durand, D., Bertrand, C. Busnel, J. P., Emery, J., Axelos, M. A. V., Thibault, J. F., Lefebvre, J., Doublier, J. L., Clark, A. H., and Lips, A. (1990). Physical gelation induced by ion complexation: Pectin–calcium systems. *In* "Physical Networks" (W. Burchard and S. B. Ross-Murphy eds.), pp. 283–300. Elsevier Applied Science Publishers, New York.

Dzuy, N. Q., and Boger, D. V. (1983). Yield stress measurement for concentrated suspensions. *J. Rheol.* **27,** 321–349.

Ferry, J. D. (1980). "Viscoelastic Properties of Polymers." Wiley, New York.

Gordon, M., and Ross-Murphy, S. B. (1975). The structure and properties of molecular trees and networks. *Pure Appl. Chem.* **43,** 1–26.

Gross, M. O., Rao, V. N. M., and Smith, C. J. B. (1982). Direct stress–strain dynamic characteristics of low-methoxyl pectin gels. *J. Texture Stud.* **13,** 97–114.

Kawabata, A. (1977). Studies on chemical and physical properties of pectin substances from fruits. Viscoelasticity of fruit pectin gels. *Memoirs Tokyo Univ. Agric.* **19,** 166–178.

Kertesz, Z. I. (1951). "The Pectin Substances." Interscience, New York.

Komatsu, H., and Sherman, P. (1974). A modified rigidity modulus technique for studying the rheological properties of w/o emulsions containing microcrystalline wax. *J. Texture Stud.* **5,** 97–104.

Lin, K. S. C., andAklonis, J. J. (1980). Evaluation of the stress-relaxation modulus of materials with rapid relaxation rates. *J. Appl. Phys.* **51,** 5125–5130.

May, C. D. (1990). Industrial pectins: Sources, production and applications. *Carbohydr. Polym.* **12,** 79–99.

Mitchell, J. R. (1976). Rheology of gels. *J. Texture Stud.* **7,** 313–339.

Mitchell, J. R. (1984). Rheological techniques. *In* "Food Analysis: Principles and Techniques" (D. W. Gruenwedel and J. R. Whitaker, eds.), pp. 151–220. Marcel Dekker, New York.

Morris, E. R. (1984). Rheology of hydrocolloids. *In* "Gums and Stabilisers for the Food Industry 2 (Proceedings of the 2nd International Conference, Wrexham, Wales)" (G. O. Phillips, D. J. Wedlock, and P. A. Williams, eds.), pp. 57–78. Elsevier Science Publishers, New York.

Morris, E. R., Cutler, A. N., Ross-Murphy, S. B., Rees, D. A., and Price, J. (1981). Concentration and shear-rate dependence of viscosity in random coil polysaccharide solutions. *Carbohydr. Polym.* **1,** 5–21.

Oakenfull, D. (1984). A method for using measurements of shear modulus to estimate the size and thermodynamic stability of junction zones in noncovalently cross-linked gels. *J. Food Sci.* **49,** 1103–1104 and 1110.

Oakenfull, D., and Scott, A. (1984). Hydrophobic interaction in the gelation of high-methoxyl pectins. *J. Food Sci.* **49,** 1093–10998.

Oakenfull, D. G., Parker, N. S., and Tanner, R. I. (1989). Method for determining absolute shear modulus of gels from compression tests. *J. Texture Stud.* **19,** 407–417.

Plashchina, I. G., Fomina, O. A., Braudo, E. E., and Tolstoguzov, V. B. (1979). Creep study of high-esterified pectin gels. I. The creep of saccharose-containing gels. *Colloid Polym. Sci.* **257,** 1180–1187.

Plaschina, I. G., Gotlieb, A. M., Braudo, E. E., and Tolstoguzov, V. B. (1982). Creep study on high-esterified pectin gels. II. The creep of glycerol-containing gels. *Colloid Polym. Sci.* **261,** 672–676.

Qiu, C-G., and Rao, M. A. (1988). Role of pulp content and particle size in yield stress of apple sauce. *J. Food Sci.* **53,** 1165–1170.

Qiu, C-G., Rao, M. A., and Walter, R. H. (1990). Creep-compliance and yield behavior of food-grade pectin gels. Unpublished data.

Rao, M. A. (1986). Viscoelastic properties of fluid and semisolid foods. *In* "Physical and Chemical Properties of Food" (M. R. Okos, ed.), pp. 14–34. American Society of Agricultural Engineers, St. Joseph, MI.

Rao, M. A., Cooley, H. J., Walter, R. H., and Downing, D. L. (1989). Evaluation of texture of pectin jellies with the Voland-Stevens texture analyser. *J. Texture Stud.* **20,** 87–95.

Ring, S. G., and Stainsby, G. (1985). A simple method for determining the shear modulus of food dispersions and gels. *J. Sci. Food Agric.* **36,** 607–613.

Ross-Murphy, S. B. (1984). Rheological methods. *In* "Biophysical Methods in Food Research" (H. W.-S. Chan, ed.), pp. 138–199. Blackwell Scientific Publications, London.

Saunders, P. R., and Ward, A. G. (1954). An absolute method for the rigidity modulus of gelatine gel. *In Proceedings of the Second International Congress on Rheology,* (V. G. W. Harrison, ed.), pp. 284–290. Academic Press, New York.

Sherman, P. (1970). "Industrial Rheology."Academic Press, New York.

Stainsby, G., Ring, S. G., and Chilvers, G. R. (1984). A static method for determining the absolute shear modulus of a syneresing gel. *J. Texture Stud.* **15,** 23–32.

Tanglertpaibul, T., and Rao, M. A. (1987). Intrinsic viscosity of tomato serum as affected by methods of determination and methods of processing concentrates. *J. Food Sci.* **52,** 1642–1645 and 1688.

Thom, D., Dea, I. C. M., Morris, E. R., and Powell, D. A. (1982). Interchain associations of alginate and pectins. *Prog. Food Nutr. Sci.* **6,** 97–108.

Watson, E. L. (1966). Stress-strain and creep relationships of pectin gels. *J. Food Sci.* **31,** 373–380.

CHAPTER 12

Nonfood Uses of Pectin

H.-U. Endress

Herbstreith and Fox KG
Pektin-Fabrik
Neuenbürg, Germany

I. INTRODUCTION

The numerous properties of pectin make it virtually universal in its applicability. Established uses in humans have been, for example, in blood plasma substitution, detoxication, and dietary-fiber fortification (Kertesz, 1951). Medical and nonmedical applications alike have expanded abreast of modern technology into new areas in which multifunctional substances predominate.

II. BIOCHEMISTRY AND PHYSIOLOGY

A. Medicine and Health

Pectin appears to exercise its beneficial effects on humans by its action on cholesterol absorption, on the exocrine pancreatic hormones and enzymes, on the ilea-hepatic and the ceca-hepatic circulation, on colon transit time, and through its intestinal microbial degradation products. Colon transit times have implications for tumorigenesis (Ink and Hurt, 1987). The action of the pectic substances, in some cases, may depend on the chemistry of galacturonic acid (Suzuki and Kajuu, 1983).

Recent benefits from applications of pectin in medicine include its effect on blood cholesterol reduction, manufacture of pharmaceuticals, salves, and emolients, etc. Pectin has been reported to have antiviral activity (Bender, 1969).

1. Hemostasis

Pectic substances are often a component in hemostatic formulations (Deuel, 1945; Kertesz, 1951; Tomic, 1981; Raskai and Szijjarto, 1984, 1985; Thiele, 1987b). Their contribution is apparently a function of the galacturonic acid, the blood concentration of which governs coagulation time (Bock *et al.*, 1964). A lyophilized sponge containing pectin (10–50%) is a medically useful hemostatic device (Lavia *et al.*, 1981). Sulfated degradation products of pectin are potent anticoagulants (Alburn and Seifter, 1956). Pulver (1961) formulated a synergistic anticoagulant composition of heparin and the polysulfuric acid ester of oxidatively degraded pectin. In the three-stage wound-treatment procedure of Alvarez (1989), 35–50% pectin is applied in the first stage and 5–10% pectin, in the second stage. Freeman and Pawelchak (1985) developed an occlusive, multilayered wound dressing containing a water-dispersible hydrocolloid, e.g., pectin (45–65%). For this purpose, the pectin particle size should be 10–40 mesh (Freeman and Pawelchak, 1988).

For burns and other wounds, Nambu (1985) composed a water-insoluble gel from polyvinyl alcohol and 2% pectin, and freeze–dried it to 95% soluble solids content. Wound dressings made of polyurethane foams, 5–50% pectin and/or other water dispersible hydrocolloids (Cilento and Freeman 1988a,b) have been patented, as well as an air-permeable, deodorizing, multilayer bandage for special open-skin disorders (Mathews and Steer, 1980). Cilento *et al.* (1984) developed a microporous adhesive tape for surgical purposes, in which a rubber material and a hydrocolloid, e.g., pectin (20–65%), composed the adhesive layer.

After some kinds of surgery, prosthetic devices must be sealed in the stoma. Sealants for this purpose may include pectin (Cilento *et al.*, 1979; Chen *et al.*, 1980 a,b, 1981, 1987; Larsen and Sorensen, 1980; Doehnert and Hill, 1985).

2. Cholesterol

A significant reduction of blood cholesterol was reported in studies with a wide variety of subjects and experimental conditions (Keys *et al.*, 1961; Palmer and Dixon, 1966; Durrington *et al.*, 1976; Jenkins *et al.*, 1976 a; Kay and Truswell, 1977; Mietinen and Tarpila, 1977; Kay et al., 1978 Jenkins *et al.*, 1979; Ginter *et al.*, 1979; Stasse-Wolthuis *et al.*, 1980; Nakamura *et al.*, 1982; Schwandt *et al.*, 1982; Judd and Truswell, 1982; Challen *et al.*, 1983; Hundhammer and Marshall, 1983; Schuderer, 1986). At least 6–15 g/day seems to be necessary. Blood-cholesterol reduction was almost nonsignificant, when less than 6 g/day was consumed (Palmer and Dixon, 1966; Raymond *et al.*, 1977; Delbarre *et al.*, 1977). Better results were obtained if pectin was completely hydrated in a food product compared to powdered pectin mixed with food (Behall and Reiser, 1986). The cholesterol-lowering effect was proportional to the amount of dietary cholesterol in the blood (Ullrich, 1987). In two studies, pectin had no influence on blood cholesterol (Fahrenbach *et al.*, 1965; Hillman *et al.*, 1985). In the study by Miettinnen and Tarpila (1977), serum cholesterol was reduced by 13% within 2 weeks. In a long-term study, they further lowered blood cholesterol by 15%.

Cerda *et al.* (1988) observed, from patients at risk of coronary heart disease, that pectin supplementation decreased blood cholesterol by 7.6%, concluding that a pectin-supplemented diet can significantly reduce blood cholesterol without a change in lifestyle.

In the study of Schwandt *et al.* (1982), pectin was applied in combination with cholestyramine, a bile-acid sequestrant, and it was shown that pectin augmented the reduction of cholesterol by 20%, compared to cholestyramine alone.

Apparently, the degree of esterification (DE) of pectin has no influence on the cholesterol-lowering effect of pectin. Using 15 g high-methoxyl (DE 71%), low-methoxyl (DE 37%), and amidated pectins, Judd and Truswell (1982) obtained similar results from each in reducing serum cholesterol. These results could not be confirmed by Miettinnen and Tarpila (1977). They found no effect on serum cholesterol in hypercholesterolemic patients who consumed 6 g/day low-methoxyl citrus pectin. According to Schuderer (1986), only high-methoxyl pectins have a cholesterol-lowering effect, but in studies with rats, Ershoff and Wells (1962) showed that pectins with an approximately 10% methoxyl content counteracted the increment of rat-liver cholesterol induced by cholesterol feeding, whereas pectic substances with approximately 5% methoxyl content were ineffective.

Pectic substances are involved in bile-acid metabolism (Eastwood *et al.*, 1980). An addition of 15 g pectin to the meal of ileostomy patients increased bile-acid excretion by 35% and net cholesterol excretion by 14% (Bosaeus *et al.*, 1986). The increased bile-acid excretion seemed to be independent of the DE of the pectins (Pfeffer *et al.*, 1981; Judd and Truswell, 1982). Pectic substances not only accelerated the bile-acid excretion, but they also changed the bile-acid profile.

A 15 g pectin/day diet in man increased fecal excretion of neutral steroids by 17%, of bile acids by 35% (Kay and Truswell, 1977) and of acid steroids by 11% (Ross and Leklem, 1981). A pectin-fortified regimen increased neutral steroids excretion, observed for male but not for female humans (Stasse-Wolthuis *et al.*, 1980).

Pectin degradation products are claimed to have an influence on blood cholesterol. While pectin cannot be digested by human enzymes, the colon microflora produce deesterifying and depolymerizing enzymes.

3. Lipids

Ershoff and Wells (1962) showed that pectin counteracted the increment of liver total lipids in rats. At an intake rate of 2 to 50 g/day, pectic substances reduced liver and blood lipids, low-density (LD) and/or very low density (VLD) lipo-protein (L), and the LDL:HDL ratio. High-density lipoprotein (HDL) is only little affected by pectin. Cerda *et al.* (1988) also observed that pectin supple-mentation decreased LDL by 10.8%, and the LDL:HDL ratio by 9.8%. The other plasma–lipid fractions showed no significant change. Richter *et al.* (1981), Nakamura *et al.* (1982), and Schwandt *et al.* (1982) showed that 18 g, 9 g, and 12 g pectin per day reduced LDL and VLDL by 5, 11, and 35%, respectively. HDL was increased by 4% (Richter *et al.*, 1981).

Pectin forms complexes with LDL in the gut and so hinder LDL absorption. The results of Baig and Cerda (1981) indicate that the interaction is of an electrostatic nature, and perhaps hydrogen bonding. Falk and Nagyvary (1982) found that one part of high-methoxyl pectin bound 4 parts of LDL. Low-methoxyl pectins bound less LDL, meaning that the binding mechanism depended on the DE of the pectic substances. Pectin consistently failed to reduce the level of blood triglycerides (Reiser, 1987).

4. Enzymes and Hormones

Schuderer (1986) discussed the possibility of the influence of pectin on some exocrine pancreatic hormones. Morgan *et al.* (1979) and Levitt *et al.* (1980) reported that a pectin supplement reduced the production of gastric inhibitory polypeptide. This hormone reduces gastric motility and insulin secretion, which could result in a lowered activity of the enzyme, α-hydroxy-α-methylglutaryl-(HMG) CoA-reductase, because its activity depends on the insulin concentration. Serum cholesterol can also be reduced, because HMG-CoA-reductase is involved in an early step of the endogenous cholesterol synthesis.

There are two possibilities by which pectic substances might have an influence on exocrine pancreatic enzymes. Pectic substances are able to increase the vis-cosity of the stomach contents and prohibit contact between food components and digestion enzymes. They can also reduce the enzymes' activities by forming complexes with them (Schuderer 1986).

Isaksson (1982), Isaksson *et al.* (1982 a,b, 1984), Dutta and Hlasko (1985) reported a reduction of amylase activity by 10 to 40%, of lipase activity by 40 to 80%, and of trypsin activity by 15 to 80%, caused by pectin. This effect was accompanied by an increase in viscosity of the digestion fluids.

Cummings *et al.* (1979) reported a significantly greater excretion of fatty acids (80%) and bile acids (35%), compared to a control period, when healthy volunteers consumed 36 g/day of pectin.

5. Glucose Metabolism

Several studies with insulin-dependent and insulin-nondependent diabetics have shown that pectic substances lower blood glucose and insulin levels after a carbohydrate meal (Jenkins et al., 1976b; Monnier *et al.*, 1978; Vaaler *et al.*, 1980; Poynard *et al.*, 1980; Williams *et al.*, 1980; Kanter *et al.*, 1980; Levitt *et al.*, 1980; Tunali *et al.*, 1990). The response was comparable in studies with healthy volunteers (Jenkins *et al.*, 1977, 1978; Holt *et al.*, 1979; Gold *et al.*, 1980; Kanter *et al.*, 1980; Sahi *et al.*, 1985, Sandhu *et al.*, 1987).

6. Weight Reduction

Pectin is thought to immobilize food components in the intestines, thereby reducing the rate of digestion (Hansen and Schulz, 1983). The result is that food absorption is concomitantly lessened. Flourie *et al.* (1984) found that with increasing pectin dosages, the absorption of food components from the stomach was reduced. This reduction was also correlated with the size of the jejunal unstirred water layer, which increases in thickness as a result of consuming pectin (Gerencser *et al.*, 1984).

The thickness of the layer is said to have an influence on absorption by prohibiting contact between the intestinal enzymes and the food, thus reducing the latter's availability (Gatfield and Stute, 1972; Wilson and Dietschy, 1974; Dunaif and Schneemann, 1981; Hansen an Schulz, 1983).

Pectin, a hydrocolloid, is able to bind a large volume of water that, in the process, bestows a feeling of satiety. The effect is again to reduce food consumption. Experiments showed a prolongation of the gastric-emptying half-time from 23 to 50 min, of a meal fortified with pectin (Holt *et al.*, 1979). The gastric-emptying half-time was doubled by the intake of 20 g apple pectin per day for 4 weeks (Schwartz *et al.*, 1983, 1988). Inasmuch as pectin delays gastric emptying and confers a feeling of satiety in obese subjects, it may be a useful adjuvant in the treatment of disorders related to overeating (Di Lorenzo *et al.*, 1988). Formulas containing apple pectin mixed with insoluble dietary fiber, certain enzymes, and proteins are sold for this purpose.

Schneemann (1985) stated that "food intake is decreased by the concurrent intake of nonnutrients". Pectin is a nonnutrient.

7. Medicaments and Pharmaceuticals

The ability of pectin to delay the absorption of drugs (Kertesz 1951) was taken advantage of in the manufacturing of encapsulated pharmaceuticals (Wuhrmann, 1939; Murray and Finland, 1946; Welch et al., 1947; Welch, 1950; Gyarmati et al., 1980; Asano et al., 1984; Ogawa et al., 1987a,b; Salatinjants, 1987; Washington et al., 1988).

Problems with the solubility and absorption of nonuniform procaine and penicillin crystals led Sumner and Grenfell (1955) to discover that their precipitation in the presence of pectin resulted in greater uniformity of shape and size. Corticotropin, used to treat rheumatism and arthritis, suffers the disadvantage of having to be frequently injected, in order to maintain a continuous therapeutic effect. To circumvent this problem, a complex with pectic acid or other polybasic acids was formed (Murphy et al., 1963). The ability of pectinic and pectic acids to form sodium, potassium, calcium, and magnesium salts is of medical benefit, when prescribing remedies that contain immunologically active substances requiring slow release (Mill et al., 1967).

Salts of pectinic acids improve the solubility of some drugs (Becher and Leya, 1946; Owens and Maclay, 1951). Welch (1952) formulated a pectin-coated penicillin salt that increased and prolonged blood levels of penicillin. Pectin particles stabilize penicillin preparations (Welch, 1950), 1-(p-nitrophenyl-sulfonamido)-thiazole retention enemas (Hoehn, 1950), and X-ray compositions (Slaybaugh, 1953). Pectin and mixtures of pectin with compatible substances have performed as drug carriers (Sackler and Sackler, 1967; Hiroshi et al., 1987).

Negrevergne (1969, 1974a,b) described the advantages of a novel pyrazolidone–pectin derivative in which the pectin molecule is bound to the number 4 position of 3,5-dioxo-1,2-diphenyl-4-N-butylpyrazolidine. This derivative is an antiinflammatory substance. Pectin in demulcents helps to alleviate gastrointestinal distress (Bender, 1969; Hill, 1976). It is a carrier of interferon (Chany et al., 1975, 1977). Aqueous dispersions (0.1–5%) are used to prepare microcrystalline beads of vitamin A acetate by wet-milling (Keller and Klaeni, 1981). Calcium salts of polygalacturonic acid are a remedy for hyperphosphatemia (Kulbe and Weber, 1984). Encapsulation of sulfamerazine was accomplished by coacervation with pectin and gelatin (McMullen et al., 1984). Bates (1989) patented a decongestant composed of vegetable oil, aloe vera, zinc, vitamins, and pectin.

For treating diarrhea (Malyoth, 1934), pectin and combination preparations of pectin with other substances like agar (Tompkins, 1938), tannic substances (Mansfield, 1940), iodine (Otto, 1941), kaolin (Kaopectate, Upjohn), kaolin, bentonite, neomycin (Bennett, 1958), bentonite and alkyl polyalcohol (Jensen, 1969), aluminum phosphate (Raudnitz, 1979), medical activated charcoal (Thiele, 1987a) or mixtures with charcoal, kaolin (*bolus alba*) and sweet whey (Laves,

1979) have been invented. Nickel–pectinate (Myers, 1941) and the combination pectin and tomato pulp were also produced.

Pectin also abolishes or alleviates the symptoms of dumping in patients suffering from dumping syndrome, because of its viscous nature (Leeds *et al.*, 1977, Lawaetz *et al.*, 1983). By increasing the viscosity of a meal, gastric emptying is reduced significantly. Zimmaro *et al.* (1989) showed that liquid stool induced by isotonic tube-feeding formula is reversed by pectin.

B. Heavy-Metal Toxicity

Since the first reports that pectin was a good antidote for heavy-metal poisoning (Kertesz, 1951), it has been discovered that the absorption of strontium into the bone structure of rats can be suppressed by pectin (McDonald *et al.*, 1952; Rubanovskaya, 1960; Patrick, 1967). The gastrointestinal tissue of rats fed a diet containing pectin was found to contain only 0.1% of strontium -90 in the diet. The binding of strontium *in vivo* is less effective of acidic than at alkaline pH (Bessubov and Hatina, 1960). Waldron-Edward *et al.* (1965) did not detect any strontium in the blood, 24 hr after their subjects ingested pectin; strontium concentration in the skeleton was significantly lower, when compared to that from a pectin-free diet.

The affinity of pectin for metals is, as follows Mg < Mn < Cr < Hg < Fe < Ni < Co < Cu < Zn < Sr < Cd < Ba < Pb (Paskins-Hurlburt *et al.*, 1977). Pectin has been evaluated as a prophylactic against lead toxicosis (Bondarev *et al.*, 1978). Pectic substances form insoluble pectinates *in vivo* that are excreted in the stool and urine (Bessubov and Khatina, 1960; Nicoulescu *et al.*, 1968; Markova *et al.*, 1976; Paskins-Hurlburt *et al.*, 1977; Stantschev *et al.*, 1979). Typical symptoms of lead poisoning disappeared when some factory workers ate 8–9 g of pectin per day (Stantschew *et al.* 1979).

Following the Chernobyl nuclear disaster, Ukrainians were fed pectin-enriched foodstuffs, in order to take advantage of the complexing ability of pectin with radioactive nuclides.

C. Mutagenicity

There was a significant reduction of nitro-compounds when Rowland *et al.* (1983) fed rats a diet fortified with pectin. Jongen *et al.* (1987) reported that pectin inhibited *in vitro* mutagenicity by fava beans treated with nitrite. The inhibition was accompanied by a diminished N-nitroso content.

III. DENTISTRY

The first toothpaste containing pectin was patented in 1932 (Deutsche Pektin-gesellschaft 1932a). Since then, dental impression materials having improved shelf-life in the dry state have been formulated to contain pectin, along with other reactive substances (Lochridge, 1951; Noyes and Lochridge, 1953a,b,c; Rabchuk, 1957, 1959). Pectin and pectinates are suitable polyanionic materials for making dental adhesives (Bohne, 1942; Gidwani et al., 1974, 1975a,b; Keegan et al., 1975; Beachner, 1968). A barrier of pectin was designed in a dry toothpaste to protect a bicarbonate and a peroxide from premature wetting (Bohm et al., 1989).

IV. OTHER PRODUCTS

A. Skin-Care Products

Contemporary recommendations for pectin include shaving (Deutsche Pektin-gesellschaft, 1932b) and skin lotions (Kröper, 1934; Anonymous, 1935; Bates, 1987), soaps (Kertesz, 1951), hair pomades (Lesser, 1939; Rae, 1944), exci-pients (Toni, 1946), emolients (Okuyama et al., 1981), and deodorants (Holzner, 1989). For treating acne vulgaris, Ciesla et al. (1982) patented an aqueous gel containing 1–3% pectin, salicylic acid, and/or benzoylperoxide.

B. Cigarette Manufacture

In the tobacco industry, pectic substances are used to make self-extinguishing cigarettes (Simon, 1984), as a binder for reconstituted tobacco sheets and films (Hind and Seligman, 1968a,b, 1969; Hind and Hopkins, 1968, 1978; Hind, 1970; Deszyck, 1973 Anonymous, 1977; Perkins and Bale 1979; Ohashi et al., 1986), as a humectant (Georgiev, 1979), in tobacco substitutes (Hind and Hop-kins, 1968; Anonymous, 1971; Hedge, 1972; Hind and Kelly, 1972; Deszyck, 1974; Perkins and Bale, 1979), in wrappers (Anonymous, 1977; Hind and Hop-kins, 1978) and in flavors (Nichols et al. 1987; Denier et al., 1989; Tateno and Masuko, 1989).

C. Microbiological Culture Media

It is possible to identify pectolytic enzyme activity by the transparent zones created around colonies of enzyme-producing microorganisms growing in pectin-containing (turbid) culture media. Jones (1946) and Roth (1980, 1981) prepared culture media with low-methoxyl pectin. A sodium ammonium pectate gel was described by Baier and Wilson (1941). Youssef (1981) designed a self-sterilizing medium with pectin, containing a chlorine-liberating compound.

The production of a highly potent interferon is enhanced on culture media by inclusion of polyanions, e.g. pectin, in the medium (Iizuka *et al.*, 1984).

D. Soil Conservation

Van Leuven (1967) used the binding power of pectin (0.05–7.5%, dry mass) in a mixture to prevent soil erosion on newly planted areas, an application accommodated by soil calcium (Hirsbrunner, 1988).

In a related application, an aqueous pectin dispersion was injected into reservoirs constructed with highly permeable earth, and when the pectin came in contact with brine containing divalent cations, it precipitated and effectively sealed the pores (Christopher and Clauset, 1980).

E. Feeding Animals

Sawadogo *et al.* (1988) discovered that pectin as well as oligo-galacturonic acid in the ration stimulated blood prolactin and growth hormone secretion in ewes. During the growth period of pigs, a 2% pectin addition to the feed caused a decrease in weight gain. During the finishing period it resulted in an increase, but did not affect the feed-conversion rate. Pectin decreases the back-fat thickness of pigs (Lagreca and Marotta, 1985). Some pellets for marine fish feeding are coated with pectin (Cox, 1985).

F. Miscellaneous Uses

Given the multifunctionality of pectin, it has been viewed, not surprisingly, as an industrial chemical and as an intermediate in chemical synthesis. Insecticides (Baier, 1940; Wilson, 1940) including plant virucides (Kasugai *et al.*, 1975) and a plant antifreeze (Woods, 1988) have been formulated with pectin.

In the enzymatic synthesis of dehydrogalacturonosylgalacturonates, pectin is the substrate (Bock *et al.*, 1987). It has been put in some therapeutic gels (Raudnitz, 1979), sanitary napkins (De Merre, 1967), and contraceptives (Gero, 1987a,b). More recently, Kasten (1989) invented biodegradable drinking straws in which coloring and flavoring substances in a pectin layer are released when liquid passes through the straw. Sausage coatings are made from gelatin and pectin (Childs, 1957; Julius, 1967). In various other ways, pectic substances have been involved in the manufacture of sausage coatings, as, for example, in dewatering baths in collagen extrusion (Higgins, 1979) and edible casings with water-resistant printing ink (Winkler, 1973).

Fruit juice decolorization is facilitated by activated carbon coated by calcium pectate (Wilson, 1956). A 1% potassium pectate solution has performed as a fining agent (Baker, 1976; Strohm *et al.*, 1987) without appreciably changing juice composition (Wucherpfennig *et al.*, 1988). This pectate increased by 50% the ultrafiltration flux rate of a continuous process for apple-juice making (Endress, 1988; Wucherpfennig, 1990). In berry juice, no color was removed (Baumann, 1989). In wineries, the removal of copper, iron, and lead from musts and wines has been attempted (Schlemmer, 1986).

Pectic substances have been assigned an important role in making paper substitutes (Schoepe, 1979), carbonless copy paper (Matsukawa and Saeki, 1975, 1976; Matsukawa *et al.*, 1976; Sankar and Arun, 1989) photographic films (Land, 1956; Salminen and Weyerts, 1961; Schmidt *et al.*, 1962), lead storage batteries (Beste *et al.*, 1966; Ryhiner *et al.*, 1967a), homeoporous electrodes (Ryhiner, 1967b), ceramics (Ruben, 1988), porous silicates (Robinson, 1988), gas filters (Moroni and Kalbow, 1978), foams (Kennedy, 1985) including flame-extinguishing foams (Chiesa, 1973; Hiltz *et al.*, 1987; Pless and Ullmann, 1990), catalytic silver (Ramirez *et al.*, 1975), rust removers (Reghin *et al.*, 1951), lubricants (Morway and Mikeska, 1954; Lanini *et al.*, 1989) and, finally, in plasticizers (Fetzer and von Rex, 1986).

References

Alburn, H. E., and Seifter, J. (1956). U.S. Patent 2,729,633 *Chem. Abstr.* **50,** 8144

Alvarez, O. M. (1989) U.S. Patent 4,813,942

Anonymous (1935) Ger. Patent 620 843

Anonymous (1971) Ger. Offen DE 20 12 070

Anonymous (1977) Brit. Patent GB 147 1943

Asano, M., Kaetsu, I., and Yoshida, M. (1984). U.S. Patent 4,483,807

Baier, W. E. (1940) U.S. Patent 2,207,694

Baier, W. E., and Wilson, C. W. (1941). Citrus pectates—properties, manufacture, and uses. *Ind. Eng. Chem.* **33,** 287.

Baig, M. M., and Cerda, J. J. (1981). Pectin: Its interaction with serum lipoproteins. *Am. J. Clin. Nutr.* **34,** 50–53.

Baker, R. A. (1976). Clarification of citrus juices with polygalacturonic acid. *J. Food Sci.* **41,** 1198–1200.

Bates, H. L. (1987). U.S. Patent 4,704,280.

Bates, H. L. (1989) U.S. Patent 4,826,683.

Baumann, G. (1989). Neue Wege der Buntsaftbehandlung. *Confructa Studien* **33,** 137–151.

Beachner, C. E. (1968). U.S. Patent 3,410,704. *Chem. Abstr.* **70,** 40644.

Becher, R., and Leya, S. (1946). Löslichkeitserhöhung von Sulfonamiden durch Pektin. *Experientia* **2,** 459.

Behall, K., and Reiser, S. (1986). Effects of Pectin on Human Metabolism *In* "Chemistry and Function of Pectins" (Fishmen and Yen, (eds.), pp. 248–265. ACS Symposium Series 310, Washington, D.C.

Bender, W. A. (1969). U.S. Patent 3,485,920. *Chem. Abstr.* **72,** 59069.

Bennett, M. E. (1958). U.S. Patent 2,828,242. *Chem. Abstr.* **52,** 11366.

Beste, H., König, A., Ryhiner, G., and Voss, E. (1966). U.S. Patent 3,271,199. *Chem. Abstr.* **60,** 15443.

Bessubov, A. D., and Khatina, A. I. (1960). [Pectin as a prophylactic agent in strontium intoxication.] russ. Vsesojuzni naucno-issledowatelskij institut konditerskoj promislennosti: *Postupila v. redakciju* **19**/XII, 39–42.

Bock, W., Pose, G., and Augustat, S. (1964). Über die Resorption und hämostatische Wirkung von Pektin bei der Ratte. *Biochem. Z.* **341,** 64–73.

Bock, W., Anger, H., Schneider, E., Flemming, C., and Henninger, H. (1987). Ger. East Patent 24 79 23.

Bohm, P. D., Denholtz, J. R., Denholtz, M., and Rudy, J. B. (1989). U.S. Patent 4,837,008.

Bohne, A. (1942). Ger. Patent 728 003.

Bondarev, G. I., Anisova, A. A., Alexeewa, T. E., and Sisranzew, J. K. (1978). [Evaluating pectin with a low degree of esterification as prophylactic agent in lead poisoning.] russ. Laboratorija profilakticeskogo pitanija, Instituta pitanija AMN SSSR, Moskwa, Postupila 1/III, 65–67.

Bosaeus, I., Carlsson, N. G., Sandberg, A. S., and Andersson, H. (1986). Effect of wheat bran and pectin on bile acid and cholesterol excretion in ileostomy patients. *Hum. Nutr. Clin. Nutr.* **40,** 429–440.

Cerda, J. J., Robbins, F. L., Burgin, C. W., Baumgartner, T. G., and Rice, R. W. (1988). The effects of grapefruit pectin on patients at risk for coronary heart disease. *Clin. Cardiol.* **11,** 589–594.

Challen, A. D., Branch, W. J., and Cummings, J. H. (1983). The effect of pectin and wheat bran on platelet function and haemostatis in man. *Human Nutr. Clin. Nutr.* **37,** 209–217.

Chany, C., Galliot, B., Chevalier, M. J., and Ankel, H. (1975). Ger. offen 24 36 172. *Chem. Abstr.* **82,** 175224. U.S. Patent 4,041,152 (1977).

Chen, J. L., Cilento, R. D., Hill, J. A., and Lavia, A. L. (1980 a,b). U.S. Patent 4,204,540. U.S. Patent 4,192,785.

Chen, J. L., Cilento, R. D., Hill, J. A., Lavia, A. L. (1981). U.S. Patent 4,253,460.

Chen, J. L., Cilento, R. D., Hill, J. A., and Lavia, A. L. (1987). Ger. Patent 28 25 196 *Chem. Abstr.* **90,** 110009.

Chiesa, P. J. (1973). Ger. Offen. 23 57 281.

Childs, W. H. (1957). U.S. Patent 2,811,453 *Chem. Abstr.* **52,** 3199.

Christopher, C. A. Jr., and Clauset, A. O., Jr. (1980). U.S. Patent 4,210,204.

Ciesla, P. F., Kligman, A. M., and McKenzie, W. L. (1982). U.S. Patent 4,355,028.

Cilento, R. D., Lavia, A. L., Chen, J. L., and Hill, J. A. (1979). U.S. Patent 4,166,051.

Cilento, R. D., Lavia, A. L., and Riffkin, C. (1984). U.S. Patent 4,427,737.

Cilento, R. D., and Freeman, F. M. (1988 a,b). U.S. Patent 4,773,408. U.S. Patent 4,773,409.

Cox, J. P. (1985). PZT-Gazette 851121.

Cummings, J. H., Southgate, D. A. T., Branch, W. J., Wiggins, H. S., Houston, H., Jenkins, D. J. A., Jivraj, T., and Hill, M. J. (1979). The digestion of pectin in the human gut and its effect on calcium absorption and large bowel function. *Br. J. Nutr.* **41,** 477–489.

Delbarre, F., Roudier, J., and de Géry, A. (1977). Lack of effect of two pectins in idiopathic or gout-associated hyperdyslipidemia hypercholesterolemia. *Am. J. Clin. Nutr.* **30,** 463–465.

Denier, R. F., Litzinger, E. F., and Alford, A. D. (1989). U.S. Patent 4,825,884.

Deszyck, E. J. (1973). U.S. Patent 3,746,012.

Deszyck, E. J. (1974). U.S. Patent 3,796,222.

Deutsche Pektingesellschaft (1932 a,b). Ger. Patent 551 888 Ger. Patent 554 084.

Di Lorenzo, C., Williams, C. M., Hajnal, F., and Valenzuela, J. E. (1988). Pectin delays gastric emptying and increases satiety in obese subjects. *Gastroenterology* **95,** 1211–1215.

Deuel, H. (1945). Pektin, seine Reaktionen mit dem Blut, besonders die hämostatische Wirkung. *Schweiz. Med. Wochschr.* **75,** 661.

Doehnert, D. F., and Hill, A. S. (1985). U.S. Patent 4,505,976.

Dunaif, G., and Schneeman, B. O. (1981). The effect of dietary fiber on human pancreatic enzyme activity *in vitro. Am. J. Clin. Nutr.* **34,** 1034–1035.

Durrington, P. N., Bolton, C. H., Manning, A. P., and Hartog, M. (1976). Effect of pectin on serum lipids and lipoproteins, whole-gut transit time and stool-weight. *Lancet* **21,** 394–396.

Dutta, S., and Hlasko, J. (1985). Dietary fiber in pancreatic disease: Effect of high fiber on fat malabsorption in pancreatic insufficiency and *in vitro* study on the interaction of dietary fiber and pancreatic enzymes. *Am. J. Clin. Nutr.* **41,** 517–525.

Eastwood, M. A., Brydon, W. G., and Tadesse, K. (1980). *In* Medical aspects of dietary fiber "Effect of fiber on colonic function." G. A. Spiller, and R. M. Kay (eds), pp. 1–26. Plenum Press, New York.

Endress, H.-U. (1988). Erste praktische Erfolge mit der Pektinsäure-Schönung. *Flüss. Obst, Liquid fruit* **55,** 645–650.

Ershof, B. H., and Wells, A. F. (1962). Effects of methoxyl content on anti-cholesterol activity of pectic substances in the rat. *Exp. Med. Surg.* **20,** 272–276.

Fahrenbach, M. J., Riccardi, B. A., Saunders, J. C., Lourie, I. N., and Heider, J. G. (1965). Comparative effects of guar gum and pectin on human serum cholesterol levels. *Circulation* **31/32** (Suppl. II), 0.

Falk, J. D., and Nagyvary, J. J. (1982). Exploratory studies of lipid–pectin interactions. *J. Nutr.* **112,** 182–188.

Fetzer, H., and von Rex, W. G. (1986). U.S. Patent 4,617,162.

Flourie, B., Vidon, N., Florent, C. H., and Bernier, J. J. (1984). Effects of pectin on jejunal glucose absorption and unstirred layer thickness in normal man. *Gut* **25,** 936–941.

Freeman, F. M., and Pawelchak, J. M. (1985). U.S. Patent 4,538,603.

Freeman, F. M., and Pawelchak, J. M. (1988). U.S. Patent 4,728,642.

Gatfield, I. L., and Stute, R. (1972). Enzymatic reactions in the presence of polymers. *FEBS Lett.* **28,** 29–31.

Georgiev, S. (1979). Effect of some substances on the moisture content of tobacco. *Bulg. Tyutyun* **24,** 10–14 through *Chem. Abstr.* **92,** 125257.

Gerencser, G. A., Cerda, J., Burgin, C., Baig, M. M., and Guild, R. (1984). Unstirred water layer in rabbit intestine: Effects of pectin. *Proc. Soc. Exp. Biol. Med.* **176,** 183.

Gero, I. B. (1987 a,b). U.S. Patent 4,692,143 U.S. Patent 4,693,705.

Gidwani, R. N., Keegan, J. J., and Rubin, H. (1974). U.S. Patent 3,833,518.

Gidwani, R. N., Keegan, J. J., and Rubin, H. (1975 a,b). U.S. Patent 3,868,260 U.S. Patent 3,868,340.

Ginter, E., Kubec, E. J., Vozár, J., and Bobek, P. (1979). Natural hypocholesterolemic agent: Pectin plus ascorbic acid. *Intl. J. Vit. Nutr. Res.* **49,** 406.

Gold, L. A., McCourt, J. P., and Merimee, T. J. (1980). Pectin: An examination in normal subjects. *Diabetes Care* **3,** 50.

Gyarmati, L., Plochy, J., Racz, I., and Szentmiklosi, P. (1980). U.S. Patent 4,199,560.

Hansen, W. E., and Schulz, G. (1983). *In* Lösliche und Fixierte Inhibitoren der Amylase in Guar und Anderen Ballaststoffen. K. Huth, ed., "Pflanzenfasern—Neue Wege in der Stoffwechseltherapie." pp. 144–150. S. Karger, Basel, Switzerland.

Hedge, R. W. (1972). Ger. Offen DE 21 50 388.

Higgins, T. E. (1979). U.S. Patent 4,154,857.

Hill, W. H. (1976). U.S. Patent 3,946,110. *Chem. Abstr.* **84,** 155714.

Hillman, L. C., Peters, S. G., Fisher, C. A., and Pumare, E. W. (1985). The effects of the fiber components pectin, cellulose, and lignin on serum cholesterol level. *Am. J. Clin. Nutr.* **42,** 207.

Hiltz, R. H., Greer, J. S., and Friel, J. V. (1987). U.S. Patent 4,713,182. *Chem. Abstr.* **108,** 206709.

Hind, J. D., and Hopkins, W. C. (1968). French Patent FR 15 35 520.

Hind, J. D., and Seligman, R. B. (1968 a,b). U.S. Patent 3,409,026 U.S. Patent 3,411,515.

Hind, J. D., and Seligman, R. B. (1969). U.S. Patent 3,420,241.

Hind, J. D. (1970). U.S. Patent 3,499,454.

Hind, J. D., and Kelley, M. F., Jr. (1972). Ger. Offen. DE 21 40 507.

Hind, J. D., and Hopkins, W. C. (1978). U.S. Patent 4,129,134.

Hiroshi, I., Toshiaki, O., Kiyohide, S., Joshio, U., and Toshinobu, U. (1987). U.S. Patent 4,690,822.

Hirsbrunner, P. (1988). U.S. Patent 4,743,288.

Hoehn, W. M. (1950). U.S. Patent 2,508,388.

Holt, S., Heading, R. C., Carter, D. C., Prescott, L. F., and Tothill, P. (1979). Effect of gel fiber on gastric emptying and absorption of glucose and paracetamol. *Lancet* **24**, 637–639.

Holzner, G. (1989). U.S. Patent 4,803,195.

Hundhammer, K., and Marshall, M. (1983). Wirkung niedriger Dosen Apfelpektin auf die Blutlipide, ihre Verträglichkeit und Akzeptanz bei ambulanten hypercholesterinämischen Patienten. *Akt. Ernähr.* **8**, 222–225.

Iizuka, M., Kubota, H., and Sano, E. (1984). U.S. Patent 4,433,052.

Ink, S. L., and Hurt, H. D. (1987). Nutritional implications of gums. *Food Technol.* **41**, 77–81.

Isaksson, G. (1982). *In vitro* inhibition of pancreatic enzyme activities by dietary fiber. *Digestion* **24**, 54–59.

Isaksson, G., Lundquist, I., and Ihse, I. (1982a). Effect of dietary fiber on pancreatic enzyme activity *in vitro*. *Gastroenterology* **82**, 918–924.

Isaksson, G., Lundquist, I., Akesson, B., and Ihse, I. (1982b). Influence of dietary fiber on intestinal activities of pancreatic enzymes and on fat absorption in man. *Digestion* **25**, 39.

Isaksson, G., Lundquist, I., Akesson, B., and Ihse, I. (1984). Effects of pectin and wheat bran on intraluminal pancreatic enzyme activities and on fat absorption as examined with the triolein breath test in patients with pancreatic insufficiency. *Scand. J. Gastroenterol.* **19**, 467–472.

Jenkins, D. J. A., Leeds, A. R., Gassull, A., Houston, H., Goff, D., and Hill, M. (1976a). The cholesterol-lowering properties of guar and pectin. *Clin. Sci. Mol. Med.* **51**, 8–9.

Jenkins, D. J. A., Goff, D. V., Leeds, A. R., Alberti, K. G. M. M., Wolever, T. M. S., Gassull, M. A., and Hockaday, T. D. R. (1976b). Unabsorbable carbohydrates and diabetes: Decreased postprandial hyperglycaemia. *Lancet* **2**, 172–174.

Jenkins, D. J. A., Leeds, A. R., Gassull, M. A., Chochet, B., and Alberti, K. G. M. M. (1977). Decrease in postprandial insulin and glucose concentrations by guar and pectin. *Ann. Intern. Med.* **86**, 20–23.

Jenkins, D. J. A., Wolever, T. M. S., Leeds, A. R., Gassull, M. A., Haisman, P., Dilawari, J., Goff, D. V., Metz, G. L., and Alberti, K. G. M. M. (1978). Dietary fibres, fibre analogues, and glucose tolerance: Importance of viscosity. *Br. Med. J.* **1**, 1392–1394.

Jenkins, D. J. A., Reynolds, D., Leed, A. R., Walker, A. L., and Cummings, J. H. (1979). Hypocholesterolemic action of dietary fiber unrelated to fecal bulking effect. *Am. J. Clin. Nutr.* **32**, 2430–2435.

Jensen, E. H. (1969) U.S. Patent 3,449,492. *Chem. Abstr.* **71**, 53586.

Jones, D. R. (1946). A medium for investigating the breakdown of pectin by bacteria. *Nature* **158**, 625.

Jongen, W. M., van Boekel, M. A., and van Broekhoven, L. W. (1987). Inhibitory effect of cheese and some food constitutents on mutagenicity. *Food Chem. Toxicol.* **25**, 141–145.

Judd, P. A., and Truswell, A. S. (1982). Comparison of the effects of high- and low-methoxyl pectins on blood and faecal lipids in man. *Br. J. Nutr.* **48**, 451–458.

Julius, A. (1967). U.S. Patent 3,329,509 *Chem. Abstr.* **67**, 74267.

Kanter, Y., Eitan, N., Brook, G., and Barzilai, D. (1980). Improved glucose tolerance and insulin response in obese and diabetic patients on a fiber-enriched diet. *Israel J. Med. Sci.* **16**, 1–6.

Kasten, H. (1989). Ger. Offen DE 37 31 058.

Kasugai, H., Motojima, S., and Inoue, K. (1975). U.S. Patent 3,891,756.

Kay, R. M., and Truswell, A. S. (1977). Effect of citrus pectin on blood lipids and fecal steroid excretion in man. *Am. J. Clin. Nutr.* **30**, 171–175.

Kay, R. M., Judd, P. A., and Truswell, A. S. (1978). The effect of pectin on serum cholesterol. *Am. J. Clin. Nutr.* **31**, 562–563.

Keegan, J. J., Patel, G., and Rubin, H. (1975). U.S. Patent, 3,919,138. *Chem. Abstr.* **84**, 35369.

Keller, H. E., and Klaeni, H. (1981). U.S. Patent 4,254,100.

Kennedy, R. B. (1985). Europ. Pat. Appl. EP 0 169 580.

Kertesz, Z. I. (1951). "The Pectic Substances." Interscience Publishers, New York, London.

Keys, A., Grande, F., and Anderson, J. T. (1961). Fiber and pectin in the diet and serum cholesterol concentration in man. *Proc. Soc. Exp. Biol. Med.* **106**, 555–558.

Kröper, H. (1934). Ger. Patent 601 475.

Kulbe, K. D., and Weber, H. (1984). Ger. Patent DE 32 28 231.

Lagreca, L., and Marotta, E. (1985). Efecto nutricional de la pectina en cerdos en crecimiento y terminacion. *Arch. Latinoam. Nutr.* **35**, 172–179.

Land, E. H. (1956). U.S. Patent 2,759,825.

Lanini, M., Periard, J., and Staub, H. R. (1989). U.S. Patent 4,808,324.

Larsen, H.-O., and Sorensen, E. L. (1980). U.S. Patent 4,231,369.

Laves, H. G. (1979). U.S. Patent 4,162,306. *Chem. Abstr.* **89**, 220903.

Lavia, A. L., Pawelchak, J. M., and Wang, J.-C. J. (1981). U.S. Patent 4,292,972.

Lawaetz, O., Blackburn, A. M., Bloom, S. R., Aritas, Y., and Ralphs, D. N. L. (1983). Effect of pectin on gastric emptying and gut hormone release in the dumping syndrome. *Scand. J. Gastroenterol.* **18**, 327–336.

Leeds, A. R., Ralphs, D. N., Boulos, P., Ebied, F., Metz, G., Dilawari, J. B., Elliott, A., and Jenkins, D. J. A. (1977). Pectin and gastric emptying in the dumping syndrome. *Proc. Nutr. Soc.* **37**, 23 A.

Lesser, M. A. (1939). *Drug and Cosmetic Ind.* **45**, 549.

Leuven van, J. W. (1967). U.S. Patent 3,346,407.

Levitt, N. S., Vinik, A. I., Sive, A. A., Child, P. T., and Jackson, W. P. U. (1980). The effect of dietary fiber on glucose and hormone responses to a mixed meal in normal subjects and in diabetic subjects with and without autonomic neuropathy. *Diabetes Care* **3**, 515–519.

Lochridge, E. H. (1951). U.S. Patent 2,568,752. *Chem. Abstr.* **46**, 4280.

Malyoth, G. (1934). Das Pektin der Hauptträger der Wirkung bei der Apfeldiät. *Klin. Wochenschr.* **13**, 51–54.

Mansfeld, G. (1940). Ger. Patent 692 938.

Markova, M., Grescheva, G., Nikolova, K., Koen, K., and Angelova, R. (1976). Third Sci. Pract. Symp. HEI, Varna 21. to 22.4.1976.

Mathews, H., and Steer, P. L. (1980). U.S. Patent 4,341,207.

Matsukawa, H., and Saeki, K. (1975). U.S. Patent 3,869,406 Ger. Patent 22 25 274.

Matsukawa, H., and Saeki, K. (1976). U.S. Patent 3,965,033 Ger. Patent 21 37 574 (1977).

Matsukawa, H., Katayama, S., and Kiritani, M. (1976). U.S. Patent 3,970,585 Ger. Patent 20 27 819 (1979).

McDonald, N. S., Nusbaum, R. E., Ezmirlian, F., Barbera, R. C., ALexander, G. V., Spain, P., and Rounds, D. E. (1952). Gastrointestinal absorption of ions I.—Agents diminishing absorption of strontium. *J. Pharmacol. Exp. Ther.* **104**, 348–353.

McMullen, J. N., Newton, D. W., and Becker, C. H. (1984). Pectin gelation complex coacervat II: Effect of microencapsulated sulfamerazine on size, morphology, recovery, and extraction of water-dispersible microglobules. *J. Pharm. Sci.* **73**, 1799–1803.

Merre de, L. J. (1967). U.S. Patent 3,329,145.

Miettinen, T. A., and Tarpila, S. (1977). Effect of pectin on serum cholesterol fecale-bile acids and biliary lipids in normolipidemic and hyperlipidemic individuals. *Clin. Chim. Acta.* **79**, 471–477.

Mill, P. J., Cresswell, M. A., Feinberg, J. G. (1967). Brit. Patent 1,174,854 *Chem. Abstr.* **72**, 101121. U.S. Patent 4,003,792 (1977) Ger. Patent 17 68 841 (1980).

Monnier, L., Pham, T. C., Aguirre, L., Orsetti, A., and Mirouze, J. (1978). Influence of indigestible fibers on glucose tolerance. *Diabetes Care* **1**, 83–88.

Morgan, L. M., Goulder, T. J., Tsiolakis, D., Marks, V., and Alberti, K. (1979). The effect of unabsorbable carbohydrates on gut hormones. *Diabetologia* **17**, 85–89.

Moroni, R., and Kalbow, H. (1978). U.S. Patent 4,070,300.

Morway, A. J., and Mikeska, L. A. (1954). U.S. Patent 2,694,683. *Chem. Abstr.* **49**, 3528.

Murphy, H. W., Probst, G. W., and Stephens, V. C. (1963). U.S. Patent 3,108,042. *Chem. Abstr.* **60**, 377.

Murray, R., and Finland, M. (1946). Pectin adjuvant for oral penicillin. *Proc. Soc. Exp. Med.* **62**, 242.

Myers, P. B. (1941). U.S. Patent 2,259,767.

Nakamura, H., Islukawa, T., Tada, N., Kagami, A., Koudo, K., Miyazami, E., and Takeyama, S. (1982). Effect of several kinds of dietary fibres on serum and lipoprotein lipids. *Nutr. Rep. Int.* **26**, 215–221.

Nambu, M. (1985). U.S. Patent 4,524,064.

Negrevergne, G. (1969). U.S. Patent 3,487,046. *Chem. Abstr.* **72**, 79414.

Negrevergne, G. (1974 a,b). U.S. Patent 3,787,389 U.S. Patent 3,790,558. *Chem. Abstr.* **80**, 108817.

Nichols, W., Newsome, R., Thesing, R., and Houck, W. (1987). Eur. Pat. Appl. EP 223 454.

Nicoulescu, T., Rafaila, E., Eremia, R., and Balasa, E. (1968). *Igiena* **7**, 421–427.

Noyes, S. E., Lochridge, E. H. (1953 a,b,c). U.S. Patent 2,628,153. *Chem. Abstr.* **47**, 4655. U.S. Patent 2,631,081. *Chem. Abstr.* **47**, 5586. U.S. Patent 2,631,082. *Chem. Abstr.* **47**, 5586.

Ogawa, Y., Okada, H., and Yashiki, T. (1987 a,b). U.S. Patent 4,652,441 U.S. Patent 4,711,782.

Ohashi, J., Furuya, N., Kataoka, S., and Watanabe, M. (1986). PCT Int. Appl. WO 86/05366.

Okuyama, G., Otani, Y., Tezuka, K., and Tokunaga, K. (1981). U.S. Patent 4,278,657. *Chem. Abstr.* **95**, 138395.

Otto, R. (1941). Ger. Patent 706 630.

Owens, H. S., and Maclay, W. D. (1951). U.S. Patent 2,555,354. *Chem. Abstr.* **45**, 7753.

Palmer, G. H., and Dixon, D. G. (1966). Effect of pectin dose on serum cholesterol levels. *Am. J. Clin. Nutr.* **18**, 437–442.

Paskins-Hurlburt, A. J., Tanaka, Y.. Skoryna, S. C., Moore, W., Stara, J. F., and Stara, J. R. (1977). The binding of lead by a pectic polyelectrolyte. *Environ. Res.* **14**, 128–140.

Patrick, G. (1967). Inhibition of strontium and calcium uptake by rat duodenal slices: Comparison of polyuronides and related substances. *Nature* **216**, 815–816.

Perkins, P. R., and Bale, C. R. (1979). Au. Patent AU 499 651 Ger. Patent DE 25 05 149 (1982).

Pfeffer, P. E., Doner, L. W., Hoagland, P. D., and McDonald, C. G. (1981). Molecular interactions with dietary fiber components. Investigation of the possible association of pectin and bile salts. *J. Agric. Food Chem.* **29**, 455.

Pless, G., and Ullmann, L. (1990). Einsatz von Schäumen auf Basis synthetischer Netzmittelschaumbildner bei schaumzerstörenden brennbaren Flüssigkeiten. brandschutz/*Deutsche Feuerwehr-Zeitung* **44**, 467–470.

Poyanard, T., Slama, G., Delage, A., and Tchobroutsky, G. (1980). Pectin efficacy in insulin-treated diabetics assessed by the artificial pancreas. *Lancet* **1**, 158.

Pulver, R. (1961). U.S. Patent 3,247,063. *Chem. Abstr.* **61**, 543.

Rabchuk, A. (1957). U.S. Patent 2,816,040.

Rabchuk, A. (1959). U.S. Patent 2,878,129. *Chem. Abstr.* **53**, 10598.

Rae, J. (1944). *Mfg. Chem.* **15**, 363.

Ramirez, E. G., Thomas, L. C., and Fry, W. E. (1975). U.S. Patent 3,887,491. *Chem. Abstr.* **83**, 85618.

Raskai, M., and Szijjarto, E. (1984). U.S. Patent 4,466,961.

Raskai, M., and Szijjarto, E. (1985). U.S. Patent 4,503,037.

Raudnitz, J. P. (1979). U.S. Patent 4,139,612. *Chem. Abstr.* **79**, 9888.

Raymond, T. L., Connor, W. E., Lin, D. S., Warner, S., Fry, M. M., and Connor, S. L. (1977). The interaction of dietary fibers and cholesterol upon the plasma lipids and lipoproteins, sterol balance, and bowel function in human subjects. *J. Clin. Invest.* **60,** 1429–1437.

Reghin, A. J., Hamberg, P. F. Jr., and Smith, H. E. (1951). U.S. Patent 2,558,167. *Chem. Abstr.* **45,** 8964.

Reiser, S. (1987). Metabolic effects of dietary pectins related to human health. *Food Technol.* **41,** 91–99.

Richter, W. O., Weisweiler, P., and Schwandt, P. (1981). Therapie der Hypercholesterinämie mit Apfelpektin. *Deutsche Med. Wochschr.* **106,** 19.

Robinson, E. (1988). U.S. patent 4,752,458.

Ross, J. K., and Leklem, J. E. (1981). The effect of dietary citrus pectin on the excretion of human fecal neutral and acid steroids and the activity of α-dehydroxylase and β-glucuronidase. *Am. J. Clin. Nutr.* **34,** 2068–2077.

Roth, J. N. (1980). U.S. Patent 4,241,186.

Roth, J. N. (1981). U.S. Patent 4,282,317.

Rowland, I. R., Mallet, A. K., Wise, A., and Bailey, E. (1983). Effects of dietary carrageenan and pectin on the reduction of nitro-compounds by the rat caecal microflora. *Xenobiotica* **13,** 251–256.

Rubanovskaya, A. A. (1960). [The influence of pectin on the radio strontium absorption from the gastrointestinaltract under experimental conditions.] russ. Postupila v redakciju 16/XI, 43–47 Institut gigieny truda i profzabolevanij AMN SSSR

Ruben, G. C. (1988). U.S. Patent 4,755,494. *Chem. Abstr.* **109,** 133876.

Rubenchik, B. L., Karpilovskaia, E. D., Tiktin, L. A., Gorban, G. P., and Pliss, M. B. (1985). [Food Inhibitors of the formation of carcinogenic nitroso compounds]. russ. *Vopr. Pitan* **1,** 48–51.

Ryhiner, G., Ryhiner, A., Voss, E., and König, A. (1967a). U.S. Patent 3,328,208. *Chem. Abstr.* **62,** 1333.

Ryhiner, G., Döhren von, H. H., and Jung, M. (1967b). U.S. Patent 3,350,232. *Chem. Abstr.* **60,** 15443.

Sackler, R. R., and Sackler, M. D. (1967). U.S. Patent 3,325,472. *Chem. Abstr.* **65,** 6998.

Sahi, A., Bijlami, R. L., Karmarkar, M. G., and Nayar, U. (1985). Modulation of glycemic response by protein, fat, and dietary fiber. *Nutr. Res.* **5,** 1431.

Sandhu, K. S., el Samahi, M. M., Mena, I., Dooley, C. P., and Valenzueala, J. E. (1987). Effect of pectin on gastric emptying and gastroduodenal motility in normal subjects. *Gastroenterology* **92,** 486–492.

Sankar, P., and Arun, S. (1989). U.S. Patent 4,822,770.

Salatinjants, A. (1987). U.S. Patent 4,708,952. *Chem. Abstr.* **108,** 62507.

Salminen, W. M., and Weyerts. W. J. (1961). U.S. Patent 2,968,582.

Sawadogo, L., Houdebine, L. M., Thibault, J. F., Rouan, X., and Ollivier-Bousquet, M. (1988). Effect of pectic substances on prolactin and growth hormone secretion in the ewe and on the induction of casein synthesis in the rat. *Reprod. Nutr. Dev.* **28,** 293–301.

Schlemmer, U. (1986). Verminderung des Gehaltes von Blei und anderen Elementen in Wein mit Hilfe von Pektinsäure. *Z. Lebensm. Unters. Forsch.* **183,** 339–343.

Schmidt, M. P., Neugebauer, W., and Rebenstock, A. (1962). U.S. Patent 3,046,131. *Chem. Abstr.* **60,** 477.

Schneeman, B. O. (1985). Effects of nutrients and nonnutrients on food intake. *Am. J. Clin. Nutr.* **42** (5 Suppl.), 966–972.

Schoepe, E. (1979). Ger. Offen 27 39 180.

Schuderer, U. (1986). Wirkung von Apfelpektin auf die Cholesterin- und Lipoproteinkonzentration bei Hypercholesterinämie. Dissertation, University of Giessen, Germany.

Schwandt, P., Richter, W. O., Weisweiler, P., and Neureuther, G. (1982). Cholestyramine plus pectin in treatment of patients with familial hypercholesterolemia. *Atherosclerosis* **44,** 379–383.

Schwartz, S. E., Levine, R. A., Singh, A., Scheidecker, J. R., and Track, N. S. (1983). Sustained pectin delays gastric emptying. *Gastroenterology* **83**, 812–817.

Schwartz, S. E., Levine, R. A., Weinstock, R. S., Petokas, R. S., Mills, C. A., and Thomas, F. D. (1988). Sustained pectin ingestion: Effect on gastric emptying and glucose tolerance in noninsulin-dependent diabetic patients. *Am. J. Clin. Nutr.* **48**, 1413–1417.

Simon, E. (1984). U.S. Patent 4,489,738.

Slaybaugh, E. H. (1953). U.S. Patent 2,659,690. *Chem. Abstr.* **48**, 2993.

Stantschev, S., Kratschanov, C., Popova, M., Kirtschev, N., and Marteschev, M. (1979). Anwenduug von granuliertem Pektin bei Bleiexponierten. *Z. ges. Hyg.* **25**, 585–587.

Stasse-Wolthuis, M., Albers, H. F. F., van Jeveren, J. G. G., de Jong, J. W., Hautvast, J. G. A. J., Hermus, R. J. J., Katan, M. B., Brydon, W. G., and Easwood, M. A. (1980). Influence of dietary fiber from vegetables and fruits, bran, or citrus pectin on serum lipids, fecal lipids, and colonic function. *Am. J. Clin. Nutr.* **33**, 1745–1756.

Strohm, G., Wucherpfennig, K., and Otto, K. (1987). Ger. Patent 36 14 656 C1.

Sumner, O. R., and Grenfell, T. C. (1955). U.S. Patent 2,725,336. *Chem. Abstr.* **50**, 4463.

Suzuki, M., and Kajuu, T. (1983). Suppression of hepatic lipogenesis by pectin and galacturonic acid orally fed at the separate timing from digestion–absorption of nutrients in rat. *J. Nutr. Sci. Vitaminol. (Tokyo)* **29**, 553–562.

Tateno, A., and Masuko, K. (1989). Manufacture of granules containing fragrant substances for cigarettes. *Jpn. Kokai Tokkoyo Koho JP* 01027461 through Chem. Abstr. 111:229281.

Thiele, H. (1987a). Ger. Offen. 35 25 440 A1.

Thiele, H. (1987b). Ger. Offen. 35 27 410.

Tomic, D. (1981). U.S. Patent 4,277,463.

Tompkins, C. A. (1938). U.S. Patent 2,139,139.

Toni, G. (1946). *Bull. Chim. Farm.* **85**, 3–21 through: *Chem. Abstr.* **40**, 6755.

Tunali, G., Stetten, D., Schuderer, U., and Hofmann, H. (1990). Wirkung von Pektin—in Form von Apfelpektin Extrakt—auf die postprandiale Serumglucose- und Serum-Insulin-Konzentration bei Probanden mit Diabetes Mellitus Typ II. 27. Wiss. DGE-Kongress, *München Ernährungs-Umschau aus Forschung und Praxis* **37**, 141.

Ullrich, I. H. (1987). Evaluation of a high-fiber diet in hyperlipidemia: A review. *J. Am. Coll. Nutr.* **6**, 19–25.

Vaaler, S., Hanssen, K. F., and Aagenaes, O. (1980). Effect of different kinds of fiber on postprandial blood glucose in insulin-dependent diabetics. *Acta Med. Scand.* **208**, 389–391.

Waldron-Edward, D., Paul, T. M., and Skoryna, S. C. (1965). Suppression of intestinal absorption of radioactive strontium by naturally occurring nonabsorbable polyelectrolytes. *Nature* **205**, 1117–1118.

Walker, A. R. P., Walker, B. F., and Segal, I. (1979). Faecal pH-value and its modification by dietary means in South African black and white school children. *Med. J.* **55**, 495–498.

Washington, N., Wilson, C. G., Graeves, J. L., and Danneskiold-Samsoe, P. (1988). An investigation into the floating behavior of a pectin-containing antireflux formulation. *Scand. J. Gastroenterol.* **23**, 920–924.

Welch, H., Hirsch, H. L., and Taggart, S. R. (1947). quoted from Kertesz (1951). *Science News Letter* **52**, 210.

Welch, H. (1950). U.S. Patent, 2,518,510. *Chem. Abstr.* **45**, 311.

Welch, H. (1952). U.S. Patent, 2,603,583. *Chem. Abstr.* **46**, 9810.

Williams, D. R. R., James, W. P. T., and Evans, I. E. (1980). Dietary fiber supplementation of a "normal" breakfast administered to diabetics. *Diabetologia* **18**, 379–383.

Wilson, C. W. (1940). U.S. Patent 2,207 185.

Wilson, C. W. (1956). U.S. Patent 2,764,512. *Chem. Abstr.* **51**, 3867.

Wilson, F., and Dietschy, J. (1974). The intestinal unstirred water layer: Its surface area and effect on active transport kinetics. *Biochim. Biophys. Acta* **363**, 112–126.

Winkler, B. (1973). U.S. Patent 3,961,082. Ger. Patent DE 22 12 398 (1977).

Woods, D. (1988). U.S. Patent 4,735,737.

Wucherpfennig, K., Otto, K., Strohm, G., and Weber, K. (1988). Pektinsäure—ein pflanzliches Schönungsmittel. *ZFL—Int. Z. Lebensm.technol. Verfahrenstechn.* **39**, 383–393.

Wucherpfennig, K. (1990). Kontinuierliches Arbeiten—Herstellung von Apfelsaftkonzentrat. *Die Getränkeindustrie* **44**, 290–300.

Wuhrmann, F. (1939). Klinische Erfahrungen mit einem Dauer-Insulin ohne Eiweiß-Zusatz; Decurvon—ein Pektin-Insulin. *Schweiz. Med. Wochschr.* **69**, 1275.

Youssef, K. A. (1981). U.S. Patent 4,248,971.

Zimmaro, D. M., Rolandelli, R. H., Koruda, M. J., Settle, R. G., Stein, T. P., and Rombeau, J. L. (1989). Isotonic tube-feeding formula induces liquid stool in normal subjects reversal by pectin. *J. Parenter. Enter. Nutr.* **13**, 117–123.

INDEX

FOOD SCIENCE AND TECHNOLOGY
A Series of Monographs

Maynard A. Amerine, Rose Marie Pangborn, and Edward B. Roessler, PRINCIPLES OF SENSORY EVALUATION OF FOOD. 1965.

Martin Glicksman, GUM TECHNOLOGY IN THE FOOD IINDUSTRY. 1970

L. A. Goldblatt, AFLATOXIN. 1970

Maynard A. Joslyn, METHODS IN FOOD ANALYSIS, second edition. 1970

A. C. Hulme (ed.), THE BIOCHEMISTRY OF FRUITS AND THEIR PRODUCTS. Volume 1—1970. Volume 2—1971.

G. Ohloff and A. F. Thomas, GUSTATION AND OLFACTION. 1971

C. R. Stumbo, THERMOBACTERIOLOGY IN FOOD PROCESSING, second edition. 1973

Irvin E. Liener (ed.), TOXIC CONSTITUENTS OF ANIMAL FOODSTUFFS. 1974

Aaron M. Altschul (ed.), NEW PROTEIN FOODS: Volume 1, TECHNOLOGY, PART A—1974. Volume 2, TECHNOLOGY, PART B—1976. Volume 3, ANIMAL PROTEIN SUPPLIES, PART A—1978. Volume 4, ANIMAL PROTEIN SUPPLIES, PART B—1981. Volume 5, SEED STORAGE PROTEINS—1985.

S. A. Goldblith, L. Rey, W. W. Rothmayr, FREEZE DRYING AND ADVANCED FOOD TECHNOLOGY. 1975.

R. B. Duckworth (ed.), WATER RELATIONS OF FOOD. 1975

Gerald Reed (ed.), ENZYMES IN FOOD PROCESSING, second edition. 1975.

A. G. Ward and A. Courts (eds.), THE SCIENCE AND TECHNOLOGY OF GELATIN. 1976.

John A. Troller and J. H. B. Christian, WATER ACTIVITY AND FOOD. 1978

A. E. Bender, FOOD PROCESSING AND NUTRITION. 1978.

D. R. Osborne and P. Voogt, THE ANALYSIS OF NUTRIENTS IN FOODS. 1978.

Marcel Loncin and R. L. Merson, FOOD ENGINEERING; PRINCIPLES AND SELECTED APPLICATIONS, 1979.

Hans Riemann and Frank L. Bryan (eds.), FOOD-BORNE INFECTIONS AND INTOXICATIONS, second edition. 1979.

N. A. Michael Eskin, PLANT PIGMENTS, FLAVORS AND TEXTURES: THE CHEMISTRY AND BIOCHEMISTRY OF SELECTED COMPOUNDS. 1979.

J. G. Vaughan (ed.), FOOD MICROSCOPY. 1979.

J. R. A. Pollock (ed.), BREWING SCIENCE, Volume 1—1979. Volume 2—1980.

Irvin E. Liener (ed.), TOXIC CONSTITUENTS OF PLANT FOODSTUFFS, second edition. 1980.

J. Christopher Bauernfeind (ed.), CAROTENOIDS AS COLORANTS AND VITAMIN A PRECURSORS: TECHNOLOGICAL AND NUTRITIONAL APPLICATIONS. 1981.

Pericles Markakis (ed.), ANTHOCYANINS AS FOOD COLORS. 1982.

Vernal S. Packard, HUMAN MILK AND INFANT FORMULA. 1982.

George F. Stewart and Maynard A. Amerine, INTRODUCTION TO FOOD SCIENCE AND TECHNOLOGY, second edition. 1982.

Malcolm C. Bourne, FOOD TEXTURE AND VISCOSITY: CONCEPT AND MEASUREMENT. 1982.

R. Macrae (ed.), HPLC IN FOOD ANALYSIS. 1982.

Héctor A. Iglesias and Jorge Chirife, HANDBOOK OF FOOD ISOTHERMS: WATER SORPTION PARAMETERS FOR FOOD AND FOOD COMPONENTS. 1982.

John A. Troller, SANITATION IN FOOD PROCESSING. 1983.

Colin Dennis (ed.), POST-HARVEST PATHOLOGY OF FRUITS AND VEGETABLES. 1983.

P. J. Barnes (ed.), LIPIDS IN CEREAL TECHNOLOGY. 1983.

George Charalambous (ed.), ANALYSIS OF FOODS AND BEVERAGES: MODERN TECHNIQUES. 1984.

David Pimentel and Carl W. Hall, FOOD AND ENERGY RESOURCES. 1984.

Joe M. Regenstein and Carrie E. Regenstein, FOOD PROTEIN CHEMISTRY: AN INTRODUCTION FOR FOOD SCIENTISTS. 1984.

R. Paul Singh and Dennis R. Heldman, INTRODUCTION TO FOOD ENGINEERING. 1984.

Maximo C. Gacula, Jr., and Jagbir Singh, STATISTICAL METHODS IN FOOD AND CONSUMER RESEARCH. 1984.

S. M. Herschdoerfer (ed.), QUALITY CONTROL IN THE FOOD INDUSTRY, second edition. Volume 1—1984. volume 2 (first edition)—1968. Volume 3 (first edition)—1972.

Y. Pomeranz, FUNCTIONAL PROPERTIES OF FOOD COMPONENTS. 1985.

Herbert Stone and Joel L. Sidel, SENSORY EVALUATION PRACTICES. 1985.

Fergus M. Clydesdale and Kathryn L. Wiemer (eds.), IRON FORTIFICATION OF FOODS. 1985.

John I. Pitt and Ailsa D. Hocking, FUNGI AND FOOD SPOILAGE. 1985.

Robert V. Decareau, MICROWAVES IN THE FOOD PROCESSING INDUSTRY. 1985.

S. M. Herschdoerfer (ed.), QUALITY CONTROL IN THE FOOD INDUSTRY, second edition. Volume 2—1985. Volume 3—1986. Volume 4—1987.

F. E. Cunningham and N. A. Cox (eds.), MICROBIOLOGY OF POULTRY MEAT PRODUCTS. 1986.

Walter M. Urbain, FOOD IRRADIATION. 1986.

Peter J. Bechtel, MUSCLE AS FOOD. 1986.

H. W.-S. Chan, AUTOXIDATION OF UNSATURATED LIPIDS. 1986.

Chester O. McCorkle, Jr., ECONOMICS OF FOOD PROCESSING IN THE UNITED STATES. 1987.

Jethro Jagtiani, Harvey T. Chan, Jr., and William S. Sakai, TROPICAL FRUIT PROCESSING. 1987.

J. Solms, D. A. Booth, R. M. Dangborn, and O. Raunhardt, FOOD ACCEPTANCE AND NUTRITION. 1987.

R. Macrae, HPLC IN FOOD ANALYSIS, second edition. 1988.

A. M. Pearson and R. B. Young, MUSCLE AND MEAT BIOCHEMISTRY. 1989.

Dean O. Cliver (ed.), FOODBORNE DISEASES. 1990.

Marjorie P. Penfield and Ada Marie Campbell, EXPERIMENTAL FOOD SCIENCE, third edition. 1990.

Leroy C. Blankenship, COLONIZATION CONTROL OF HUMAN BACTERIAL ENTEROPATHOGENS IN POULTRY. 1991.

Yeshajahu Pomeranz, FUNCTIONAL PROPERTIES OF FOOD COMPONENTS, second edition. 1991.

Reginald H. Walter, THE CHEMISTRY AND TECHNOLOGY OF PECTIN. 1991.

Printed and bound by CPI Group (UK) Ltd, Croydon, CR0 4YY

03/10/2024

01040422-0007